Value Practices in the Life Sciences and Medicine

Value Practices in the Life Sciences and Medicine

edited by

Isabelle Dussauge,

Claes-Fredrik Helgesson,

and Francis Lee

Great Clarendon Street, Oxford, OX2 6DP,
United Kingdom

Oxford University Press is a department of the University of Oxford.
It furthers the University's objective of excellence in research, scholarship,
and education by publishing worldwide. Oxford is a registered trade mark of
Oxford University Press in the UK and in certain other countries

First Edition published in 2015

Impression: 1

Published in the United States of America by Oxford University Press
198 Madison Avenue, New York, NY 10016, United States of America

British Library Cataloguing in Publication Data
Data available

Library of Congress Control Number: 2014941823

ISBN 978–0–19–968958–3

Printed and bound by
CPI Group (UK) Ltd, Croydon, CR0 4YY

ACKNOWLEDGEMENTS

Most endeavours worth pursuing take time and involve many people. This volume is no exception. It began as an ambition of ours together with Boel Berner and Sven Widmalm to initiate a conversation about values, economization, life science, and medicine. This evolved in April 2011 into a workshop on the 'Moral Economy of Life Science' at the Vadstena Monastary in Sweden (with gracious support from Riksbankens Jubileumsfond). This meeting became the starting point for this book.

We have benefited from presenting drafts of the introductory (Chapter 1) and concluding chapter (Chapter 14) at many meetings and conferences, such as at the 'Value, Values and Valuation' sub-theme at Egos 2012 in Helsinki; at a 'Bio-objects and their Boundaries' COST Action meeting in Madrid; at the 'Dimensions of Value and Values in STS' symposium at University of Edinburgh; and at 4S 2013 in San Diego. At these meetings and in other conversations this volume has benefited from comments, critique, discussions, and encouragement from numerous colleagues.

In addition to thanking the contributors to this volume for making it possible, we would especially like to thank Luis Araujo, Cecilia Åsberg, Patrik Aspers, Boel Berner, Kean Birch, Ingemar Bohlin, Anna Bredström, Michel Callon, Alberto Cambrosio, Franck Cochoy, Barbara Czarniawska, Liliana Doganova, Aant Elzinga, Lena Eriksson, Karin Fernler, John Finch, Emma Frow, Masato Fukushima, Susi Geiger, Rebecca Herzig, Andrew Hoffman, Klaus Høyer, Ericka Johnson, Petra Jonvallen, Anne Kerr, Hans Kjellberg, Lene Koch, Jenny Lee, Javier Lezaun, Donald MacKenzie, Liz McFall, Alexandre Mallard, Jessica Mesman, Tiago Moreira, Shai Mulinari, Fabian Muniesa, Nicole Nelson, Vololona Rabeharisoa, Morten Sager, Kerstin Sandell, Ebba Sjögren, Mikaela Sundberg, Karin Svedberg Helgesson, Pascale Trompette, Anna Tunlid, David Tyfield, Andrew Webster, Marianne Winther Jørgensen, and four anonymous OUP reviewers, for fruitful exchanges and comments.

The unit Technology and Social Change in the Department of Thematic Studies at Linköping University was a good home for this project. We are not least indebted to past and present participants of the research programme ValueS and our quirky and dynamic weekly seminars: Rèka Andersson, Lotta Björklund Larsen, Mats Bladh, Baki Cakici, Maria Eidenskog, Vasilis Galis, Anders Hansson, Andrew Hoffman, Linus Johansson Krafve, Lisa Lindén, Oscar Javier Maldonado Castañeda, Anna Morvall, Johan Nilsson, Mikael Ottosson, Karin Thoresson, Anna Wallsten, Sven Widmalm, Steve Woolgar, and Teun Zuiderent-Jerak. The staff have also helped out in countless ways; a special thanks goes to Eva Danielsson and Josefin Frilund for all assistance. Hans Winberg and Annemarie Brandt at the academic think tank Leading Health Care have also provided support, not the least through finding space for editorial meetings. Our commissioning editor David Musson has, together with Emma Booth, Clare Kennedy, Patricia Baxter, and Subramaniam Vengatakrishnan, provided invaluable guidance and help in the process from book proposal to this finished volume.

The support from Riksbankens Jubileumsfond for the project 'Trials of Value' (Helgesson and Lee) has been instrumental in the conceptual development of the introductory and concluding chapters. Finally, although unconventional, we cannot refrain from acknowledging that our own collaboration in this endeavour has been rewarding for each of us. Well aware that the ordering of author names has become a common valuation practice, we have found no reason to deviate from a plain alphabetical ordering of our names.

Isabelle Dussauge,
Claes-Fredrik Helgesson,
and Francis Lee

Linköping,
Stockholm,
Uppsala,
December 2014

TABLE OF CONTENTS

LIST OF ILLUSTRATIONS

Figures

Tables

LIST OF CONTRIBUTORS

Kristin Asdal is an Historian and Professor of science, technology, and culture at the Social Science Faculty, University of Oslo. In her research she has been particularly interested in the ways in which politics of nature emerge in exchanges and confrontations between the natural sciences, economics, and economic reasoning. She is also profoundly interested in questions related to method in science and technology studies.

Isabelle Dussauge is a Researcher at the Center for Gender Research at Uppsala University. Her primary research interests are in the science and politics of the body, at the intersection of science and technology studies, gender studies, and the history of medicine. She has worked with visualization in medicine; the early computerization of health care; and the place of the brain in contemporary culture. She is currently concluding a research project entitled 'Brain Desires', a critical inquiry into the contemporary neurosciences of sexuality and pleasure.

Carrie Friese is an Associate Professor in the Sociology Department at the London School of Economics and Political Science. Her research interest concerns the constitutive practices of reproductive and genetic sciences in social life, which includes human and other animal species. She examines these areas through the perspectives of medical sociology, science and technology studies, and animal studies. Her book *Cloning Wild Life: Zoos, Captivity and the Future of Endangered Animals* was published by New York University Press in 2013.

Kor Grit is an economist and philosopher. He received his PhD in philosophy from the University of Groningen in 2000. He has written books in Dutch about the advance of economic discourse and dilemmas of health-care managers. He is currently Assistant Professor at the Department of Health Policy and Management, Erasmus University, Rotterdam. His current research includes studies of patient participation, solidarity, and ethical aspects of health-care markets. Recent publications examine the role of patient organizations in the Dutch neo-corporatist model of participation (*Health Expectations*), undocumented migrants (*Journal of Health Politics, Policy and Law)* (*JHPPL*), performance indicators (*Public Administration*), and the possibilities of strengthening the position of patients with the aid of new financial instruments, such as personal health budgets (*Journal of Medical Ethics*).

Claes-Fredrik Helgesson is Professor in Technology and Social Change at Linköping University. His research interest concerns in broad terms the intertwining of economic organizing, science, and technology. The theoretical inspiration comes primarily from economic sociology and social studies of science and technology (STS). His current project, 'Trials of Value' together with Francis Lee, investigates the designing of controlled medical experiments as a site where scientific, medical, and economic aspects are deliberated upon when establishing what knowledge is worth pursuing.

Helgesson is co-founder and co-editor of *Valuation Studies*, a new open access journal, which published its first issue in spring 2013.

Linus Johansson Krafve is a PhD Candidate in the Department of Thematic Studies—Technology and Social Change, Linköping University. He is interested in economization of the public sector and its effects on the relations and practices in and between politics, administration, and citizens. His dissertation is about the design of a market for primary health care in Swedish county councils.

Francis Lee is Assistant Professor at the Department of Thematic Studies—Technology and Social Change, at Linköping University. His primary research interests are in the practices, politics, and technologies of knowledge. His work has dealt with the valuation of knowledge in the biosciences, epistemic standards in education, and exclusion in socio-technical processes. He is currently studying the valuation of biomedical knowledge in the project 'Trials of Value'. He has co-edited *Posthumanistiska nyckeltexter* [Posthumanist Key Texts], a Swedish reader on posthumanist theory (Studentlitteratur, 2011).

Ilana Löwy is a Senior Researcher at INSERM (Institut National de la Santé et de la Recherche Scientifique), Paris. She is interested in the history of bacteriology and immunology, tropical medicine, history of oncology, and the intersection between gender studies and biomedicine. Among others, she has published *Virus, moustiques et modernité: Science, politique et la fièvre jaune au Brésil* (Archives d'Histoire Contemporaine, 2001) and *Preventive Strikes: Women, Precancer and Prophylactic Surgery* (Johns Hopkins University Press, 2009). Her most recent book is *A Woman's Disease: A History of Cervical Cancer* (Oxford University Press, 2011). She is now studying the history of prenatal diagnosis.

Daniel Neyland is Professor of Sociology at Goldsmiths, University of London. His research interests cover issues of governance, accountability, and ethics in forms of science, technology, and organization. He draws on ideas from ethnomethodology, science and technology studies (STS) (in particular forms of radical and reflexive scepticism, constructivism, actor–network theory, and the recent STS turn to markets and other forms of organizing); his research is ethnographic in orientation. Dan's substantive interests are focused on: traffic management, waste, airports, biometrics, parking, signposts, malaria and the utility of social science, ideas of equivalence, parasitism, the mundane, market failures, problems and solutions, deleting, value, and publicity. These ideas form themselves into a variety of publications and research projects (currently including ERC project: MISTS <http://www.gold.ac.uk/sociology/research/researchprojects/mists/> and EU FP7 project: ADDPRIV <http://www.addpriv.eu/>).

Christer Nordlund is Professor of History of Science and Ideas at the Department of Historical, Philosophical, and Religious Studies at Umeå University. He is an historian focusing on the cultural, material, and intellectual history of science, technology, medicine, and the environment from 1800 onwards, mainly in a Scandinavian context. He has also taken an interest in current science policy and theories of research organization and cooperation. Nordlund has been a Visiting Scholar at Uppsala

University, the University of Cambridge, and Max Planck Institute for the History of Science (MPIWG) in Berlin. He was Torgny Segerstedt Pro Futura Scientia Fellow at the Swedish Collegium for Advanced Study (SCAS) in Uppsala 2007–14 and is currently directing the interdisciplinary research group Umeå Studies in Science, Technology and Environment (USSTE) at Umeå University. Nordlund's most recent books are *Vägar till vetenskapen* (ed.) (Santérus Förlag, 2013), and *Motorspriten kommer!* (co-authored) (Gidlunds, 2014). He is currently exploring the intellectual history of creativity.

Philip Roscoe is Reader in Management at the School of Management, University of St Andrews. His research focuses on the formation and organization of markets and on processes of economization; empirical sites include online dating, retail investing, organ transplantation, and alternative economies. He has a PhD in management from Lancaster University, an MPhil in medieval Arabic thought from Oxford University, and a BA in theology from the University of Leeds. He has published in *Accounting, Organization and Society*, *Business History*, *Organization*, and *The Sociological Review*. In 2011 he was one of the winners of the inaugural *AHRC BBC Radio 3 'New Generation Thinkers'* scheme; his book '*I Spend Therefore I Am*' was published by Penguin Viking in February 2014.

Elena Simakova is Lecturer in Innovation at the University of Exeter Business School. Her current research mainly concerns post-positivist perspectives on governance of science and technology, especially in the postcolonial and post-Cold War context of the contemporary techno-science. She has researched university, industry, and policy settings in the UK, USA, and France, and is author of *Marketing Technologies: Corporate Cultures and Technological Change* (Routledge, 2013). From 2012 until 2017 engaged as Scientific Director of the Policy Analysis and Studies of Technologies (PAST) Centre at the National Research Tomsk State University, Russian Federation, supported by the George Soros Open Society Foundations (OSF) under the Higher Education Support Programme (HESP). She has recently co-edited (with Christopher Coenen) a special issue of *Science, Technology & Innovation Studies* on governance of visionary techno-sciences <http://www.sti-studies.de/ojs/index.php/sti>.

Sergio Sismondo is a Professor in the Departments of Philosophy and Sociology at Queen's University, Ontario. His current research is on the political economy of pharmaceutical knowledge, looking at relations between research and marketing in a variety of areas running from clinical trials through medical education. He is co-editor with Jeremy Greene of *The Pharmaceutical Studies Reader* (Wiley-Blackwell, 2015). He also is the author of *An Introduction to Science and Technology Studies* (2nd ed., Wiley-Blackwell, 2010) and a number of other general and philosophical works in STS. He is Editor of the journal *Social Studies of Science*.

Tom van der Grinten is Professor Emeritus of Health Policy and Organization at the Department of Health Policy and Management of Erasmus University, Rotterdam. He studied economics and sociology at the University of Amsterdam. His research is focused on the institutional factors influencing policymaking in public health and health care, with particular interest in the significance of the public–private–professional dependencies. He is the author of many articles and contributions to books on the subjects of health policy and organization. Tom van der Grinten sat on the editorial board of various scientific journals and of national and international bodies for research, development, and policymaking in the field of public health and health care. Among other responsibilities, he chairs the advisory council for the Dutch Inspectorate on Health Care, is a member of the advisory council for the Health Care Insurance Board (CVZ), and the scientific council for Health Forecast 2014.

Sven Widmalm is Professor at the Department of History of Science and Ideas, Uppsala University. His research is in the social history of science from the eighteenth century onwards and has focused on e.g. military cartography and astronomy, the academic culture of nineteenth-century physics, and relations between academic research and industry in twentieth-century biochemistry. Current research concerns include the 'scientific family', relations between Swedish and German science during the Third Reich, the history of intellectual property, and post-war research policy.

Steve Woolgar is Professor of Science and Technology Studies at Linköping University and Professor of Marketing at Saïd Business School, University of Oxford. He has published widely in social studies of science and technology, social problems, and social theory. His books include *Laboratory Life: The Construction of Scientific Facts* (with B. Latour, Princeton, 1986); *Science: The Very Idea* (Routledge, 1988); *Knowledge and Reflexivity* (Sage, 1988); *The Machine at Work: Technology, Organisation and Work* (with K. Grint, Polity, 1996); *Virtual Society? Technology, Cyberbole, Reality* (Oxford University Press, 2002); *Mundane Governance: Ontology and Accountability* (with D. Neyland, Oxford University Press, 2013); and *Representation in Scientific Practice Revisited* (with C. Coopmans, J. Vertesi and M. Lynch, MIT Press, 2014); *Globalization in Practice* (with N. Thrift and A. Tickell, Oxford University Press, 2014); and *Visualisation in the Age of Computerization* (with A. Carusi, S. Hoel, and T. Webmoor, Routledge, 2014). His work has been translated into Chinese, Dutch, French, Greek, Japanese, Portuguese, Spanish, and Turkish. His main current research interests include the influence of the neurosciences on social sciences and the humanities; and the nature and dynamics of provocation and intervention.

Teun Zuiderent-Jerak is Linköping University Research Fellow and Docent (~Senior Lecturer) at the Department of Thematic Studies—Technology and Social Change at Linköping University. He obtained his PhD in Science and Technology Studies (STS) from the Erasmus University, Rotterdam, in 2007. His research focuses on standardization and innovation practices in health care, the construction of markets and public values, and STS research that explicitly aims to 'intervene' in the practices it studies. He has published articles in the academic journals *Social Studies of Science, Science, Technology, and Human Values, Social Science & Medicine, Human-Computer Interaction, Science as Culture*, and the *British Medical Journal*. He has also co-edited a special issue of 'Science as Culture on Unpacking "Intervention"' in *Science and Technology Studies*. His book *Situated Intervention; Sociological Experiments in Healthcare* is forthcoming at The MIT Press.

1 On the omnipresence, diversity, and elusiveness of values in the life sciences and medicine

Isabelle Dussauge, Claes-Fredrik Helgesson, Francis Lee, and Steve Woolgar

Valuations of life are intermingled with values such as scientific reputation, profitability, fairness, economic efficiency, and accessibility of care. Reciprocally, the practices of the life sciences themselves produce values, for instance: public health, the preservation of endangered species, profitability of tamed animals for farming, usability of clinical data, or bodily autonomy.

Scientists, regulators, analysts, and publics regularly strive to define what counts as proper conduct in science and health care, what is economically and socially valuable, and what is known and worth knowing. The life sciences, and the biomedicalization of human life more generally, are an increasingly important site for the production of life forms and of culturally available meanings and orderings of life—not least of social life (Clarke et al. 2003). In the crucible of day-to-day practice, we face decisions that assemble the value of things, humans, and animals in certain ways.

So we need to ask: How are values made and ordered? What determines what comes to count as important values in a given setting? What actors are allowed to make values and yardsticks, and with what means? How are metrics agreed upon? Which values are put into play? How are a multitude of possible divergent values coordinated or separated in practice?

The social and material orderings of the life sciences and health care are permeated with the making, attribution, and performance of different values: cultural, economic, legal. Our aim in this volume is to examine the ongoing composition of these values in the life sciences. We do this using a broad approach that does not treat values as stable and predefined, but rather as something grappled with, articulated, and made in concrete practices. By looking at dissonances, discordances, and ruptures between values, we find situations of the explicit assembling, articulation, coordination, and negotiation of values. These practices illuminate the various yardsticks, different technologies, and matters of concern that inform these mundane, but fundamental, activities.

In this volume we use a broad definition of the domain under study, the life sciences, to allow for an inclusive investigation into different value practices across and between contexts. The empirical domains range from medical science and health-care markets, through zoos, to cod farming. What sets this book apart from many of the dominant theoretical approaches to values (in economics, anthropology, sociology, philosophy, history) is that it takes values as enacted: in actions, in technical practices, and in practices of valuation. Our exercise is rooted in an ambition to consider 'values' as things to be explained and explored rather than as given entities with explanatory power.

Quod vide value: price × quantity, irreducible, and mystified

When it comes to values, two inevitable and recurring questions are 'what are values?', or 'how do we define values?' Many discussions involve a struggle to define, delineate, or even to reconcile different notions of value. What is a value? How do you know when something is a value? How are economic values different from cultural values? How do you know if you are studying values? We believe that these questions are posed in the wrong manner. Here and throughout this volume we explore what would happen if we stopped asking 'What is a value?' and started asking 'How are values made?' That is, what happens if we shift from an etic register, attempting a priori to delineate and define what proper values are, to posing the emic question: 'How does something come to count as a value?'

This move has consequences for how we think and write about values. How can we talk about different sets of values without using readily available categories? In this introduction, we deploy a few clumsy and provisional placeholder terms to give an indication of the sets of values that recur across the empirical fields of interest to us. We use words like 'economic', 'medical', and 'cultural' to point to the diverging registers of value that this volume explores. These placeholder terms stem from empirical work, and are not an appeal to analyse values using these words as the only categories. Hence, a few words of caution are necessary: these temporary placeholders are dangerous. They shape the way we approach values. They filter our analysis and they divide the world. It follows that they cannot be an innocent point of departure. They are always already implicated in world-making. What would happen if we moved beyond these (and other) prefixed notions of values? What would happen if we instead explored the multifaceted, shifting, and entangled making of values?

To underscore the futility of attempting to answer the question 'What are values?' we wish to play a short game of *Quod Vide Value*, or looking up value: Many analytical efforts have been made analyse values, or to use the concept of value for various analytical purposes. A glance at several recent social science reference works provides rich insight into the multiple ways in which the concept of value has figured in various disciplines. First, under the word 'value', the *Encyclopedia of Semiotics* affirms that 'disciplines as varied as economics, philosophy, aesthetics, logic, linguistics, and via linguistics, semiotics all employ a concept of value, though with varying applications and meanings.' (Bouissac 1998 q.v. value). The *Encyclopedia of Aesthetics* concurs, asserting that 'value is one of the weightiest, most indispensable, and perhaps most mystified concepts in aesthetics and, beyond that, in formal thought.' (Kelly 1998 q.v. value).

We learn from *A Dictionary of the Social Sciences* that '[c]lassical economics distinguished among "use value" (or how useful an item is to a given person or situation), "exchange value" (which reflects its price on the market), and "labour value" (which reflects the amount of human effort invested in its production)' as well as that 'sociologists and anthropologists often have a completely different understanding of value' (Calhoun 2002 q.v. value). *A Dictionary of Sociology* informs us that 'distinctions are often drawn between values, which are strong, semi-permanent, underlying, and sometimes inexplicit dispositions; and attitudes, which are shallow, weakly held, and highly variable views and opinions.' It further states that certain sociological scholars, 'and Talcott Parsons in particular, overemphasize the importance of shared values in maintaining social order' (Scott and Marshall 2009 q.v. value).

A Dictionary of Economics is succinct in its entry, giving no hint of scholarly debates about the term. As a first entry it defines 'Value = price × quantity' and as a third entry it defines values as a '. . . general term of praise, used in a phrase such as "good value". Value in this sense refers to something similar to price but more important and more permanent' (Black et al. 2012 q.v. value). The latter refers to marketing, and *A Dictionary of Marketing* defines value as 'the benefit that a good or a service provides, as perceived and measured by the beholder' and underlines that '[w]hat one customer finds of value can be different from what another finds of value' (Doyle 2011 q.v. value).

On the other hand, in philosophy, *A Dictionary of Critical Theory* defines value as a 'measure for distinguishing the absolute and relative worth of a thing (an object or a service) both to its owner and to others' (Buchanan 2010 q.v. value). It further asserts that value 'is theorized in two main ways, as an ethical problem and as an economic problem.' *The Oxford Dictionary of*

Philosophy declares that acknowledging 'some feature of things as a value is to take it into account in decision making.' (Blackburn 2008 q.v. value) This dictionary further distinguishes between those 'who see values as "subjective"' and 'those who think of values as objective.' The former, it is posited, think of values 'in terms of a personal stance, occupied as a kind of choice', whereas those who hold values as objective, according to the dictionary, 'suppose that for some reason—requirements of rationality, human nature, God, or other authority—choice can be guided and corrected from some independent standpoint' (Blackburn 2008 q.v. value).

In our game above, it seems that value is Very Important, but also that it can pertain to use, exchange, labour, semi-permanent dispositions, praise, price × quantity, a benefit of a good or service, relative or absolute worth of a thing; it can pertain to economics or ethics, sociology, or philosophy; it can be objective or subjective. When does this game of looking up value end?!

Perhaps, in the search for a definition of value we find ourselves in the same unfortunate position as Socrates in trying to find an answer to what virtue is:

> How fortunate I am, Meno! When I ask you for one virtue, you present me with a swarm of them, which are in your keeping. Suppose that I carry on the figure of the swarm, and ask of you, What is the nature of the bee? and you answer that there are many kinds of bees, and I reply: But do bees differ as bees, because there are many and different kinds of them; or are they not rather to be distinguished by some other quality, as for example beauty, size, or shape? How would you answer me? (Meno by Plato)[1]

Figure 1.1 graphically summarizes the kaleidoscope of definitions of value in the social sciences. The notion of value is evidently at the heart of several different avenues of inquiry and distinctions as well as a source of scholarly disagreements.

What to look for when looking for values?

> Do you know what that is asking? You are asking, 'Could you tell me, without knowing what kind of world we are in, what a ... [value] would look like?'
>
> (Lynch 1993: 144 citing Harvey Sacks. Our insertion of value.)

Alas, we do not offer a substantive theory of values: we do not propose an account of what values are, or a definition of them. If this is what you need, other approaches apparently furnish a multitude of answers (a starting point

[1] Thanks to Sergio Sismondo who suggested the similarity between Plato's discussion of virtues and our discussion of values.

Quod vide Value, or looking up value

Ferdinand de Saussure introduces the concept of linguistic value in his Course in General Linguistics (1916), in which it acts as the fulcrum of his general thesis about language.

Encyclopedia of semiotics (Bouissac 1998)

Classical economics distinguished among use value ..., exchange value ..., and labor value .... // Sociologists and anthropologists often have a completely different understanding of value. They generally speak of social or cultural values as mechanisms of solidarity and collective identity.

A dictionary of the social sciences (Calhoun 2002)

Ethics uses value as a means of determining the difference between the various ideas and concepts impacting on everyday life, such as the notions of freedom and life which come into conflict over issues like the right to life of the unborn foetus versus the right to decide of the mother...

A dictionary of critical theory (Buchanan 2010)

Regarding values as a type of social data, distinctions are often drawn between values, which are strong, semi-permanent, underlying, and sometimes inexplicit dispositions; and attitudes, which are shallow, weakly held, and highly variable views and opinions.

A dictionary of sociology (3rd ed.) (Scott and Marshall 2009)

1 Value = price × quantity. ...
2 A synonym for price. The theory of value is concerned with explaining the relative prices of goods.
3 A general term of praise, used in a phrase such as 'good value'. ...

A dictionary of economics (4th ed.) (Black et al. 2012)

[T]he term [value] seems to name an aspect of the world so fundamental to our thinking —so elementary and at the same time so general—as to be both irreducible and irreplaceable.

Encyclopedia of aesthetics (Kelly 1998)

To acknowledge some feature of things as a value is to take it into account in decision making, or in other words to be inclined to advance it as a consideration in influencing choice and guiding oneself and others.

The Oxford dictionary of philosophy (2nd rev. ed.) (Blackburn 2008)

The benefit that a good or a service provides, as perceived and measured by the beholder.

A dictionary of marketing (3rd ed.) (Doyle 2011)

Excerpts of entries q.v. value

Figure 1.1. Quod vide value, or looking up value

for finding a definition could be *Quod vide value* above). We attempt to offer a different approach to the study of values; an approach that attempts taking the making of values in practice seriously. We are asking the analyst to suspend his/her preconceived notions of what values are and how values are maintained.

We believe that articulations, choices, exchanges, hierarchizations, sortings, displacements, and commensurations of values lend themselves to understanding how values are made. We argue that values should not be seen as intrinsic properties of objects, people, or cultures. Like ethnomethodologists study members' methods, 'ethnomethods', without recourse to structure as explanatory resource we request that the researcher pay attention to members' construction of values, that is, 'ethnovalues' (cf. Lynch 1993: 148). We argue that the making of values happens in practice and that therefore a study of values must pay attention to the bricolage of making them. Our injunction is to attempt an escape from familiar conceptualizations of value in our own cultures—and fashion emically sensitive descriptions of the exotic and mundane practices of valuation (On emics, see M. Harris 1976).

Regretfully then, the only answer we provide to the question of what to look for when looking for values is: look for *the practices of making value* (see following section entitled Values as enacted). Wherever. Whenever. However.

Values as enacted

Instead of helping tie tighter the Gordian knot of a definition of value, we ask what would happen if we pose the question of values differently. What would happen if we take value as something which happens in practices, rather than as a prefixed entity which explains action? What would happen if we saw the genesis, articulation, dispute, and settling of what comes to count as values as matters for empirical investigation and explanation? For instance, the above definitions would function well as entry points—as textual practices—for a study of values in different disciplinary cultures. Thus we would move from an etic register attempting to bolt down what a value is, to an emic register asking how values are made.

We suggest that this more pragmatic, practice-based definition of values allows us to see the world differently. Rather than thinking of values as belonging to different domains—price as economic value, fairness as an ethical principle, efficiency as an organizational ideal—we study them in connection with each other. This breaks down disciplinary barriers and taken-for-granted definitions of what types of values are commensurable, or even possible to study beside one another.

By studying the making of values traditionally seen as belonging to different domains we can see power struggles over which values are to be dominant, the making of boundaries between values (that may become made as separate), and when different values are made commensurable. We can see how coordination is achieved between systems of classification and valuation. How valuations and values shift over time: at one moment value is tied to a biomedical platform, at another it becomes linked to a patent application, at yet another it is connected to the possibility of medical intervention. We can also get a glimpse of how different values are made to occur in opposition to each other, for example, in a commercial ambulance dispatch that must weigh the economic costs against the medical needs. Or when values are tied to organizations and people we can ask: whose values are important?

This leads us to embrace an analytical perspective where values are seen as the outcome of work, where they are the upshot of a wide range of activities. Fairness, to take an example that has frequent currency in the provision of health-care services, is thus seen as a value inseparable from the work of evoking, implementing, and organizing it. This emphasis on practices is very much inspired by the pragmatist stance of John Dewey which has been understood a 'flank movement' in approaching values (Dewey 1913; Muniesa 2012). Dewey emphasized the role of valuation as an activity, asking whether values are prefixed or whether they are inseparable from activities of valuation: 'Do values antecede, or do they depend upon valuation—understanding by valuation a process of reflective estimation or judgment?' (Dewey 1913: 269). His own orientation was clearly towards emphasizing the process of valuation (Dewey 1939). Our answer to the question is along similar lines: we need to see values as depending upon and being enacted by a wide range of activities, where valuations of various forms are part of such activities.

Our use of the term 'values' in the plural relates to the traditional sociological sense of the word, according to which values are seen as conceptions of what is good, proper, and desirable (cf Graeber 2001: 1–2). Anthropologist David Graeber (2001) explains that this use of values in the plural has traditionally been distinct from a notion of value in the singular where it has been used in an economic manner to signify 'the degree to which objects are desired, particularly, as measured by how much others are willing to give up to get them' (cf. Graeber 2001: 1). Graeber insists on relating conceptions of what is good, desirable, or important to economic processes and other processes of meaning-making:[2]

> When anthropologists nowadays speak of 'value'—particularly, when they refer to 'value' in the singular when one writing twenty years ago would have spoken of 'values' in the plural—they are at the very least implying that the fact that all these things should be called by the same word is no coincidence. That ultimately, these are all refractions of the same thing. But if one reflects on it at all, this is a very challenging notion. (Graeber 2001: 1–2)

Our use of the notion of 'values' in the plural joins previous efforts made to break down the distinction between social and cultural *values* and economic value (see, for instance, Appadurai 1986; Aspers and Beckert 2011; Stark 2000; Zelizer 1978, 1981, 1989, 2005). However, we argue that Graeber's problem is posed in the wrong manner. While he suggests that different manifestations of value ultimately refer to the same thing, we instead argue that we need to attend to how economic value and social/cultural values are *made* separate in

[2] For Graeber the challenge furthermore entails understanding the limitations of how anthropologists have addressed this challenge and proposing a better way to untie the complex conceptual knitted piece of values.

practice. When the distinction between values (social, cultural) and value (economic) is dissolved, we cannot use one concept to explain the arrangement of the other. Economic value can no longer be seen as a determinant of social and cultural values. Conversely, social and cultural values can no longer be seen as determinants of economic value (as for example in traditional economic sociology and marketing thought; see Baker 1984). What is gained is an opening up of a space to explore recurrent intersections between different values, for instance, between matters pertaining *simultaneously* to the ethical and the economic.

There is thus a strong case for studying the enmeshing of economic and other values. This stance is also reflected in an influential paper where economic sociologist David Stark (2000) advocated reconsidering what he called Parson's pact. From the mid-twentieth-century this pact specified that value was the object of study for economists while values in the plural were for economic sociologists to study. The pact had had its benefits, Stark admitted, but it relied on the limiting distinction between value and values. Inspired by the work of Luc Boltanski and Laurent Thévenot (2006), Stark (2000) advocated for a sociology of worth that did not recognize the strong distinction between economic value and other values, permitting investigation of many relations between various forms of worth. This in turn opens up the topicalization of practices in the life sciences as summoning and producing both economic and other values, and avoids reifying economic practices as different in kind from social or scientific practices.[3]

The proposition that we examine *all kinds of values* as upshots of practices may appear utterly degrading to any notion of fundamental moral values. Can one really avoid making any strong a priori distinctions between the qualities of values? Are not some values clearly incommensurable with others? Are we, to put it bluntly, selling out ethics and aesthetics by proposing to understand them as potentially no different from say the value of a cupcake?

The answer is 'no'. We are not trying to fit ethics into an economic equation: we instead consider how different values are articulated and sometimes considered commensurate (or incommensurate) in different situations (on commensuration see Espeland and Stevens 1998.) We observe, and indeed want to highlight, that this is what actors, algorithms, and institutions do in many contemporary practices (see a conspicuous example of this in the allocation of donated organs analysed by Philip Roscoe in Chapter 5). Again, this is why the pragmatist stance, and in particular its focus on valuation as a social practice, is so helpful. Such an approach allows us, for instance, to

[3] Such reifications are, incidentally, what happens either when the economic is ignored or when it is foregrounded as something preferentially studied in peculiar 'economic' settings such as those of financial markets.

examine competing processes of, and orders of, valuation as a means to examine the articulation (literally) of competing sets of values. It also opens up questions about the technologies, devices, and organization of valuations, which in effect constitute assemblages enacting values. (For more extensive discussions about valuation as a topic of study, see the overviews in Kjellberg et al. 2013; Lamont 2012.)

So our point is *not* to argue that ethics has been economic all along, or that morals can be reduced to social or economic arrangements; instead, we believe that people and institutions deal with ethical and economic concerns in complex ways, and that they try to arrange these values in acceptable combinations or 'matches' (Zelizer 2005), among others by assigning meaning and boundaries to situations, actions, and relationships. Whereas what counts as ethical does matter for what counts as economic, and vice-versa, these values are often composed differently in practice, and thus also remain in tension with each other. In this, we wish to caution against the complete analytical naivety proposed by certain pragmatists (cf. Latour 2005) who tend to disregard the historical regularities of society, as well as against the complete and utter reification of values as being decidedly economic/social/cultural.

By narrowing the gap between social and cultural values and economic value we are acknowledging the commonalities between them. They both denote the desirability of certain acts over others; and they both refer to the collective production of that desirability and its governing effect on individual actions. This reminds us that values are not simply a cognitive concept of ordering. Values also have an emotional valence. The desirability of valuable acts and things is not reducible to a hierarchy (of categories of actions, things, etc.) of meaning-making—cognitive knowledge. Rather, value/s refer to orderings that are simultaneously of meaning/knowledge and emotional. Whereas detailed explications of desirability fall beyond the scope of the present book (and would require going into psychoanalytically inspired theories, for instance), we shall stay with the observation that value/s denote and produce the desirable. Desirability must then become plural, as competing orders of desirabilities: different values are made beside each other.

The present book approaches the life sciences with a provisional and agnostic definition of values. We purposefully leave as an empirical question what comes to count as a relevant order of value in given situations, practices, socio-technical systems, institutions, and professional cultures. While this may appear somewhat disorienting to those committed to definitive versions of values, our intention is to use this agnostic position to allow us fruitfully to address a few well-grounded matters of concern.

From matters of concern to stakemaking

Recurringly, scholarly work and media coverage highlight central areas of conflict and tension in relation to values in life sciences: money, fairness, efficiency, and priority are constantly put into play in complex negotiations over value. Accordingly we echo Bruno Latour's call for a renewed critique addressing matters of concern rather than matters of fact (Latour 2004). We believe that inquiries into the workings of values afford us a critique that *adds* to the concerns it addresses—for instance commodification processes in the life sciences—rather than merely debunking them. Latour writes 'Give me one matter of concern and I will show you the whole earth and heavens that have to be gathered to hold it firmly in place' (Latour 2004: 246).

Maria Puig de la Bellacasa's work with matters of care—the emotive dimension of matters of concern—is also of particular significance in this regard. In her work she grapples with questions of commitment and attachment (de laBellacasa 2011). For de la Bellacasa this means that we need 'to engage properly with the becoming of a thing, we need to count all the concerns attached to it, all those who care for it' (de laBellacasa 2011: 90). But rather than just counting concerns or attending to those who care, we wish also to call for attention to the making of concerns; the production—in practice—of what comes to count as valuable, desirable, or otherwise worth caring for. To emphasize a shift in attending to the ready-made, a connotation which is present in both matters of concern or care, we wish to call attention to the making of such matter—the making of stakes, or as we choose to call it, stakemaking.

We argue that value practices are crucial practices by which people and things make stakes, matters of concern, or matters of care—or displace them. Sometimes, values and their practices are salient matters of concern themselves—salient things that have to be brought and kept in place through complex compositional processes. Below we introduce the three main areas of concern for the present book, and our endeavour.

HOW ARE MATTERS OF CONCERNS MADE? VALUATION PRACTICES AND STAKEMAKING

> Restrictions covering sheep movements after the Chernobyl nuclear disaster have finally been lifted from all farms in England and Wales after 26 years. After the 1986 disaster, the Food Standards Agency (FSA) placed controls on 9,800 UK farms, but these were gradually removed. The final eight in Cumbria and the last 327 in Wales are now free of them. Adam Briggs, from the National Farmers Union, said it meant an end to the 'sorry situation' of the Chernobyl legacy.
>
> (Anonymous 2012)

In the wake of the Chernobyl accident, in April 1986, the sale of sheep from Cumbrian hill-sheep farmers was severely restricted due to the presence of high levels of cesium. For some farmers there was great uncertainty as to whether the nuclear fallout affecting the Cumbrian sheep resulted from the proximity to the Sellafield nuclear plant. However, the British government's culture of reassuring the public and projecting certainty and control permeated government communication. The stakes were high: government officials wanted to handle risks of nuclear fallout and reassure the public, while hill-sheep farmers' livelihoods were put at risk (cf. Wynne 1989). This situation of unsettled stakes—unsettled matters of concern—repeated itself almost uncannily after the Japanese Fukushima disaster on 11 March 2011. The uncertainty about radiation damage produced widely diverging interpretations of what were supposed to be the matters of concern:

> The divergence of opinion has led to divisions among families, generations and communities. 'Should I stay or should I go?' is a question that weighs heavy on countless minds. It is why hotels in north-eastern Japan are struggling to attract tourists. It explains the rash of postponed visits by foreign dignitaries to Tokyo. And it is a particular worry for those whose DNA is most vulnerable to change: expectant mothers and young children. (Watts 2011)

The Chernobyl and Fukushima catastrophes can both be interpreted as struggles over matters of concern or care. What is at stake is not a settled matter. People struggle to make sense of knowledge, companies struggle to survive, the government struggles to provide guidance and aid to those who need it. In both cases, the economic and epistemic work of central government is pitted against local work, knowledge, and experience. The concerns vary widely. Valuations of life, knowledge, and money become matters of concern. Whose assessments of radioactive fallout are valid? Whose valuation of lives, quality of life, and livestock? Assessments of different values are intertwined: economic value, risk values, health values, quality of life.

In addition, in the making of concerns, the emergent expectations around science and technology have been shown to be of crucial importance (Adams et al. 2009; Borup et al. 2006; N. Brown and Michael 2003; Fukushima 2012; Simakova and Coenen 2013; Tutton 2011). The promissory rhetorics of science and technology perform certain worths, are involved in making matters of concern or care. The promises of lower radiation levels, of health, safety, and a good life are all part and parcel of performing certain values around which matters of concern revolve.

In this volume we pose a number of questions related to *stakemaking*—the production of stakes, concern-making, and care-making. These questions move orthogonally to knowledge practices, but are not separate from them. They are about the production of concerns through the production of values:

- How are matters of concern performed through economic valuation practices (e.g. Clarke et al. 2003)?
- How are matters of concern performed with technical, classificatory, and institutional systems (Bowker and Star 1999; Star 1991)?
- How are matters of concern performed through the use of scientific theories, reflecting world views that include what/who is worth living and/or taking into consideration (Fujimura 2013)?
- Whose voices, worldviews, needs, and interests come to shape matters of concern (Callon and Rabeharisoa 2003, 2008; Galis 2006; Galis and Lee 2014)?
- How are performances of the values of objectivity, impartiality, non-bias, and calculation mobilized to make matters of concern (Haraway 1988)?
- How are subjectivities and affect co-produced with matters of concern (de laBellacasa 2011; Skeggs and Loveday 2012)?

Our theoretical interest in the production of matters of concern stems from an interest in the relation between valuation practices and questions of power, social order, and the production of subjectivity and affect. However, as we have already hinted, we do not want to treat values as an external force that explains power, order, or subjectivity. We instead call for a sensitive exploration of how valuation practices are intertwined with stakemaking in the life sciences—and more broadly, in science, society, and technology. How are valuations part of making stakes? How are matters of concern held in place? Valuation practices which implicate knowledge, expertise, and economy have far-reaching effects for who is in a position to produce stakes, to know, to influence, to gain—and who is not.

WHAT IS WORTH KNOWING? VALUES AND THE EPISTEMIC

> What happens if I get placebo and TroVax® is then shown to work? Although the five-year survival rate for metastatic renal cancer is less than 5 percent, the answer [in the patient pamphlet] to the question is given as: 'If the study shows TroVax® prolongs survival and you received the placebo [in the trial], you will be given the opportunity to be treated with TroVax®, following regulatory approval.'
>
> (Jain 2010: 89)

A randomized controlled trial (RCT) testing a new cancer drug might be valued as an invitation to both patients and oncologists to 'live in a space organized through both hope and progress' (Jain 2010: 90). Patient and oncologist hopes for at least a prolongation of days of living are entangled with the large-scale production of data for purposes of producing scientific knowledge of the efficacy of a new drug. The clinical trial entails the enactment of many values; in the recruitment, patients and their oncologists enact values of 'hope and progress', while the overall trial enacts values relating to the prospect of acquiring new scientific evidence of the efficacy of the drug.

In short, as Lochlann Jain (2010: 90) clearly demonstrates, the trial exists in a nexus of valuations that enact different and even divergent values. Most poignantly in this case, the patient pamphlet understates the temporal scope of the trial as well as the need for a certain number of trial subjects to die for the trial to provide viable knowledge. Setting these particular misrepresentations aside, the example vividly illustrates how many different and divergent values may be in play in relation to knowledge and knowledge production in the life sciences. Values of life and living are intertwined with the values of a particular scientific result and indeed with a particular mode of knowing, such as the calculability of life (and death).

It is therefore urgent to address questions such as:

- Who and what comes to decide what is worth knowing, and what it is worth (cf. Helgesson and Lee 2012)?
- What values are articulated and given weight when deciding what is worth knowing?
- With what means do data, methods, and objectives acquire value?
- How are valuations pertaining to the economic intertwined with the articulation of values in choosing subjects, objects, and methods for pursuing knowledge?

Knowledge and its means of production are linked to values and valuations in numerous ways. The presentation of scientific awards such as the illustrious Nobel prize, or the more mundane granting of grants (cf. Lamont 2009), are but two expressions of the array of valuations performed as part of science. Valuations are also deeply ingrained in the very practices of scientific work (see the contributions by Dussauge, Helgesson and Johansson Krafve, Lee, and Widmalm, in this volume).

There is thus a strong case for taking an interest in the links between values and the epistemic. To the extent that the classic sociology of scientific knowledge (SSK, see, e.g. Bloor 1976) was interested in these links, it was primarily to investigate how the values of certain social groups became imprinted in knowledge claims: values (and interests) were treated as explaining the stabilization of scientific knowledge. Our question is rather how values are made in, and in relation to, science.

Our theoretical interest is directed towards the enactment and stabilization of values in relation to the epistemic. Briefly stated, values are central to what urgently needs to be explained. This means that our primary focus is not so much on how values might guide the scientific gaze, but instead on how values are formed and articulated in, for instance, processes of valuation related to the epistemic. Even more urgently, we need to investigate the practices for dealing with shifting and conflicting values.

HOW MUCH IS IT WORTH? VALUES AND THE ECONOMIC

> On Saturday 2nd December Pfizer decided to end clinical trials of torcetrapib, a drug that promotes the creation of the good variety of cholesterol, known as HDL. Initial results had shown that the drug was associated with an unacceptably high rate of death among users. Only two days earlier Pfizer's research chief, John LaMattina, had described the drug as 'the most important new development in cardiovascular medicine in years.' On Monday investors, who had hoped that torcetrapib would be a commercial blockbuster, knocked more than $25 billion off Pfizer's stockmarket capitalisation, reducing the firm's value by one-eighth.
>
> (The Economist 2006)

The fate of the Illuminate trial indicates how financial values can be entwined in many ways in medical research. The decision to end the trial followed the release of information from a monitoring body that eighty-two trial subjects receiving torcetrapib had died versus only fifty-one subjects in the control group (Berenson 2006). The premature ending of the trial meant the loss of almost $1 billion invested in the drug and reduced the company's market capitalization even more. This is a striking and large-scale example of the links between medical research and financial markets. At one point, the trial involving some 15,000 subjects represented an investment based on promises of future revenues derived from knowledge about how torcetrapib might improve treatments for preventing heart disease. Following the news about the disturbing extent of deaths among trial subjects, the trial meant financial losses as well as a realization that the sought-after knowledge would not materialize. Not only does the very existence of the trial certify that something had been considered worth knowing, but the valuations pertaining to both its realization and premature closing clearly engaged and enacted a variety of both economic and non-economic values.

The life sciences do not only entail important intermingling between the economic and non-economic values, as illustrated by the Illuminate case. The emergence of biobanks provides an important illustration of how endeavours in medicine and the life sciences mix with various economic domains, such as commercially driven research and the public sector. For instance Catherine Waldby and Robert Mitchell (2006: 80) describe the configuration of the UK Stem Cell Bank, in which depositors agree that cell lines must not be sold for financial gain and where public sector researchers will pay marginal costs for cell lines while commercial users pay full costs. This ensures that 'the knowledge and therapies generated will be available to fellow citizens' (Waldby and

Mitchell 2006: 80). This particular initiative aims, through the use of specially designed rules and agreements, to create an economy for the exchange of cell lines which operates quite differently from a more commercially driven tissue economy. While being an economy including economic values—marginal costs, full costs, etc.—it clearly aims to enact a different set-up of values than would another form of tissue economy.

In attempting to analyse the making of economic values alongside other values it is thus urgent to address questions such as:

- How are knowledge and life transformed when they become described and treated as primarily subject to economic goals and constraints?
- How and with what means are economic values made in practice?
- How are boundaries and links made between notions of economic, epistemic, and cultural values?
- How are different values (economic, cultural/social) performed as distinct, intertwined, homogenous, or heterogeneous (Brown 2013)?

Financial losses and gains are intimately linked to what is known and unknown, to what trials do and do not demonstrate. However, as we can begin to anticipate from the above examples, economic values, in the plural, can go beyond plain profit, market capitalization, and differentiated pricing: for instance, values might involve maximized public health, or the efficiency of systems and interventions, and so on. In the above examples, economic value is complexly entangled with non-economic values: commercial decisions about new drugs are made alongside evaluations of the risks of patients dying from side effects. People wanting to contribute their bodies to the possible generation of therapies that aim to help fellow citizens might also generate profit for manufacturers.

What makes these values economic is that actors articulate them as such: through explicit categorizations; through the attribution of expertise on an issue to economic experts, etc. It is in this sense that we use the concept of economic values in the following. A crucial question is therefore how science, knowledge, and health-care practices are routinely assigned economic value, not only when pricing a drug or assessing the marginal or full costs of supplying a cell line. The making of economic values is deeply ingrained in the practices of research and health-care provision and moreover enacted with/against other values. Our main concern here is how economic value is continuously constituted as a steering mechanism (and object of critique) in biomedicine and health services (see, for instance, Johansson Krafve 2012; Moreira 2012; Sjögren and Helgesson 2007), and how the assembling of economic value tends to dominate the assembling of other non-economic values (cf. Chapter 6 by Zuiderent-Jerak et al. in this volume).

We wish to articulate a perspective which takes very seriously the composition of values and the making of boundaries between them—be they fashioned as economic, cultural, or epistemic. But we also wish to emphasize that it would be dangerous not to pay attention to the repertoires in which different values are articulated. One danger in particular would be to a priori reduce them to effects of one another—or to give analytical precedence to one set of values. Just as we cannot simply replace one set of values with another, Latour (2004) has warned against the danger of simply replacing certain matters of fact with other facts. The question is how to take seriously the politics of the boundaries between different values that are created in practice.[4] Attending to matters of concern implies attending to the complex, concrete, local, and global relations between value and values.

Two precursors of the study of values in the life sciences

As outlined above, this volume proposes a vantage point from which to observe how different values in the life sciences are enacted in actions, technical practices, and valuation practices. Looking at how such 'value practices' enact values provides a route to investigate the entwinement of, and conflict between, diverse values, be they economic, moral, legal, etc. This overall theoretical starting point represents a pragmatic understanding of values which distinguishes it from the main strands of research in the area, centring around the notions of 'biocapital' and 'moral economy'.

BIOCAPITAL: THE ENMESHING OF BIOLOGY AND ECONOMY

The enmeshing of capitalism and biotechnology has been a growing concern for science studies, ever since Edward Yoxen (1981) brought attention to the increasing operation of capital at the molecular level of the biotic: '[A] specific mode of the appropriation of living nature—literally capitalizing life' (Yoxen 1981: 112). The move to analyse these processes has enjoyed many labels

[4] In its traditional manifestations, science and technology studies (STS) has made few contributions concerning economic orderings and valuations. The emergence of STS-influenced studies of markets (Callon 1998; Callon 2007), most notably financial markets (MacKenzie 2006); MacKenzie and Millo 2003), can be seen as a move to take the economic seriously. However, this shift of attention to markets and in particular to finance actually underlines the traditional inattention in STS to economic practices in the domains of science and technology. Alas, it is only in the 'biocapitalism' literature that we find the most contributions to theorizing the relationship between economic and other values in biomedicine (see Part III on biocapital).

ranging from biovalue, genetic/-omic capital, to bioeconomy and biocapital. The common issue is the attention to the labour and commodification processes which distinguish the production and promotion of biotechnological products (Helmreich 2008).

The analysis of biocapital brings together a disparate cluster of perspectives ranging from Marxist and Weberian to post-Foucauldian analyses, whose common denominator is awareness of a change in the relationship between the business world and the modes of organizing the life sciences. Two works, Kaushik Sunder Rajan's *Biocapital* (2006) and Nikolas Rose's *The Politics of Life Itself* (2007), have emerged as key texts to come to grips with these developments.

Sunder Rajan (2006) compares the US and Indian biotechnology industries in seeking to understand the capitalist practices and subjectivities that emerge in the porous relationship between the academic and the corporate world. He hones in on the promissory rhetorics of biotechnology, the colonialist moulded flows of inequality, and genomics as information science. Rose's (2007) approach is different, but with similar theoretical predecessors. Rose's main interest is to explore the emerging 'somatic ethics' of biocapital and the processes of subjectification in relation to life-science technology. The 'biocapitalism' literature thus proposes one way to theorize the relationship between life sciences, power, and social orderings in late capitalist culture.

However, with Helmreich (2008) we might ask how we could look beyond capitalism as the stable signifier towards other complex arrangements of values. How do we move beyond re-inscribing capitalism as the stable entity around which all other values are ordered? What would happen if we with Kristin Asdal (2011) start asking questions on processes of biocapitalization instead? Another issue which sets our approach apart from that of the biocapitalism literature is our point of expanding the Marxist flavoured analysis of value as accruing from labour (see, for example, Mitchell and Waldby 2010) by paying attention to the enactment of value in different actions and practices. By broadening the issues that the biocapitalism literature outlines, we emphasize values, not as a static category revolving around capital and labour, but as a heterogeneous phenomenon that must be understood in its many guises.

VALUES IN THE MAKING OF SCIENCE: THE MORAL ECONOMY PERSPECTIVE

The notion of moral economy, originally introduced by E. P. Thompson (1971), was reintroduced in the 1990s into the history of science as a way to capture the various norms and values that regulate scientific activities and structure communities. As used in recent work on scientific practices, *moral economy* refers to a system of exchange based on certain principles of fairness and means of control regarding access to the resources scientists need. The moral economy concept, resurgent in an increasing number of publications (Atkinson-Grosjean and Fairley 2009; Daston 1995; Kohler 1994; Lock 2001; McCray 2000; Strasser 2011), encompasses notions of moral values and notions of how a scientific community should conduct various kinds of exchange, including what rewards are appropriate for productive work:

> Moral conventions regulate access to tools of the trade and the distribution of credit and rewards for achievement. As the moral economy of eighteenth-century English laborers was rooted in concrete, historical systems of agricultural production and marketing, so are the moral economies of experimental scientists rooted in specific configurations of materials, literary, and social technology. (Kohler 1994: 12)

The notion of moral economies in science is an interesting attempt to capture and characterize a given scientific domain in terms of a particular set of values. Lorraine Daston (1995) proposes a somewhat different conceptualization with a much wider focus on how cultural values shape the epistemic aspects of scientific practice: 'Moral economies . . . are integral to science: to its sources of inspirations, its choice of subject matter and procedures, its sifting of evidence, and its standards of explanation' (Daston 1995: 6).

This conceptualization opens up for examination the intertwining of values in scientific practice in which economic, epistemic, and methodological values are in practice seen as facets rather than distinct categories (cf. also Atkinson-Grosjean and Fairley 2009; Rasmussen 2004). Daston takes pain to emphasize how the notion differs from that of the Mertonian norms, stressing precisely the historically contingent nature of moral economies: 'In contrast to Mertonian norms, moral economies are historically created, modified, and destroyed: enforced by culture rather than nature and therefore both mutable and violable; and integral to scientific ways of knowing' (Daston 1995: 6).

Yet, despite this emphasis on contingency, the notion of moral economy seems often to be used to focus attention on how values *regulate* action. In other words, it is put to greater use in characterizing what appears to *guide* valuations and other behaviours in a given scientific field, rather than focusing on how such composites of values are shaped and sustained. What we have here is a notion that is frequently used to highlight the regulation of expectations and behaviour, and the historical contingency of such guiding regulations (values), but does not tell us how these very values are produced in the first place. The traditional use of moral economy thus carries with it the risk of becoming not only a concept for depicting the (actual) set of values manifest in a domain of scientific practice, but also for backgrounding the very making of such values.

The rudiments of a 'value practices' approach: The four themes of this volume

The two approaches discussed in the two preceding sections, The Moral Economy Perspective and The Biocapital Perspective, provide an influential contribution to the analysis and problematization of values in life science. However, we have argued that there is a pressing need to move beyond these conceptualizations of value in life science, in favour of a perspective which analyses the diversity of value practices in the life sciences.

First, we argue that there is a need for more pluralistic analyses of value. Rather than fixing economic notions of 'capital' (Franklin and Lock 2003; N. Rose 2007; Sunder Rajan 2006, 2012) or more recently 'assets' (Birch and Tyfield 2012) as the stable reference points for analysing value we wish to reorder the analysis to allow for analyses based on understanding values in the plural (Helmreich 2008). Thus, we wish to nurture empirically sensitive accounts about which values come to count as important at any given time or place. The question of which values determine (or corrupt) action are replaced with questions as to how values are made in practice. We thus wish to shift our analytical awareness towards a sensitivity to processes and away from a sensitivity to how values guide behaviour in different settings. Rather than taking an interest in which values are predominant in a given period or place (Daston 1995), we would urge for investigating the processes and practices that constitute borders, commensurations, or orders of values. We argue that values should be seen as always already constituted in practices, not as static entities which exist outside of action.

In terms of a pragmatist take on values, a suggestion from a number of authors in this volume would be to ask a novel set of questions that stresses the performative aspects of moral economization or biocapitalization. Here questions of how capital (or assets) shapes biomedical practice or objects are replaced by an interest in how emic notions of economy are made beside other notions of value in biomedical practice. Questions on the topography of specific moral economies are replaced by questions on how such topographies are constituted. How are values (in the plural) made? Thus, rather than leaving behind the important work done by scholars on biocapital or moral economy we wish to make an orthogonal move which takes as point of departure the making of biocapital or moral economies. With Kristin Asdal or Carrie Friese (both in this volume) we wish to inquire into biocapitalization (Asdal, Chapter 9) or moral economization (Friese, Chapter 8).

Second, the present volume aims to develop an approach that engages with the question of making multiplicities of values. In summary, the pragmatist approach we have begun to outline above investigates how values are made, rather than taking values as stable predefined entities. Values should be seen as the outcome of work (they are made, assembled, enacted in practice). This necessitates an analytical interest in what comes to count as value, and the processes that lead to this fleeting state of affairs. A corollary of this is that values cannot be seen as an explanatory factor, but rather something that needs to be explored and explained.

Through these pragmatist moves we wish to foster understanding of the composition of multiplicities of values: the establishing of relationships, the drawing of boundaries, commensurations and incommensurations, the ordering of hierarchies, and the production of desirabilities. In an effort to outline a 'value practices' approach to the concerns raised above, we aim to focus on how values are enacted: in actions, in technical practices, and in practices of valuation. In this we will try to account for both how peoples' actions draw on values, and how agents' actions and reactions come to enact values.

Third, our pragmatist approach to values crucially allows us to address three concerns in the fields we address.

Our first concern relates to stakemaking: the production of matters of concern or care. We wish to inquire into how the assembling of values is part of producing stakes. In the production of values, people, practices, and devices produce a transient fixation of what counts as important, valuable, or desirable. We argue that it is crucial to understand how matters of concern are continuously accomplished and held in place through the production of values.

Our second concern deals with how the making of values is intertwined with the making of facts. Here we do not ask how values guide the scientific gaze, but instead how values are made in relation to the epistemic. How is 'valuable knowledge' established? How do controversies about 'values' implicate controversies on facts? How do values and facts mesh or diverge?

Our third concern pertains to how economic values are made alongside other values. Here we wish to counteract the tendency to give precedence to a certain repertoire of value. The pragmatist approach we propose makes the question of determining values an empirical question. To ask whether social or economic values should be the basis for analysis is to miss the pragmatist mark. Attention to the making of stakes necessitates attention to the articulation and disarticulation of different types of values, be they economic, social, or something else altogether.

INTRODUCING THE FOUR THEMES OF THE VOLUME

This book gathers together fourteen distinct chapters on the various aspects of the life sciences and medicine. These chapters encompass contemporary as well as historical contexts and probe, in different settings, what comes to count as valuable, desirable, or condoned. The chapters of this book have different entry points and take different approaches. They develop theoretical insights useful for further studies, and may well not be fully coherent with one another. In other words, the contents of this book have not been, and are not, bound to follow the routes we outline above. This is why we will not at this point further articulate our proposal for a 'value practices' approach: We do that in the concluding essay, the last chapter of this volume.

CONFLICTED 'PUBLIC' VALUES

The first part of the book deals with the production of boundaries of acceptable behaviour in the life sciences. It comprises three chapters that ask: How is proper economic, medical, and scientific behaviour made in practice? How are economic goals and personal ambitions made compatible (or not) with the life-saving purposes and ethical practices of health care and medicine? How is the production of medical knowledge affected by economic interests and 'publication planning' by big pharma companies for profit maximization made acceptable? The chapters in this part examine the boundary work, the negotiations, and moral quandaries of delimiting and defining proper conduct in the biosciences, medicine, and the pharmaceutical industry. The chapters chart an uneasy terrain of value—where scientists and their actions are constantly evaluated by different yardsticks—monetary, moral, objective. These yardsticks become crucial in ordering and creating hiearchies. Understanding the measure of 'good science' involves including significantly different metrics, enacted in widely varying practices.

The first chapter of this part, by Sergio Sismondo (Chapter 2), explores how key opinion leaders (KOLs) in medicine delimit what is considered acceptable medical, scientific, and economic behaviour. At stake is the handling of economic and scientific 'conflicts of interests' in the marketing and publication of pharmaceutical products. Sismondo demonstrates how the KOLs construct boundaries between science/medicine and pharmaceutical marketing practices by referring to the objectivizing practices of the US Food and Drug Administration (FDA) and the randomized clinical trial. Here Sismondo uses the example of the KOLs to explore the conflicts of value between the pharmaceutical industry and medical research. He asks how KOLs work to reconcile their medical/research role with their close collaboration with the pharmaceutical industry. The chapter shows how KOLs use a belief in the mechanical objectivity of medical science to justify their actions. A belief in mechanical objectivity counteracts a fear of biased research, as objectivity is achieved through an impersonal FDA-machine, even making interested research a goal to strive for. In the words of one KOL quoted in the chapter: 'No conflict, no interest.'

Chapter 3, by Christer Nordlund, also examines the reciprocal exchanges between the pharmaceutical industry and scientists. However, Nordlund focuses on exploring the shared interest in collaboration at different levels: the material exchanges of resources such as laboratory equipment or urine, the exchange of laboratory services, as well as the exchange of reputation. Nordlund argues that the long-term exchange relationship between pharma and science must be constantly hidden from the public eye. The free exchange of resources between the two spheres is deemed highly problematic in terms of a medical code of conduct. Thus, rather than focusing on the justificatory repertoires of scientists, the chapter explores the mutual benefit that the exchanges between the pharmaceutical industry and science bestow on each party. The chapter shows how mutual interests in developing hormones directed the actions of both the pharmaceutical industry and researchers. However, the chapter also demonstrates how the quite different moral economies of science and of the pharmaceutical industry caused the public to perceive of the collaboration as less than ideal, making it necessary to downplay collaboration and stress the independence of scientists from industry.

Chapter 4, by Sven Widmalm, analyses the practices of rule enforcement in matters of priority, publication, and plagiarism. Here, analysis of a conflict over publication priority and its adjudication by a 'court of peers' provides insight into how scientific rules were negotiated and interpreted in the international biochemical elite. This process focused on the specific virtue, character, and conduct of scientists rather than on higher order social norms. Enacting the importance of the fact/value distinction, the process of judgement attempted to separate the establishment of factual statements from the evaluation of the same statements. However, the process of establishing factual publication priority was intimately intertwined with the evaluation of the character of scientists, and ultimately with upholding the legitimacy of the scientific system.

These chapters explore how actors struggle to define and construct ethical behaviour in practice. What are acceptable courses of action in science and medicine? What is a virtuous clinician? What is an upstanding scientist? How does s/he behave? Values do not precede behaviour but rather are constituted in practice.

MARKETS AS CARERS FOR HEALTH

The second part of the book examines the fostering of health services and medical science through market arrangements and the parallel making of the values of justice, utility, and efficiency. The three chapters in this part treat issues that are crucial in light of the contemporary trend in which markets are seen as means to create public value. How is competition made in practice?

How can the scarce supply of transplant organs be distributed fairly? How is quality counterpoised against to the cost of care in health-care provision? By what means and to what effects are market arrangements deployed when organizing the provision of health services? How are economic valuations performed as part of or alongside other valuations? How are economic values intertwined with epistemic and cultural values?

The studies in the second part examine the precarious work of fostering markets catering to the many problems and values of health care—be it in terms of generating competition among potential developers of a malaria vaccine (Chapter 7 by Neyland and Simakova); securing a good supply of transplant organs (Chapter 5 by Roscoe); or providing health-care services more broadly (Chapter 6 by Zuiderent-Jerak et al.). These studies contribute to the growing and diverse literature on the markets and economization in health care (e.g. Greener 2003; Johansson Krafve 2012; Jost 2007; Kurunmäki 1999; Kurunmäki and Miller 2008; Le Grand 2007; Mol 2008; Moreira 2012; Porter and Teisberg 2006; Porter 2000; Sjögren and Helgesson 2007).

A recurring subtext in conversations about markets in the provision of health care is that a number of diverging values are at stake in this area. Each of these three contributions critically and closely examines the efforts to organize health services and medical science with markets. Yet, instead of remaining focused on abstract principles of markets and health care, the merit of these contributions is their impertinence in looking more closely at the challenges faced, the efforts made, and their multiple and uncertain consequences. While stopping short of concluding that health-care markets do not work in principle, these contributions emphasize the uncertainty of their outcomes in practice.

The first chapter of this part, Chapter 5, by Philip Roscoe, deals with the precarious construction of organ allocation algorithms, which entails integrating different values, such as justice and fairness, utility, and exchange value, into prosaic algorithms. 'Who should get an organ?' is the crucial issue at stake. Here, Roscoe examines how, in the case of protocols for allocating transplant organs, surgeons integrate moral valuations with prosaic notions of utility and exchange value. Research into the plurality of value regimes has shifted attention from a subjective, marginal utility approach, in which commodities have value only in as much they are valued by buyers; the chapter argues that non-financial arrangements for exchange may still maintain some of the characteristics of a marginalist economy, in which exchange value is a product of scientific coding. It invokes the concept of a moral economy to frame the negotiations between exchange value, and considerations of right and wrong, of inherent worth, scientific knowledge, and the rights and responsibilities of patients and practitioners.

The next chapter (Chapter 6), by Teun Zuiderent-Jerak, Kor Grit, and Tom van der Grinten, examines the valuation practices by which multiple values (such as quality and cost) are integrated into market practices of the Dutch healthcare system. The chapter demonstrates how, despite the availability of a multiplicity of 'valuemeters', quality of care is still framed in cost-saving terms. The authors thus provide an alternative to the usual polarizing discussion about the merits of markets in biomedical domains. By examining valuation practices Zuiderent-Jerak et al. investigate how experimental valuemeters might bring multiple worths into market practices. Empirically the chapter scrutinizes the relationship between markets and public values in Dutch hospital care. The authors demonstrate that although the policy aim of the financial instrument analysis was to create care products that would stimulate competition over quality, and ensure affordability and accessibility of care, this market device profoundly influenced how public values such as quality were defined in practice, i.e., quality was shaped in cost-saving terms. This happened not in the absence of, but despite the wide availability of precisely the valuemeters (Latour and Lépinay 2009) that according to both health economists and social studies of markets scholars claimed would enable the articulation of non-financially defined quality. The authors conclude that a narrow focus on either 'devizing valuemeters' (as proposed by authors from social studies of markets) or 'reducing information asymmetry' (a key strategy in health economics) is insufficiently politically sensitive. To account for a wider range of public values in health-care markets, value practices need to focus on the political process of articulating and disarticulating certain values.

Daniel Neyland and Elena Simakova have written the final chapter (Chapter 7) of this part. It focuses on attempts to introduce a market for malaria vaccine, and how this market is framed as an ontologically singular means to incentivize vaccine research. However, the markets' ontological singularity slips away, becoming messy, unsettled, and uncertain amid market experimentation. Despite this disjunction between the ontological singularity of the market and the multiplicity of market experimentation, the market is still claimed to be the best means to produce malaria vaccines. Here, Neyland and Simakova focus on how those involved in attempts to produce a malaria market continually work to manage slippages between apparent ontological singularity in market framing (that the nature of things are settled and well known) and multiplicity in market experimentation (in which the nature of things appears to become messy, unsettled, subject to new questions and assessment). In research, these slippages are noted by economists, scientists, and policymakers. However, despite any counter-expectation that such slippages should lead to criticism or even abandonment of the market as a focus for managing malaria, the market (broadly construed) continues to be heralded as the solution to the problem of malaria. The chapter critically examines how health-care concerns around malaria are made amenable to a 'world of worth' (Boltanski and Thévenot 2006) manifested in the market.

Here we explore the how the topic of markets as carers for health involves the enactment of several multiplicities of values. How is the market made to relate to other values, such as public health, justice, fairness, quality, or public image? The production of values around efforts to make markets produces stakes: what is and is not important?

VALUING HUMAN AND NON-HUMAN BODIES

The third part of the book examines how the worth of various lives and life forms are crafted focusing on the management of human and non-human populations through health technologies and biotechnology. The part grapples with questions such as: What lives come to count as being worthy of existence? What life do we—experts and laypeople—endow with the value of living? The chapters of this part address widely different practices through which life-worthiness is made. This part highlights transformation, as nineteenth- and twentieth-century institutions for managing human and non-human populations for modern purposes adapt to the twenty-first.

The first chapter in this part (Chapter 8), by Carrie Friese, examines 'genetic value' in the contemporary breeding practices of zoos struggling to maintain their place in a new century. Practices of animal breeding for conservation purposes in zoos lead Friese to ask: What populations do zoo scientists and other experts regard as valuable, or as surplus, and within what moral and economic orders do these values emerge? In this chapter, Friese explores the logic of genetic value and its socio-historical context. The chapter addresses how zoos exploit the life sciences in pursuing changing economies in wild animal bodies. The focus here is on the production of 'genetically valuable' individuals in what appears, on one hand, to be a techno-scientific economy comprising laboratory skills and research materials. On the other hand, genetic value also expresses the ways in which the zoo asserts itself as a moral institution, one that remediates its past errors which contributed to the endangerment of many wild species. Through this moral economy, zoos intend to create, exchange, and preserve endangered animal cells and bodies in order to contribute to species preservation. Friese considers the iterative relationship between these two different kinds of zoo economies in asking how moral economies and techno-scientific economies are co-constituted.

In the second chapter of this part (Chapter 9), Kristin Asdal interrogates the Norwegian efforts to create a new food product, the farmed fresh cod. Asdal suggests that the biocapitalization of the cod requires a 'co-modification' of the living entity and the market for it. Asdal asks how life and markets are co-modified to harvest value from the sea, and what kinds of values are enacted in this process. Asdal engages with cod farming and explains how the life underwater is cultivated to produce future values. The powerful locus from which such (potential) future values spring is not only the life sciences 'in the field', but also series of policy documents and strategy plans (including research and innovation policy documents) that value and evaluate such efforts. Asdal argues that these documents are devices for producing certain regimes of worth (Boltanski and Thévenot 2006) and for the timing and taming of future values. Asdal argues that these political devices are devices for 'co-modifying' markets and biological entities, and that they also operate on the values of the science and scientists involved.

The last chapter in this part (Chapter 10), by Ilana Löwy, analyses the historical implementation of and regulations for human prenatal screening in the late twentieth century. More specifically, Löwy investigates the conflicting values driving the framing of Down's syndrome as an epidemic in the history of prenatal diagnostic technologies. Löwy asks: On the basis of what moral, scientific, and health expert values was prenatal screening driven by professional and non-professional actors? Thus Löwy also raises the political question of how the values of the less-in-power (e.g. laypeople anxious to give birth to healthy children) can be exploited to induce people to accept measures driven by an entirely different set of values (i.e., a public health framing of Down's syndrome as an epidemic to be eradicated). Löwy examines the consequences of the generalization of prenatal diagnosis (PND) and its incorporation into routine pregnancy monitoring. In industrialized countries, the expansion of the PND was closely related to the wish to 'prevent' a specific condition, Down's syndrome. In countries where abortion is legal, PND and associated testing technologies gave rise to expectations that the sum of individual decisions would reduce the burden of handicap in populations, transforming the 'risk of disability' into a privately managed public health problem. In countries where abortion is illegal, management of the 'risk of disability' remains in the private—and privatized—sphere. This chapter addresses the intersection of public values, health policies, technologies of biomedicine, ethics, and private practices.

This part illustrates that biotechnologies of human and non-human animal bodies, as well as health policy and expertise, partake in systems of value attribution that define, order, and rank which bodies are worth their lives—and what those lives are worth to others.

VALUATIONS AND KNOWLEDGE

The last part of the volume deals with values in laboratory work. Its three chapters consider several questions: How do actors construct boundaries between interested and non-interested science? How are links between interests and the valuation of scientific work linked in practice? How are data valued? How are incommensurable values performed in large and distributed scientific endeavours? How do scientific models perform specific valuations of human behaviour? How are economic issues performed alongside or separate from other values in science? By beginning to answer these questions these chapters probe the intertwining of the economic and the epistemic in the life sciences. The chapters explore the making of several topographies of values, in which the life sciences handle constant clashes and coordinations between different performances of values.

In Chapter 11, on 'Purity and interest', Francis Lee demonstrates that the dichotomy between pure and interested science is simultaneously upheld and broken down in practice, the point being to demonstrate how viewing interests as an empirical category—not an explanatory concept—can shift our attention to the links between performed interests and divergent values, valuations, and evaluations in the life sciences. To do this, the chapter displaces the traditional discussion of the (bio)capitalist corruption of epistemic work, and shows how interests are performed alongside epistemic valuations of data and methods. The chapter highlights how the actors use two modes of interest-purification, temporal and organizational, to distinguish between pure and interested science. However, this value purification collapses in the face of the simultaneous and divergent evaluations on what constitutes 'good science'. The chapter stresses how dichotomous thinking about divergent interests in the life sciences breaks down in the face of the work done to coordinate and delineate different orders of worth. Specifically, the chapter focuses on the studied actors' performance of science as a moral or political project and how epistemological yardsticks—such as antibody-specificity or experimental replicability—are intertwined and contrasted with other yardsticks—such as production efficiency or medical utility. The questions that the actors grapple with are 'What constitutes good science?' and 'What should the life sciences be concerned with?'

In Chapter 12, Claes-Fredrik Helgesson and Linus Johansson Krafve investigate the various arrangements used to gather data in three clinical registry networks. Each network is made up of several researchers and clinicians at different sites who collaborate to collect observational data into a unified registry of patients with a specific condition. Their analysis shows how the registries examined accommodate both more exchange-like arrangements—where clinics delivering data might be recompensed financially and/or with co-authorships—and relations involving other parties such as laboratories. They furthermore observed that the work to gather and transfer was related to a variety of articulations as to how participation was valuable to those involved in the gathering and transfer of data. They conclude that the apparent integrity of the large-scale registry networks examined appears not to be the result of a widely shared moral economy, but instead rests on the accommodation of a number of different and partially overlapping arrangements that furthermore enacts a variety of values.

In Chapter 13, Isabelle Dussauge explores models of the desiring brain in neuroscience. The contemporary neuroscience of sex and emotions describes the desiring brain as a valuation machine—a machine which attributes value to different scenarios or possibilities of action, and processes these values to favour expected outcomes or behaviours. What values does this cerebral valuation machine produce, handle, weigh against each other—in the neuroscientific model? In this chapter, Dussauge enquires into the economic metaphors used to describe the desiring brain and through it, desire itself. The chapter analyses the consequences of the economization of emotion and behaviour in neuroscience. It also addresses how the models either pit non-economic against economic values, or integrate them together. This text addresses a powerful scientific account of what set of values constitutes life, and furthermore, of what makes a good life: the dynamics of desire and happiness made neural.

In the concluding chapter (Chapter 14) of this volume we take a few more steps in outlining a pragmatist take on values. Here we take as point of departure the empirical work achieved in the chapters. We attempt to mobilize the points and analytical moves made in order to propose how studies of value practices can be done so as to introduce the notion of valuography to denote a broader research programme for the empirical study on the enactment of values. This notion is then grounded in a pragmatist analysis of the making of values in the plural. How is the question 'What comes to count as valuable?' answered in practice?

Part I

Conflicted 'Public' Values

2 Key opinion leaders

Valuing independence and conflict of interest in the medical sciences

Sergio Sismondo[1]

Dr T steps up to the podium, a smiling, clean-cut doctor and professor in his 50s.[2] He is wearing a light suit with a striped white and blue shirt and a yellow tie, a good outfit for very hot weather, as it is in Philadelphia on this August day. Dr T, who was introduced as having authored over 500 publications and being 'one of the brightest stars in neuroscience', gives his talk without PowerPoint, the first time he has done so in years, he says—a mishap that morning involving his cat and his laptop led him to scramble to assemble notes for this talk and one he will give later that afternoon to the same conference. Nonetheless, he is a confident speaker, comfortable giving his narrative to this audience of mostly pharmaceutical company managers. He is the representative key opinion leader (KOL) at a pharmaceutical industry conference on relationships with people like himself.

After explaining how the cat lost his presentation, Dr T gives his disclosures. This is a practised move. As he explains, with apparent pride, 'In the past decade, I have been a consultant to the manufacturer of every compound that has been developed for the treatment of depression or the treatment of bipolar disorder, and some number of other compounds that haven't made it through the multi phases stages of development'. Normally he presents this as two slides. He adds a list of six pharmaceutical companies that have paid him to give talks in the past three years, and lists another four that have recently funded research projects.

[1] Research for this paper was supported by grants from the Social Sciences and Humanities Research Council of Canada (#410-2010-1033), and from the Canadian Institutes of Health Research (#106892). My research assistant Zdenka Chloubova provided invaluable assistance. I would like to thank the editors of this volume, Ilana Löwy, and audiences at Vadstena, Harvard University, and Northwestern University for their questions and very helpful suggestions.

[2] I have anonymized many of my sources here, even when they are speaking in a public venue, as is Dr T, or are being quoted in other articles. I provide names only of people who are authors or major subjects of published works. Anonymized data here stems from a number of sources, but especially thirteen interviews with prominent KOLs, and the presentations at three industry conferences (two North American, one European) on relationships with KOLs, and at one conference for publication planners.

Dr T is the quintessential high-level researcher KOL, a nationally recognized expert who is personable and a good public speaker. Dr T started his connection with the industry in the 1980s, doing dinner speaker programmes, later giving promotional talks, serving on various advisory boards, and even once helping run a speaker training programme. But there are many different kinds of KOLs, corresponding to the many uses pharmaceutical companies have for them. Indeed, Dr T's path to this point took him through many of the common KOL roles.

As I will describe here, physician and researcher KOLs are paid to give talks for pharmaceutical companies. On the one hand, this is simply a market exchange, in the form of payment for service. Modern medicine is shot through and through with the commercial and similar interests of actors such as public and private insurers, pharmaceutical companies, hospitals and clinics, and physicians themselves, to mention a few. On the other hand, KOLs are intervening in *medicine*, a highly charged and highly morally regulated area. The Hippocratic ideal demands that physicians divorce their medical judgements and their interactions with patients from all of those interests. Physicians often remind themselves and others that lives are at stake in medicine, and thus they have a particularly acute duty to exercise care and to serve the interests of patients. It is from this duty that their high status derives, and presumably also their relatively high incomes.

As a whole, medicine is conflicted about interactions with the pharmaceutical industry, and many individual physicians are also conflicted. We might see a rough parallel to the supposed 'hostile worlds' of commerce and intimacy that Viviana Zelizer (2005) describes. Apparent hostility, though, does not prevent KOLs from interacting with the industry in a range of ways, most of which involve presenting pharmaceutical companies' data, arguments, claims, and views. My goal here is to display the parts of the moral economy of medicine that allow KOLs to resolve or ignore the hostilities between the worlds of commerce on the one hand, and science and care on the other. How do they justify their work for and with pharmaceutical companies in other than self-interested terms? What makes exchanges with companies acceptable or unacceptable to KOLs?

In answering these questions in terms of a moral economy, I am taking the notion not in the broad sense of Lorraine Daston's 'web of affect-saturated values that stand and function in well-defined relationship to one another' (Daston 1995: 4), but in the narrower sense in which E. P. Thompson introduced the term (E. P. Thompson 1971). A moral economy is the set of moral rules, normative tropes, and customary practices that regulate markets (see Atkinson-Grosjean and Fairley 2009; Kohler 1994: 11–13). Particularly relevant here, I take moral economies to normatively afford and bound market transactions, as actors hew to acceptable—or at least defensible—actions and practices. Telling, then, are participants' justifications of their actions and articulations of their limits.

Emily Martin (2006) explores the moral economy of drug researchers and marketers in much the way I do here. She frames her interviews with sales representatives and marketers, for example, in terms of the intense negative publicity that the pharmaceutical industry has faced: how did her interview subjects reconcile themselves with an industry vilified as 'rapacious and profit hungry' (Martin 2006: 157). Matching my experience with KOLs, she writes, 'Nearly every person I interviewed spent considerable time, without much prompting, telling me what makes their work meaningful to them and why' (167). There is some overlap between the meaningfulness that Martin's and my subjects described, but also a difference in emphasis. Martin's pharmaceutical marketers and salespeople most often talked about helping patients, whereas my KOLs justified their work in terms of communicating objective science or clinical experience. I will return to this later in the paper.

We might see talk of moral economies as drawing attention to some specific aspects of the regulation of political economies more generally. Markets of all different kinds need regulation in order to function, and important parts of that regulation come through informal norms, conventions, and relationships and micro-social interactions (e.g. Knorr Cetina and Bruegger 2002; C. W. Smith 1990). While not all of that informal regulation is usefully described in terms of moral economy, at least some of it is connected to a distinctly moral order.

KOLs: An overview

In their efforts to engage with physicians, pharmaceutical companies often turn to KOLs as conduits of information. The abbreviation is standard within the industry, although they are also sometimes more modestly referred to as 'opinion leaders' or 'thought leaders'. The idea of a KOL has a good sociological pedigree, stemming most directly from the work of Paul Lazarsfeld. In publications on a study of political views and voting behaviour during the United States Presidential election of 1940, Lazarsfeld (e.g. 1944) coined the term 'opinion leader'. The concept and term were extended beyond politics and public affairs to other walks of life, including fashion, movies, and marketing more generally (Katz and Lazarsfeld 2006). Application to medicine came already in the early 1950s, when Katz and co-workers, with a grant from Pfizer, studied the expansion of the prescribing of tetracycline, eventually published as a very well-read book entitled *Medical Innovation* (Coleman et al. 1966), which became important to social network theory.

The pharmaceutical industry found Lazarsfeld's idea extremely valuable, and since the 1950s has steadily increased its use of KOLs. In an article on the importance of engaging KOLs on their own terms, the medical education and communications company (MECC) Watermeadow Medical writes that the term is usually 'a convenient shorthand for those people—usually eminent, usually physicians—whom we co-opt into our development and marketing strategies' (Watermeadow 2007). This bald-faced definition is associated with a set of activities to not only identify potential KOLs, but to plan and implement campaigns to use them for particular ends.

According to industry analysts, pharmaceutical companies spend 15–25 per cent of their marketing budget on speaking events (Zuffoletti and Friere 2006). Sunshine laws in the US and France are requiring pharmaceutical companies to publicly reveal how much they pay to physicians. Earlier reports, on the basis of legal settlements and earlier versions of these laws, show that some physicians can make very large sums of money. On the basis of filed reports from only seven companies, at least 350 US physicians, representing a broad array of fields, each made more than $100,000 consulting for these pharmaceutical companies in 2009 (ProPublica 2011).

In practice, the term 'KOL' is used to describe people of widely varying influence. We can roughly divide KOLs into two classes, depending on whether they are seen primarily as physicians or primarily as researchers. Members of the first group are hired by the industry to give talks to other physicians, in more or less intimate and informal venues. What I will call the 'physician KOL' might go to a clinic and make a presentation to the group of other physicians working there, or might serve as an after-dinner speaker for a group assembled by a sales rep. Pharmaceutical companies train physicians, work with them to make them 'product champions', and pay them generously for their lectures (Moynihan 2008). Kimberly Elliott, a former pharmaceutical company sales representative, says: 'Key opinion leaders were salespeople for us, and we would routinely measure the return on our investment, by tracking prescriptions before and after their presentations... If that speaker didn't make the impact the company was looking for, then you wouldn't invite them back' (quoted in Moynihan 2008).

In the second group are already-established researchers, who might be hired to give talks at larger or more formal events, but also to serve as consultants or on advisory boards, who might be investigators on company-sponsored research, and who might become authors on manuscripts stemming from that research. These tend to be called 'national', or 'global' KOLs, and I will refer to them here as 'researcher KOLs'. There is overlap between physician and researcher KOLs, but the categories mark rough outlines of roles.

As KOLs have grown in importance within the industry, independent firms have positioned themselves to offer KOL-related services to the pharmaceutical industry. This is a worldwide activity, but even within the US market there are dozens of companies that specialize in identifying, mapping the influence of, recruiting, and managing KOLs; there are many more that specialize in other national or regional markets. One such firm, for example, is Thought Leader Select, which advertises multiple services for identifying, mapping, and planning engagement with KOLs: 'Thought Leader ID, Thought Leader Impact, Thought Leader Engage'. Other firms describe overlapping services and skills, such as specializing in KOL relationship management, and have proprietary software systems for planning and tracking interactions (e.g. InsiteResearch 2011). KOLs are key to successful pharmaceutical marketing, and so all of the work of engaging and engaging with them makes for a sizeable amount of business.

The physician KOL

In an autobiographical essay for the *New York Times*, psychiatrist Dr Daniel Carlat describes his invitation into the ranks of KOLs:

> On a blustery fall New England day in 2001, a friendly representative from Wyeth Pharmaceuticals came into my office in Newburyport, Mass., and made me an offer I found hard to refuse. He asked me if I'd like to give talks to other doctors about using Effexor XR for treating depression. He told me that I would go around to doctors' offices during lunchtime and talk about some of the features of Effexor. It would be pretty easy. Wyeth would provide a set of slides and even pay for me to attend a speaker's training session, and he quickly floated some numbers. I would be paid $500 for one-hour 'Lunch and Learn' talks at local doctors' offices, or $750 if I had to drive an hour. I would be flown to New York for a 'faculty-development program,' where I would be pampered in a Midtown hotel for two nights and would be paid an additional 'honorarium.' (Carlat 2007)

Carlat's experience is not unusual. A programme utilizing KOLs would generally begin with a training session, to ensure that the speaker is well versed in the positive aspects of the product, and able to speak effectively about them. For example, Wave Healthcare claims on its website:

> It's vital that advocates are able to communicate and influence colleagues with clarity and conviction. To ensure speakers are at the top of their game, we have developed a communication skills programme for clinicians. (Wave Healthcare 2011)

Another such firm, KnowledgePoint360, which owns Physicians World Speakers Bureau, offers programmes for training speakers, sales representatives, and medical science liaisons alike. Its promotional material appears to treat KOLs and employees in the same terms: 'Whether it is for external resources, such as speakers, or internal staff, including sales representatives and medical science liaisons, a robust training program is critical to the long-term success of any pharmaceutical, biotech, or medical device company' (KnowledgePoint360 2010).

According to one Merck study, KOLs' audience members each wrote US $623.55 worth of new prescriptions for the drug being discussed, whereas comparable sales rep audience members wrote only $165.87 worth of new prescriptions; even factoring in the extra cost of the KOL, the return on investment from KOL-led meetings with physicians was almost double the return on meetings led by sales reps (Hensley and Martinez 2005). As a result, physician KOLs have become key mediators between pharmaceutical companies and other physicians. Says Kimberly Elliott, the former sales representative, 'There are a lot of physicians who don't believe what we as drug representatives say. If we have a KOL stand in front of them and say the same thing, they believe it' (Moynihan 2008). And thus it is worth training KOLs and paying them generous honoraria.

Not surprisingly, the payments are valued by KOLs. In interviews, they acknowledge this, even while most of them downplay it: 'You know, my kids are grown up... I helped use a lot of the income to support my parents' (Dr A); 'We're paid well. But we're paid I think fairly' (Dr C); 'Well, I enjoy doing promotional talks and I actually try to do education, but when it comes down to it it's really about earning extra money' (Dr K).

Before their deployment, most physician KOLs are not pre-existing opinion leaders. They are not physicians who are already influential or who have a place in a social network that would allow them to be influential. Instead, pharmaceutical companies' hiring of them makes physician KOLs influential. They are networked with other physicians, turning them into important social nodes. In an important sense, then, pharmaceutical companies turn people into KOLs, by providing the right training, resources, and venues to make these physicians influential.

In the US, it is standard for physician KOLs to be given no flexibility in their presentations, because their talks are deemed 'promotional', and therefore must adhere strictly to the indications established for the drug on which they are speaking. Dr J says: 'When you're out there actually doing a talk, you really have to follow those rules to a T. If you don't follow those rules then... you're at risk of, you know, breaking procedure and I mean arguably I guess you're at risk of breaking the law.' Physician KOLs sometimes even gain the false impression that the Food and Drug Administration (FDA) itself has approved their slides: 'So if I am doing a promotional program for a company, I have to use the slide deck that they provide me—I am not allowed to alter it in any way and every word in that slide deck is basically reviewed by their own internal counsel in conjunction with representatives from the FDA' (Dr C).

Yet the emphases within his script did not escape Daniel Carlat. 'Yes, I was highlighting Effexor's selling points and playing down its disadvantages, and I knew it. But was my salesmanship going to bring harm to anybody?' (Carlat 2007) As time went on, the sales reps treated him increasingly like a colleague, and encouraged him to sell the drug harder, targeting physicians who weren't prescribing it. About a year after he began his stint as a KOL, though, Carlat became disillusioned. Another physician's query about hypertension started leading him to question Wyeth's spin on the data. He started wondering about the short duration of the trials that showed Effexor superior to other antidepressants, about the selection of patients for those trials, and the difficulties of withdrawal from the drug. As he presents it, Carlat developed worries about Effexor and the quality of the studies he was pushing, and as a result started questioning the congruence of his roles as a salesperson and as a presenter of objective medical science.

Carl Elliott (2010: 69–70) describes the experiences of his psychiatrist brother Hal on GlaxoSmithKline's speakers bureau. Initially, Elliott (i.e. Hal) was invited to speak about depression to a community group, for an honorarium of $1000. After a successful talk, he was invited to speak to groups of physicians on the same topic, but with every talk the sales reps who were his contacts encouraged him to insert increasing amounts of information about antidepressants. Eventually, Elliott became disturbed about the sales role he was playing, and his low status when playing that role. When a sales rep asked him to speak with a particular physician after a lunch talk, he realized: 'I was literally standing in the drug rep spot begging for a minute of this doctor's time, like a cocker spaniel begging for a leftover piece of meat from the table' (Elliott 2010: 108). At that point, he quit.

Elliott and Carlat provide two different reasons for ending their positions as physician KOLs. On Elliott's account, the major difficulty was the low status of his KOL position, at least when other physicians saw him as a sales rep or too closely connected with sales reps. Carlat, on the other hand, claims to have been concerned about the integrity of the science and medicine he was presenting, resulting in a conflict of roles. These two reasons suggest key elements of the moral economy of KOLs.

Both Elliott and Carlat quit their KOL positions before they were promoted. Had they continued, their next steps would probably have been to give more prominent after-dinner talks to groups of physicians, or possibly to provide accredited continuing medical education (CME), increasing their status and the size of their honoraria. After that, they might have graduated to national speaking circuits, where 'the real money is' (Elliott 2010: 70).

The researcher KOL

Dr T, whom I described at the opening of this paper, spent many years giving promotional talks, speaking from company slide sets. As he became established as a researcher, he continued giving talks for pharmaceutical companies, but they were generally more sophisticated scientific talks. He came to wear a hat not as a well-connected and thoughtful physician presenting other people's data but as a research scientist presenting his own. In this way physician KOLs can graduate to become researcher KOLs. Along the way, they are helped to gain influence by the platforms, networks, and resources offered them by their sponsors. But by the time they are researcher KOLs they have also established their own reputations. They have attained a certain amount of independence from individual pharmaceutical companies, because their own status is in demand.

As KOLs, these researchers influence physicians and researchers. Although that influence may work through many different media and in different settings, most often these KOLs simply are paid to speak on behalf of pharmaceutical companies. They are paid to deliver CME courses, talks to specialists and other important physician groups, and presentations at workshops and conferences—and even sessions for other KOLs. In all of these venues, researchers have moved from simple promotion to its science-based successors; for example, the rise of CME is a direct response to criticisms of the commercialism of advertising (Podolsky and Greene 2008).

For these higher-level talks, the honoraria are $2500 or more (Moynihan 2008), versus the $500 to $1000 paid to most physician KOLs for their presentations. A number of governments are in the process of regulating payments to physicians, lowering payments to the level of 'fair market value', however difficult that is to assess. Fair market value is a frequent topic of discussion at industry conferences devoted to KOLs, and there are entire industry reports devoted to the topic (e.g. Cutting Edge Information 2009). The topic is important not because companies want to save money, but because they want to avoid legally dubious payments that might be seen as creating inappropriate influence or even be seen as bribes to prescribe, recommend, and otherwie act in favour of a particular drug. For pharmaceutical companies' legal purposes, payments to KOLs need to be market transactions, reimbursements for work or expertise, and need to have a certain level of transparency.

Although researcher KOLs do not engage in the direct sales-promotion activities of their local counterparts, they influence prescribing both directly and indirectly. According to InsiteResearch (2011), 70 per cent of the U.S. physicians in a therapeutic area writing the most prescriptions are 'directly or indirectly related' to the top five opinion leaders in that area. Promotional and educational material may also be built on research or studies executed or authored by KOLs. And of course KOLs can influence physicians with whom they are not already related, by speaking directly to them, and by affecting the terrain of medical knowledge in their areas.

Moreover, KOLs can smooth the path to acceptance of diseases and drugs. Jennifer Fishman (2004) describes how researchers on female sexual dysfunction acted as mediators between pharmaceutical companies, the U.S. Food and Drug Administration (FDA), physicians, and potential consumers. For example, in 2001, researchers organized a consensus conference on 'Androgen Deficiency in Women' designed to establish the definition of and diagnostic criteria for this developing disorder. The conference was supported by grants from several companies in the process of developing testosterone products for women, and was important to the prospects for success of these products, because the FDA only approves drugs that treat established medical disorders. The conference's consensus document, then, was a key step in establishing the regulatory legitimacy of female sexual dysfunction in the form of 'female androgen insufficiency syndrome' (Fishman 2004: 193). In addition to looking at documents, the FDA turns to researchers like the conference organizers and participants in order to judge the documents: these KOLs have the relevant expertise to contribute to the agency's decisions.

Summarizing, the firm InsiteResearch (2008b) claims:

> Interacting with qualified investigators, physicians experienced in regulatory reviews, well-known and respected speakers, and highly published authors will help to efficiently manage tasks within the critical path of the product and disseminate the message of the product to the end prescribing audience.

Companies draw on KOLs' influence in a broad variety of contexts, and also put them in better positions to have that influence, making them better KOLs. As we might expect, then, companies put real effort into the recruitment of, building relationships with, monitoring, and deployment of KOLs. Most of the contact is via a company's medical science liaisons (MSLs), scientifically trained representatives who are not supposed to engage in sales. Ms B, a senior manager of MSLs at a mid-sized pharmaceutical company, emphasizes that MSLs have to have goals in all of their interactions with KOLs:

> When you go in, that might be your goal, your objective, to just continue to develop that relationship. And that's OK. It's just that at some point you need to expand on that goal . . . At the end of the day we do want something from them . . . We have needs that need to be met by KOLs, on the medical affairs side. (Author notes)

Ideally, interactions between MSLs and KOLs should be part of a general 'KOL management' plan. KOL management should spread knowledge, change opinions, and change prescribing habits. It should produce a good return on investment, although it is impossible to measure—a point much rued by people putting together and working in MSL programmes.

For researcher KOLs, relationships with pharmaceutical companies offer more than payments for advising and speaking. Most visibly, the companies offer research support to their more valuable contacts. Sometimes that comes from companies proposing trials that they want done, and offering research roles and expected authorship. And sometimes that comes in company support for trials that the researchers propose, which companies do to both further relationships and contribute to positive scientific publicity.

KOLs also may be offered authorship on company manuscripts, another relationship that handsomely serves the interests of both sides. With the rise of 'publication planning' in the past three decades, companies have started systematically treating data as important resources to be managed and marshalled (Sismondo 2007, 2009). Publication planning becomes a form of 'ghost management' of clinical research and publication when pharmaceutical companies and their agents control or shape multiple steps in the research, analysis, writing, and publication of articles, in ways opaque to readers. These companies not only fund clinical trials but also routinely design and shape them, typically using contract research organizations to run those trials. Companies propose and design multiple manuscripts around studies, by lumping and splitting data (Melander et al. 2003).

Hired medical writers produce first drafts and edit many papers, and publication planners expertly shepherd manuscripts through the publication process. Because of the commercial importance of having the right sort of author, publication planners find KOLs to serve as the nominal authors of manuscripts. This allows planners to make it seem as if the articles are by respected independent researchers, instead of by coordinated corporate teams. KOL authorship increases the perceived credibility of an article and also functions to hide features of the research process; even though they usually contribute more than the nominal authors, company statisticians and researchers, reviewers from an array of departments, medical writers, and publication planners are only rarely acknowledged in journal publications (Gøtzsche et al. 2007).

One document revealed in legal proceedings lists 85 manuscripts on Zoloft managed by a medical communications company. Those manuscripts, published between 1998 and 2001, became a significant portion of the literature on Zoloft published in that period. Moreover, they were much more prominently published, authored, and cited than were the other articles on Zoloft published in the same period (Healy and Cattell 2003). Other documents reveal similar numbers of planned articles for major drug launches, suggesting that roughly 40 per cent of medical journal articles on new treatments are ghost managed.

Medical schools place unrealistic expectations on their researchers, and so academic KOLs are keen to add to their CVs; it is not unheard of for researchers to list a thousand authored and co-authored scientific publications. Most medical science articles have multiple authors, so researchers are used to making modest contributions to published research. Publication planners further pare down the necessary work. To some KOLs, a free manuscript may feel like another perk of having good relations with a drug company, complementing the dinners, the trips to meetings and conferences, speaking and consulting fees. In some cases, academic authors may not even be fully aware, or may decide not to be aware, that they are freeloading off a drug company.

Through ghost managed publications and presentations, as well as more independent speaking and publishing engagements, KOLs are given forums that increase or support their status. As long as KOLs maintain the appearance of independence from their sponsors—and to some extent even when they do not—their talks increase their prominence. Repeatedly being billed as a leading expert gives them the status of leading expert. Pharmaceutical industry resources may even be enough on their own to create leading experts, without independent expert status: half of the U.S. KOLs known to have received more than $200,000 in 2009 had no formal affiliations with academic centres (Weber 2010). 'Anything that, you know, puts you in front of people gives you the opportunity to enhance your professional status' says Dr H, a physician KOL with a research profile.

Other forms of status come in the form of perks. Even very ordinary engagements are typically accompanied by perks that mark status, such as being treated as a VIP by being driven to a talk. Dr David Healy, a professor of psychiatry in Wales, describes the days of being a high-level KOL, and the luxurious treatment he received, before becoming a prominent critic of the pharmaceutical industry.

> It's a glittering social thing...They hire the Opera House in Copenhagen for a meal. They hire the Metropolitan Museum of Art for a meal. If you were to ask my wife, she'd say: 'I miss the days when we'd fly into Geneva airport, and get picked up by a stretch limousine, and get driven to Montreux, and have the master bedroom in the Royal Hotel, overlooking the lake'. (quoted in Elliott 2010: 94–5)

Relationships between pharmaceutical companies and researcher KOLs, then, involve multiple potential benefits for each side.

Valuing/accepting/denying conflict of interest

The pharmaceutical industry's payments to KOLs create many conflicts of interest. Within medicine, conflicts of interest by themselves are not necessarily seen as a problem. They are treated as issues that can be managed, and may even be seen as valuable: 'No conflict, no interest', says David Blake, Vice-Dean of Medicine at Johns Hopkins University (quoted in Schafer 2004: 15).

In articles from the first half of 2010 in a random sample of top-level general medical journals, the number of reported conflicts of interest per author and per article varied with the journals' impact factors—impact factors being measures of the mean number of citations to recent published articles in the

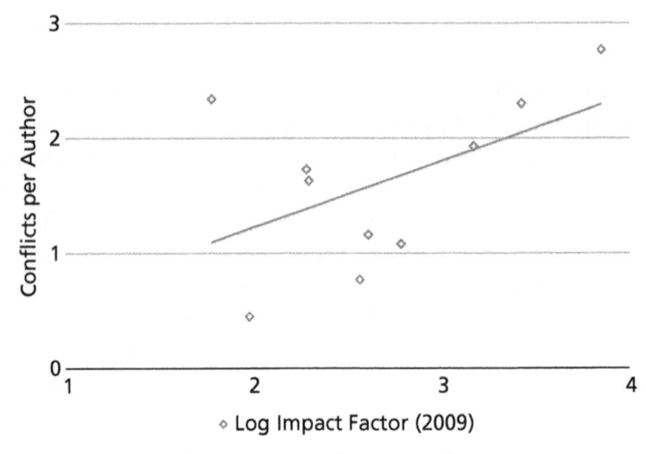

Figure 2.1. Conflicts per author

journal. The *New England Journal of Medicine*, with an astonishingly high impact factor of 47, led the pack with an equally astonishing 39.2 declared conflicts per article, and 2.77 conflicts per author. Most of those conflicts of interest represent different connections with pharmaceutical companies. While the statistical relationship (Figure 2.1) is weak, it is suggestive: Authors in prestigious medical journals clearly are likely to have multiple connections with pharmaceutical companies.

The connection is understandable. Pharmaceutical companies hold valuable resources for increasing researchers' reputations and status. Over the past 50 years, the industry has become the largest funder of medical research. It provides roughly 70 per cent of the funding for clinical trials—increasingly considered the most important studies in medicine—and also supports some basic research. Although the majority of that funding goes to contract research organizations and biotechnology firms, the total volume of industry funding is still very attractive to academic medical researchers. And studies funded by pharmaceutical companies tend to be published in higher-impact journals (e.g. Jefferson et al. 2009).

In addition, the pharmaceutical industry has the resources to facilitate career advancement even without providing research funding, by offering authorship of presentations and publications on ghost managed research. Publication planners make KOLs their authors on articles, and their speakers at conferences, workshops, and other events. In so doing, they build reputations, turning people into opinion leaders, and opinion leaders who are more key. Pharmaceutical company connections represent money, status, perks, and upward-looking careers. It is through those connections that physicians and medical researchers can become 'players' in their areas. Conflicts of interest, then, may be markers of status. For many in medicine, conflicts can be disclosed, handled, used, and valued.

Nonetheless, conflicts of interest can threaten status, when KOLs become too closely and visibly connected to companies. Within medicine, there are many people who criticize relations with industry, and disparage KOLs in particular. In interviews, every single one of my subjects took opportunities, typically many opportunities, to justify their interactions with industry. They perceived themselves as being criticized by peers and others, and often wanted to set the record straight about their uprightness and the value of what they do. They thus engaged in what Sarah Wadmann (2014) in a very similar context calls 'moral work' to justify their actions and downplay potential moral dangers.

An experienced consultant, Mr J, who has worked with the pharmaceutical industry in many roles, emphasizes how the industry needs KOLs who, even while understanding and being willing to work in the industry, will maintain their scientific integrity and independence. At the same time, he puts at the top of his list of KOLs' requirements: 'Protect my reputation'—do not create an 'appearance of [my] being an industry "sell-out."'

The companies and the KOLs, then, all value the appearance of independence even while they value their connections. KOLs are useful to the companies largely because they are not company employees. They can do commercially useful work under the guise of science and/or medical experience, effectively disguising company interests. A KOL who appears to be just an arm of the sales force will lose status and hence effectiveness among peers. One KOL, who goes on to defend his work for pharmaceutical companies, notes that paid speakers have been written about as 'hired guns, paid stooges, or "drug whores"' (McNeil 2008). And it seems reasonable to guess that any KOL who thinks that he or she is just an arm of the sales force will lose some self-esteem—most of them are, after all, successful physicians and/or researchers.

Crucial in supporting the sense of independence is an ideology of sound science. This is not just a moral economy of science, but rather a moral economy in which science provides moral force. Producing and communicating sound science is seen to be necessarily morally valuable and to provide a kind of independence from company interests. Physicians like Carlat say that they stopped being KOLs when they stopped being confident in the scientific research they were promoting. Others justify their continuing work for companies in terms of soundness. One university-based KOL is concerned about the ethics of his earning very substantial money speaking and consulting for the pharmaceutical industry, but still defends his actions in terms of science, and in terms of self-reflexive checks on his integrity:

> There is always the potential that somehow I'm getting in under the radar and then springing this very subtle and very pernicious sales message... I'm listening to myself every time I speak, and I have to ask myself the question: 'Is what I'm saying truthful?' (quoted in Weber 2010)

As one physician, challenging criticisms of medical conflicts of interest, writes, 'the content of my talks is rigorously vetted for accuracy, relevance, and fairness by the companies for sponsored programmes and the accrediting agency for CME' (McNeil 2008). A number of other KOLs echo this point, sometimes even emphasizing how their lack of agency enhances objectivity: 'In a sense the FDA really runs promotional speaking... What you are doing is you are educating physicians on the content of the FDA approved package insert' (Dr N). Or, in a similar justificatory context, Dr F says:

> Only after I've given a very regulated, scientifically backed and vetted presentation am I allowed to comment on my personal uses and preferences. Think of it as a relevant review of the scientific literature as it pertains to the use of this drug. After this review, I can comment on other aspects of the medication which may not be proven, but are my observations.

Closely related to scientific truthfulness is clinical truthfulness. KOLs who would prescribe a drug to their own patients claim that they feel perfectly justified being paid to speak on its behalf. Here is Dr F again:

> I agreed to talk about the medication because I feel it works better than the alternatives, which I have tried on my patients. Like any business, there are many who abuse the system, or are ingenuine [sic] in their statements, or who are only in it for the money. You can always find horror stories. But at least for me, I believe in the product that I recommend and won't say anything that is untrue. Drug talks are a simple way to increase my visibility with my peers as well as earn a few extra dollars recommending a product that I routinely recommend to my patients multiple times a day. (Dr F 2010)

All the interviewed KOLs insisted that they believe in the drugs for which they speak. Dr A, for example, cheerfully comments: 'My mother and father are on a lot of the drugs I speak for. I think they're terrific. So, I am not putting my parents on it because I am speaking for the company—it's the best drug.' Crucial to their understanding of the unimportance, or perhaps even the non-existence, of conflicts of interest is their sense that they are being honest about their clinical views. If they can speak with conviction they do not consider relevant any of the relationships that they might have with the company sponsoring their work.

This moral economy has a different emphasis from the one Martin (2006) found in her interviews of drug representatives and marketers. Although their stories varied, Martin found that these subjects described emotional resonances with patients that justified their work: they told stories about elderly women able to play with their grandchildren because of a painkiller, of mental illness de-stigmatized and patient advocacy groups empowered. Here is one of the drug representatives:

> So you actually don't feel like a salesman. You feel like you're educating, you know. You can take people who really can't fit into society or can't pull down jobs, and you put them under a physician's care. You work with their medication, and in two or four or six months you give them some type of a normal life back. They can function! (Martin 2006: 167)

KOLs, too, believe that they are helping patients, but they do so via the truth. They routinely argue that there is no difficulty with their being paid to act for pharmaceutical companies as long as they are communicating what they believe—on the basis of science, clinical experience, or both—and thus providing good medical education. 'Oh, it helps other patients elsewhere, it's spreading the word—it's spreading the gospel', enthuses Dr A.

Of course, KOLs and the data they present are carefully managed. Local KOLs are indoctrinated into promotional plans, given scientific data and clinical claims to explain, placed in specific prepared settings, and carefully prodded to include more commercial information in their presentations. Researcher KOLs are brought into a company's ambit through advisory boards and consultancies, their interactions with MSLs are carefully planned, they make presentations with company slide sets, are perhaps given research funds for commercially valuable projects, and may author ghosted articles, which are themselves carefully managed. Good management of them impinges on KOLs' actual independence, though they may not always be aware of it.

Only very rarely do they believe that their relationships might affect what they believe. Dr C muses, for instance,

> I'm probably biased in a pro-industry kind of way. They pay me a lot of money to do the work that I do and I really enjoy this kind of work so I always try to approach these things filtering it through those lenses of, OK... however pro-industry I think I should be... maybe I should be 20% less pro that way... um... so, but you know, we all have our blinders.

But even that level of reflexivity is not usual, or at least not usually voiced. Instead, KOLs often simply insist on their independence. For example, Dr Q, an endocrinologist who gives 200 talks per year for pharmaceutical companies, says he doesn't like reading from the company slide sets, which he understands to be a recent requirement to ensure that speakers do not discuss unapproved uses for drugs. But he claims that he is still able to give a thoughtful presentation: 'You're not just a paid monkey reading slides' (Weber 2010).

To sum up this section, conflicts of interest in medicine are remarkable things. They are ubiquitous at the highest levels of research, so much so that they may even be markers of status. Yet they can undermine pharmaceutical companies' goals and threaten status, if physicians and researchers become seen as too close to the companies. For this reason, KOLs emphasize the scientific and clinical truthfulness of what they have to say, making them not agents of the companies, but of medical science. They can thus set aside conflicts of interest, or perhaps even deny that there is any important sense in which they face conflicts of interest, because their statements and actions are guided by the truth. The moral economy that allows pharmaceutical companies to use physicians and researchers as their representatives, and to reward them handsomely, is one in which science provides a key moral force.

Conclusions

KOLs are recruited, trained, developed, engaged, and deployed by pharmaceutical companies and their agents. Those companies' interests are almost always close by, at stake in every interaction. To the KOLs' audiences, those interests are either partially obscured (in promotional talks) or entirely hidden (in CME and conference presentations, ghost managed journal articles, and other kinds of actions). KOLs serve as disguises for corporate faces, allowing pharmaceutical companies to market their products through apparently neutral representations of medical science.

In some regulatory environments, the paying of physician KOLs to speak is transparent enough that at least those physicians understand how they are acting on the company's behalf. Nonetheless, they configure their promotional talks as education, and often as education structured not by companies' interests but by regulation, science, and clinical experience. That allows them to reconcile their almost complete lack of independence in giving those talks with their roles as professionals. They have very little discretion about the content of those talks, but they see the companies themselves as having little discretion, given the demands of regulators and of science. Thus, while physician KOLs give up their freedom to say what they might want to in their talks, they nonetheless can present themselves as maintaining independence from company interests: conflicts of interest become irrelevant when KOLs are presenting what they see as sound science.

Researcher KOLs probably do not see all of the ways in which pharmaceutical companies manage them, because much of their management happens through more subtle tools than money. They are engaged in ways that further their careers while they further company interests; they may even see themselves as exploiting their industry connections. Most of the time, their engagement involves scientific research: performing it, communicating it, or taking credit for it. In this moral economy, education based on clinical experience and sound science again provides important support for independence. If a

KOL is just presenting scientific information, how could industry interests, or a conflict of interest, possibly affect him or her?

Other elements of medicine's moral economy make room for but downplay the importance of conflicts of interest—such as a belief that medicine needs the pharmaceutical industry, a belief that physicians are relatively ethical professionals, etc. But they take a backseat to the justification that KOLs are simply communicating what they take to be truths, a position that essentially denies real conflicts of interest.

Zelizer (2005) argues that the 'hostile worlds' of commerce and intimacy are not, in fact, generally hostile, but that instead intimacy does not last long without some interactions involving money or goods or some surrogate for them. However, actors engage in considerable effort to mark and account for interactions occurring between intimates that appear commercial. Similarly, medical science and care may require something like commerce (at least thought of broadly) to thrive. But that does not prevent some kinds of genuine hostility between these worlds as a whole, a hostility that KOLs can often mute with the right kinds of justifications: if they could not see themselves as maintaining some independence, and could not see themselves as promoting scientific and clinical truths, the breaking of moral norms around conflicts of interest would reveal clear hostility. KOLs do not necessarily perform extra work to mark commercial and non-commercial domains as separate, but they do moral work to account for and justify the mingling of those domains, and they do so with little prodding.

There is a more subtle hostility that the moral economy I describe largely ignores. KOLs may produce, author, and communicate real science, but we can expect that science to be interested. It is interested because we can expect systematic biases in the details of scientific results, emphases, framing, and questions (e.g. Lundh et al. 2012). And it is interested because pharmaceutical companies bring enormous resources to bear on communicating the science that they want to be dominant. Powerful and narrow interests shape the knowledge that KOLs help push. Meanwhile, there are only modest and inconsistent resources pushing in the other direction. The result we should expect, given work in science and technology studies over the past forty years, is a close coupling of much medical knowledge and pharmaceutical companies' interests.

Yet medicine, like many other domains, tends to treat good science as objective in a mechanical sense, if not a veridical sense. That is, it tends to treat science as produced through methods that, when they stand up to criticism, allow little room for choice. Stemming from a belief in mechanical objectivity is a denial of interested science, and a denial of the ways that the careful deployment of science can support commercial over patient interests. There also is an acceptance of the legitimacy of clinical experience in a political sense, although again not a veridical sense. In this acceptance, too, there is little room for interested beliefs, and little room for appreciating how the deployment of those beliefs can support commercial over patient interests. As a result, most KOLs can cheerfully take the money, status, and perks that pharmaceutical companies offer, fairly secure in the belief that they are acting in the interests of science and health.

3 The moral economy of a miracle drug

On exchange relationships between medical science and the pharmaceutical industry in the 1940s

Christer Nordlund

At last the esteemed professor ascended the rostrum. It had not been erected in a university auditorium, however, nor was the audience filled with academics. Instead, Professor Axel Westman was giving a press conference, arranged by a pharmaceutical company at a glamorous hotel in Stockholm. It was the summer of 1948 and the purpose of the event was to announce that the company, Leo, in close collaboration with Westman and his colleagues at the women's clinic of the Karolinska Hospital, had succeeded in developing a new, highly advanced hormonal preparation: Gonadex.

The audience was astonished to hear Gonadex described as a drug that possessed the unprecedented ability to help sterile women become pregnant; it could literally create life. Furthermore, thanks to its unique clinical purity, Gonadex was said to be completely free of side effects, in stark contrast to earlier hormone replacement drugs. All this made Gonadex a drug the likes of which had never before been seen, which, as the press reported the next day, amounted to nothing less than a 'Swedish triumph'. But the public was also astonished by the spectacle of a famous professor and physician conducting 'science by press conference', to use the term coined by Harry Collins and Trevor Pinch (1998: 59). At that point in time, such public appearances were still as rare in Sweden as they were elsewhere.

On the whole, the launch of the new miracle drug qualified as a success, generating a huge amount of press. At the same time, marketing Gonadex the way Leo did was a daring move. Predictably perhaps, Gonadex failed to live up to the hype. In time, interest in the much-ballyhooed preparation declined and today there are few who know that a drug named Gonadex ever existed, even though it remained on the market until 1986. Gonadex was the first original hormonal preparation developed in Sweden but it was no miracle drug and it never became an international sales sensation.

The collaboration between the company and the professor, so strongly articulated in the press, is one of several things that at once caught my interest in this case. Before familiarizing myself with the media reports from the press conference, I knew that Westman was considered a pioneer in the field of reproductive endocrinology in Europe, but not that he had hobnobbed with the business world. In the many short biographical texts written about Westman, the fact that he worked for the pharmaceutical industry for almost *three decades* was never mentioned (e.g. Borell 1994; Hamburger et al. 1960; Kottmeier 1961). I saw an opportunity to study an example of interaction between science, industry, medicine, and mass media long before it registered on the radar of research and innovation policy (Norlund 2007; Nordlund 2011). Why did Leo add Westman to its network of production, and why did Westman agree to collaborate with private enterprise, both in the lab and in the public spotlight? What values were enacted and negotiated due to this collaboration? And why was Gonadex described the way it was in the daily press? Answering these questions will shed light on the flexibility of the moral economy of pharmaceutical research and development (R&D) in the first half of the twentieth century.

From the most general perspective, moral economies of science may be defined as fusions of ethos and epistemology. It is a concept that stresses the simple but significant fact that it is impossible to separate the 'social' from the 'scientific'. Moral economies define and maintain scientific ideals, but these ideals are constituted by everyday, context-dependent practices, not abstract, stable norms like those once suggested by Robert Merton and his followers (Cf. Atkinson-Grosjean and Fairley 2009; Daston 1995). And such practices, indeed what counts as *good* practices, changes over time. In the history and social studies of the life sciences, the development of moral economies have been analysed in terms of interaction, cooperation, and reciprocity between the pharmaceutical industry, academic scientists, and physicians, as well as the arguments and expectations behind such 'exchange relationships' (Bud 2008; Rasmussen 2004), and that is the way I too use the concept.

In the interwar years, the moral economy of medical research was transformed from one in which industry and academy were largely separate (with the latter deeply suspicious of the former), to a moral economy in which collaborations between industry and academy were largely accepted (Atkinson-Grosjean and Fairley 2009: 159). A 'scientifization' of the pharmaceutical industry took place, at the same time as the pharmaceutical industry started to shape medical science. With reference to previous research into the development of biological drugs in the Western world during the period under investigation, I shall explore aspects of this process by highlighting three types of everyday practices, which gradually became central to the cooperation between private industry (Leo) and public medical science (Karolinska Hospital), and hence the history of Gonadex: 'consultation', 'procurement', and 'mediation'. In so doing, I not only analyse these exchange relationships as such; I also try to detect the main values—scientific, economic, political, medical, moral—that they challenged, created, or brought to light.

Consultation

Mass-produced medicine has a history that stretches back to the Industrial Revolution. The transition from handicraft to industry is usually described as an evolution proceeding along two parallel lines. One proceeds from big apothecaries where, with the help of mechanized pill-making machines, a surplus of certain medicaments began to be produced for sale to other retailers. The other line embarks from the chemical industry, particularly the German dye industry, which began using industrial methods to effect production of pharmaceutical substances and also found a market for its waste products (Anderson 2005). Towards the end of the nineteenth century, these two traditions were fused and institutionalized as more or less self-contained businesses, specializing in medical preparations (Liebenau 1987: 5 f.).

In Sweden, pharmaceutical production was first industrialized in the form of a small-scale chemical and instrument manufacturing industry, materializing around the turn of the century, at the same time as the industrialization process writ large had its breakthrough in the country. Faith in the financial opportunities offered by industry and technology was strong, and new businesses were encouraged and subsidized both by the government and by young, exuberant investment capitalists (see also Lundgren 1998; *Svensk Farmaceutisk Tidskrift* 1901). In the interwar period, well-established companies like Leo (founded in Helsingborg in 1914), Astra, and Pharmacia naturally had their own resources in the form of capital, technical infrastructure, and permanent staff. However, this was not enough to create and sell new pharmaceutical drugs. In order to manufacture and control the quality of the drugs, companies were compelled to enrol external competence. Visible association with recognized expert researchers was also decisive for ideological reasons, in order for the company to be deemed reputable and 'ethical' (Eriksson 1987).

In this context, pharmaceutical companies were forced to devise production networks in which university researchers and hospital physicians served as essential nodes. However, the fact that academics and physicians began to collaborate with private enterprise was not unproblematic at the beginning of the twentieth century; on the contrary, academia viewed such collaboration with some scepticism. Many, though not all, cherished the ideal that science should remain free to pursue the truth, independent of worldly interests. After all, this was an age in which the pharmaceutical business was riddled with quacks and snake-oil salesmen (Rasmussen 2004: 165–7). In the United States, pharmacists working in the industry were actually refused membership of the American Society for Pharmacology and Experimental Therapeutics. The society wished to avoid, as it stated, 'every external influence which would be inimical to the scientific interests of pharmacology' (Parascandola 1990: 30). Independence was thus highlighted as a central value.

The aim of academic science was primarily to assess and evaluate the quality and therapeutic value of new drugs, not contributing to research in order to develop them. In Sweden, it was the Academy of Pharmaceutical Sciences that took the first initiative to critically examine preparations not produced at the official pharmacies. A certain state-run pharmaceutical lab with supervisory powers was later founded, in 1915, when a special laboratory unit was set up at the Department of Forensic Chemistry at the Swedish Medical Institute. When the latter was dissolved in 1918, the lab was granted the status of independent authority and named the National Pharmaceutical Laboratory (NPL), working under the auspices of the Royal Swedish Board of Medicine.

However, during the end of the interwar period a change in the moral economy of the drug company–medical scientist collaboration took place and new values were enacted (Rasmussen 2004). Experts who could advise on or conduct technical, laboratory, and clinical research were typically hired by companies as consultants, either 'general' or 'specialist' according to the typology of John P. Swann (1988). The former arrangement meant that university scientists with high academic status, broad competence, and a comprehensive network of contacts were contracted as scientific advisors. They worked closely with management and were charged with supervising strategic choices of new research projects and products in one or several fields of medicine. They also had the task of contributing to the reputation of the company and its products by dint of their personal reputations (Swann 1988: 57 ff.). Specialist consultants were researchers hired on an ad hoc basis as experts in their particular field, be it chemistry, microbiology, or pharmacology. These individuals were particularly valuable for companies having trouble attracting enough sufficiently competent staff. But they were also hired in order to quickly impart the latest news from the 'research front' to their staff (Swann 1988: 87–93).

Pharmaceutical companies needed consultants, but the consultants also realized that they needed the companies, partly due to the fact that their own research could benefit from the financial and material resources of which the industry availed them. This was not least true for the scientist who became a central actor in the Gonadex story, Axel Westman (1894–1960). Westman was a physician specializing in gynaecology and obstetrics, but he was also an entrepreneurial life scientist. In his first inaugural address as professor, in Uppsala in 1932, Westman (1933) maintained that modern gynaecology demanded new skill sets. The gynaecologist, who was typically trained as a surgeon, needed to become an endocrinologist, and in so doing adopt experimental methods. Westman emphasized the value of research in order to produce new knowledge, not only for its own sake but in order to improve health care.

But experiments and hormone analyses required specific material prerequisites. Physicians and researchers had to have well-equipped laboratories at their disposal, something which at that point in time was only a 'future goal' in Uppsala as well as in the rest of Sweden (Westman 1933: 190). This requirement, together with the need for animals on which to experiment, industrially purified hormones, and a variety of indispensable but expensive instruments, was probably the most important reason why Westman, already in 1933—fifteen years before the launch of Gonadex—agreed to work with Leo. Because the company not only paid Westman an honorarium, they also financed the construction and maintenance of a laboratory at the women's clinic in Uppsala, which has subsequently gone down in history as 'Sweden's first hormone laboratory'. Hence, Leo was one of the first companies in Sweden to finance and materialize a laboratory study of hormones in an academic environment (Ahlin and Lundgren 2002: 75; Norgren 1989: 64–70). Inversely, government employee Westman and his consultant colleagues were among the first medical scientists in the country to actively collaborate with the private pharmaceutical industry.

Westman spent a long time as Leo's consultant, from 1933 until 1960. His main fields of research were sex hormones, reproductive endocrinology, and sterility treatment, topics of importance for involuntarily childless couples but also for the much debated 'population question' of the 1930s and 1940s. This political question was surrounded with discussions about the 'quality' of the populace but first and foremost it was based on the fact that nativity rates in Sweden (as in many any European countries) had dropped steadily since the end of the nineteenth century. In this context, Westman argued—for example in the governmental inquiry *Report on Sexuality*—that many couples were involuntarily childless due to infertility problems, and that progress in endocrinology would help them and, ultimately, Sweden as a nation (Westman 1936). Hence, Westman stressed that his work, in the end, also had political value.

Westman's interests and line of research were also of importance for Leo. Having started in the hormone business as a distributor of insulin in the 1920s, Leo soon included many types of preparations in its catalogue, mainly 'female sex hormones'. Since Westman was a practising chief physician and senior researcher with high academic status, doing both clinical and pre-clinical research, he was the ideal partner, providing medical competence to the development and supervision of Leo's products and contributing to its reputation as a science-based concern. A further bonus was the fact that Westman had many contacts in the research field, to which Leo acquired indirect access by adding him to its network.

The single most important reason for Leo to finance an external lab at this time was, however, the new quality demands imposed on pharmaceutical drugs in Sweden in 1934, when a promulgation on trade with 'pharmaceutical specialties' took effect (Svensk Författningssamling 1934: 306). Before a factory-produced medicine could be made freely available in pharmacies, it had to be registered. Applications required submission of facts including characteristics or composition, how it was manufactured, the dimensions of its packaging, and price as well as advertising material intended for circulation. Samples of the drug were to be submitted for testing, which would verify whether the drug satisfied 'appropriate pharmaceutical standards'. The NPL conducted the inspection and the Board of Medicine decided if it qualified for registration (Lönngren et al. 1999: 184–92). This new legislation made tougher demands on the industry. In order to guarantee that a new drug would be approved, companies were compelled to make sure that they delivered what they promised. One way of fulfilling this requirement was to hire experts and resort to external competence, i.e. consultants, which is exactly what Leo did.

A central component in these guidelines concerned *standardization* of the drugs, and a main task of the hormone lab was to perform such standardization tests. In the present context, standardization refers to heeding the formal compliance to use particular instruments and methods in order to facilitate testable and reproducible measurements in different laboratories. This in turn leads to the manufacture of authorized examples considered 'standards', which serve as reference objects or units (cf. Golinski 1998: 173–7; O'Connell 1993). In order to effect harmonization throughout the pharmaceutical world, scientists have collaborated since the 1920s in stipulating international standards for biological pharmaceutical substances. After the Second World War, responsibility for this work was put into the hands of the World Health Organization (WHO) and its Expert Committee for Biological Standardization (ECBS) (Cockburn 1991; Cockburn et al. 1991).

The goal has been to achieve consensus on how biological substances are defined, how their potency should be determined, and which units were to be used in order to arrive at rational, universal solutions. In practice, however, these solutions have, for better or worse, often proven temporary, pragmatic, and dynamic (Pieters 1998; Rosemberg et al. 1968; Sinding 2002). For standardization also concerns *standardized values* about what was considered reliable and what skills were considered adequate for the job at the time. Gaudillière and Löwy (1998: 10) nicely characterize the situation in the following manner: 'It is about standardized items being similarly produced, as well as judging these items to be reliable means of operation and evaluation. It also refers to judging individuals to be reliable when using these adequate means.' A pharmaceutical company intent on marketing a new drug was required to defer to these complicated conditions, but so was the body inspecting the finished product. Maintaining a good mutual relationship was in everybody's best interests.

Westman and his assistant's assignment consisted of standardizing sex hormones as they rolled off the conveyor belt, checking and confirming the strengths of the dosages in conformance with the international norms. In return, they received an undisclosed annual sum but also unlimited free samples for the clinic, along with access to lab animals, chemicals, and other laboratory utensils ('Till ledning för styrelsesammanträde', Løvens kemiske Fabrik to Genell, Leo Historical Archives 1933). The collaboration between Leo and the clinic in Uppsala did not last long, however. When Westman moved to Lund Hospital in 1935 as the new director of the women's clinic there, and later to the Karolinska Hospital in 1942 when he assumed the professorial chair at the new women's clinic there, he was able to bring the hormone lab (including assistants) with him ('Avtal mellan direktionen för Karolinska sjukhuset och aktiebolaget Leo', Leo Historical Archives 1943). In Stockholm, Leo continued to finance the lab's day-to-day expenses for another full decade.

The arrangement between Leo, Westman, and the external laboratory meant that hormone drugs could be standardized and tested in a clinical environment, which might ultimately lead to increased sales (Gaudillière 2005; Oudshoorn 1993). Westman introduced successive new drugs into his clinical work—in Uppsala, Lund, and finally Stockholm—and together with his many public appearances, his essays and popular books (e.g. Westman 1940a, 1940b) contributed to whipping up interest in hormone therapy. Furthermore, his endocrinological expertise, unparalleled in Sweden, lent indirect lustre to the same drugs he had a hand in producing (cf. Fishman 2004: 205). Westman was not unaware of this, for as he admitted in a letter, 'It was in the financial interests of Leo to subsidize the laboratory since most of the hormone drugs were manufactured by the Leo factory in Helsingborg' (Westman to the Directorate of the Karolinska Hospital, 23 January 1946, Stockholm County Council Archives). Obviously, this was acceptable due to the contemporary moral economy of the pharmaceutical company–medical scientist collaboration.

Procurement

When establishing its external laboratory, Leo followed in the footsteps of companies such as Organon in Holland and Schering in Germany, both of which had developed successful field collaborations with gynaecologists and gynaecological clinics. These networks were important for the 'scientification' of the industry, but they also had a significant material aspect, which ultimately concerned access to natural resources. Producing drugs in the first half of the twentieth century was not merely a matter of maintaining a well-stocked cabinet of chemical substances. Large-scale production of many types of hormones (so-called 'biologicals') also required subcontractors who could be counted on to deliver raw materials with prompt regularity, as these hormones were made from fresh, organic material. It was therefore imperative that the companies establish steady, reliable contacts with the places from whence this raw material came, which included slaughterhouses, farms—and clinics.

Through relationships with gynaecological clinics, companies gained access to raw material, such as glands removed from female patients and urine from pregnant women visiting the clinics. That the companies procured such material was in the interest of science as well, because gynaecologists, in the midst of conducting experimental and clinical research, required access to an uninterrupted flow of hormones that the companies were able to extract from this material. As Nelly Oudshoorn (1993) has demonstrated, this situation further intensified the necessity for more or less formal collaboration—infrastructural arrangements—between academy, industry, and wholesale suppliers. This was true already in the 1930s when Westman started to work for Leo. At that point in time Leo manufactured only one type of hormone, oestrogen, made from urine of pregnant mares (which Leo bought from farmers in the Scanian countryside). Other types of hormone drugs were imported from Leo's mother company Løvens in Denmark. But it was still true during the Second World War, when importation was difficult and Leo decided to expand its own R&D of hormone preparations.

In the autumn of 1944, Leo's managing director Carl Holtman wrote to Westman informing him of the company's plans to invest in what he called a 'line of hormones'. Westman's delight at the news and his eagerness to be involved is obvious in his reply to Holtman. 'As you understand, it greatly pleases me that the Company intends to increase production of hormone drugs, and like you I find it propitious that we continue to collaborate in the manner hitherto pursued' (Westman to Holtman, 3 November 1944, Stockholm County Council Archives). A few days after this exchange, a meeting was arranged between Westman and Leo's representatives at the Karolinska Hospital's women's clinic. Westman explained that the 'incorporation of the domestic manufacture of hormone preparations, sex hormones and others, [is] of the greatest significance'. According to Westman, Leo was the one company in Sweden suited to succeed in such an endeavour and he repeated that he was eager to participate in the work himself and recommended that collaboration between Leo and his clinic 'not only proceed according to plan but deepen and expand' (Deutsch and Kallós undated, Samtal med professor Axel Westman, Leo Historical Archives).

At a meeting with Leo's representatives in Malmö on 4 April 1945, Sune Genell, chief of medicine at Malmö General Hospital and another of Leo's consultants on hormone preparations, also applauded investing in sexual hormone drugs. Genell stated that it was of 'great significance and import' that Leo begin manufacturing its own hormones, mentioning in particular gonadotropin preparations, which could be produced from the urine of pregnant women. He also had a concrete suggestion as to how this raw material could be collected. In their report on the meeting, Leo's representatives wrote that Genell suggested that the company approached women's clinics, and that 'a small payment for the urine would certainly assure the acquisition of larger amounts of material' (Deutsch and Kallós, 'Rapport över besök hos Överläkare S. Genell', 4 April 1945, Leo Historical Archives).

Two days later, a meeting brought Westman and Leo's director Holtman together. The theme of the meeting was Leo's planned investment in new hormone preparations and the discussion focused on one specific sort—pituitary gland hormones and gonadotropins, which could be used for treating many types of female sexual dysfunction, including sterility ('Protokoll vid sammanträdet den 6 April 1945', Leo Historical Archives). As far as gonadotropins were concerned, it was pointed out that testing had shown that the biological effect could vary widely depending on whether they had been produced from the urine of pregnant women (so-called human chorionic gonadotropin, hCG) or the blood of pregnant horses, or if real pituitary glands were used as the starting material. This, replied Westman, was a matter deserving further investigation.

Later, Westman, who had been studying various types of gonadotropic hormones since the late 1930s, maintained that the observed difference between the different types of gonadotropins was due to the doses given (e.g. Westman 1937a, 1937b, 1938). This objection was not uncommon. Since the 1920s, the standard reply to the question as to why commercial sexual hormone drugs did not work was that they were simply too feeble (Oudshoorn 1993: 13). The quandary was that when treatment was intensified, so were the undesirable side effects. Similar problems had however occurred in the use of other types of hormones, and the standard solution to that problem had been to attempt to make the drug as pure as possible. In other words, theoretical opinion still insisted that side effects were caused by 'pollutants' in the drug, not the drug itself.

In light of this situation, and in combination with the fact that Leo was braced to beef up its in-house arsenal of hormone drugs, there were more than enough reasons to attempt to produce absolutely pure hCG from urine. Since they should ideally not cause any side effects whatsoever, these improved drugs could be injected in very strong doses and thus, predicted Westman, be more effective than their predecessors. But Westman also argued in this context that a pure chorionic gonadotropin would be a valuable 'scientific instrument' in endocrine research in and of itself (cf. Goodman 1998). Through clinical treatment, new and significant knowledge about the effect of chorionic gonadotropin on the human body would accrue. Hence, although the parties highlighted different values—economic, scientific, medical, political—they shared an interest in trying to produce the substance.

Upon consultation, Leo decided to create some new hormone preparations and one of these would be comprised of pure hCG, extracted from urine of pregnant woman. A research team, which included researchers from both Leo and the Karolinska Hospital, was set up. Leo's purification efforts were rewarded when a pure substance that crystallized in the shape of delicate, transparent needles was produced. This crystalline substance was tested with electrophoresis, a relatively new method for separating charged molecules from one another, and the research group concluded that the substance was completely pure and homogeneous. The group had achieved its first goal, one that no one had achieved before (Ahlin and Lundgren 2002: 113; Claesson et al. 1948: 2–5).

As required by law, the launching of Gonadex was preceded by tests to measure its strength in comparison with the international standard for hCG, established in 1938. These tests were conducted at the hormone laboratory at Karolinska. The biological activity of the substance, tested on baby female mice, showed that the purified preparation caused the expected effect on the animals' sex glands and that its strength was proportionately potent. However, the results of these biological trials were not entirely unequivocal—the sex glands of the animals were apparently not affected the same way each time (Claesson et al. 1948: 6–7). Although standardization proved difficult, data from this experiment were used when Gonadex was registered.

The medical value of these purified preparations was still an open question, at least in practice. In order to determine this, Leo had to rely not only on Westman but also on his patients at the women's clinic. Westman selected approximately thirty women on whom to test the effect and possible toxicity of the drug. The results of the initial experiments with the new, crystalline hCG were discussed at a meeting held on 17 January 1948. Though the experimental phase was not yet complete, it appeared clear that no side effects manifested themselves, despite the fact that the preparation had been injected in 'very large doses' ('P. M. för sammanträde i Hälsingborg lördagen den 17 januari 1948', Leo Historical Archives). According to the article 'Crystalline Human Chorionic Gonadotrophin and Its Biological Action', in which the results of the science–industry teamwork were presented, the substance was also very potent.

On the other hand, the medical effects were presented as only preliminary, and as far as the medical value of the drug was concerned, the tone was cautious (Claesson et al. 1948: 16). The authors maintain that menstruation can be affected but say nothing about the possibility of successfully treating sterility. Nor do they mention in the article that the research was conducted in order to refine a substance that would be used in the new drug Gonadex.

Instead, the name 'Gonadex' first appeared in a Swedish-language article published by Westman in the medical journal *Svenska Läkartidningen* shortly after the Stockholm press conference. This article, which more or less recycles the content of the first article, was entitled 'Klinisk prövning av choriongonadotropinet "Gonadex Leo"' [Clinical trial of the chorionic gonadotropin 'Gonadex Leo'] (Westman 1948). But no evidence is provided here either that Gonadex could have a positive effect on sterile women. It is only stated that tests are currently being conducted (Westman 1948: 1159).

In any case, after the meeting in January 1948, Leo entered into a period of intense work culminating in the public launch of Gonadex at the press conference five months later. This work consisted partly in preparing a marketing strategy for the new product and partly in rapidly organizing a campaign to collect urine from pregnant women in order to scale up production of hCG. As we shall see, Westman was enrolled in both activities.

Leo needed access to urine from pregnant women, and its network of consultant contacts proved its value. The first urine used in the development of the new drug probably came from Genell's clinic in Malmö, suggested by directives he issued in 1945. However, when production of Gonadex began in earnest, the campaign focused on Stockholm, mainly the clinic Westman oversaw ('P. M. för sammanträde i Hälsingborg lördagen den 17 januari 1948', Leo Historical Archives). When this was not considered enough, a second campaign was launched in Helsingborg were the company was situated. In April 1948 the company calculated that the collection could surely be multiplied several times over if the company ran a 'systematic advertising campaign' ('Kalkyl över framställning av chorion-gonadotropin', 6 April 1948, Leo Historical Archives 1948 f.). The catchment area was later expanded to include prenatal clinics in the Gothenburg region, thus covering all of south-west Sweden.

Pregnant women visiting any one of the clinics belonging to Leo's network were asked to provide a urine sample in the service of the pharmaceutical industry. The fact that a private business entered into the public medical realm does not seem to have raised any eyebrows, perhaps because the women thought that they were participating in a scientific project. They also received a fee commensurate with the amount of urine delivered. Some preserved receipts show a fee of one crown per litre, but the amount may have varied from clinic to clinic ('Redovisning för tiden den 11–16 december 1950', Leo Historical Archives). However, the success of the system was contingent on the willingness of nurses and midwives to oversee the procedure on top of their regular duties. When Leo realized their irreplaceable value, they also introduced a commission for those willing to get their patients to sign up for the campaign ('Rapport från sammanträde den 14 juli 1950 beträffande urininsamling', Leo Historical Archives). The urine was subsequently shipped to Leo's industrial park in Helsingborg ('Konto nr 304', Leo Historical Archives undated). While it was a quite expensive and complicated arrangement, it worked and ensured a steady supply of raw material necessary for the large-scale production of Gonadex.

Mediation

Marketing departments and public relations firms have comprised other important nodes in the network of pharmaceutical companies. The bigger companies had their own, in-house 'information departments' that made sure that the physicians were regularly called on and convinced of the excellence of their wares. And in turn, salesmen, or 'pharmaceutical consultants', gathered opinions and information from their clients for the company; communication was a two-way street. By distributing ads directly to physicians, pharmacists, and scholarly periodicals, they were considered *ethically acceptable*, as opposed to advertisements in the daily papers.

In this picture, ordinary medical scientists and physicians are absent. There is also a standard to which physicians are held, dictating that they ought not to participate in the marketing of pharmaceuticals, especially if they themselves have a financial interest in any of the products. The World Medical Association International Code of Medical Ethics of 1949 states specifically that doctors 'shall not receive any financial benefits or other incentives solely for referring patients or prescribing specific products' and 'should use due caution in divulging discoveries or new techniques or treatment through non-professional channels'. This standard has largely been upheld insofar as physicians seldom participate directly in marketing campaigns. However, that does not mean that they have not participated in other ways.

In practice, scientific consultants have played a role in marketing and public relations. They have in fact been deeply involved in introducing new drugs, not only through consultant work and clinical trials but also by actively collaborating with industry in identifying and creating new markets. From ethnographic studies of the practices of developing and introducing drugs to treat so-called 'female sexual dysfunction', Jennifer R. Fishman (2004) declares that public medical experts act as *mediators* between pharmaceutical companies and regulatory agencies, the professional medical community, and potential consumers of new drugs.

As an example, Fishman states that medical researchers relay information about clinical trials at a variety of medical gatherings, and by increasing public awareness about new diagnoses and treatments via internet websites. At the same time, they pass information and knowledge from clinics to companies; this channel of communication is also multidirectional. Medical researchers acting as mediators between pharmaceutical companies and consumers (doctors and patients) when, and even before, new drugs are introduced is however nothing new, even if the precise form of mediation has changed over time (e.g. Nelkin 1995; Pieters 2002, 2005: 134–57).

When Gonadex was about to be launched, Leo distributed academic articles and advertising flyers to the medical corps, who were expected to recommend the drug to their patients. But that was only one part of the campaign Leo orchestrated, according to a strategic document entitled 'Memo re: Publicity for Gonadex' composed in mid-February 1948, three months before the Board of Medicine approved the new drug. The document lists several bullet points concerning a planned press conference, including suggestions that an 'interview' with Prof. Westman should be arranged, that pictures of Westman and the drugs be taken, and that a venue for the conference should be reserved. As soon as the necessary preparations were complete, a personalized letter should be delivered to the editors-in-chief of the Stockholm papers, which clearly stated that it was 'a truly revolutionary preparation with the full weight of Professor Westman behind it'.

Leo and Westman generally followed these instructions, culminating in the much-anticipated press conference. Thus, Leo made a conscious decision to spread information and propaganda with the aid of their consultant via various news media, which gave them wide channels to the general public. In this Leo was ahead of the curve, but it was not alone. Rather, the company had probably been inspired by the latest public relations strategies from the US, which had gained a foothold on Swedish soil after the Second World War. The Swedish Association of Advertising Agencies had just arranged its first course in 'Corporate Public Relations' and press departments were springing up in government agencies, private organizations, trade associations, and some of the heavy industries. As media scholar Larsåke Larsson points out, public relations at this time had served different needs in different sectors, but as far as the private sector was concerned, this need usually involved the launch of single products (Larsson 2005: 49–76). In Leo's case, the goal was 'publicity for Gonadex'.

Two drafts of an invitation to the press were produced. Both state that AB Leo had the pleasure of inviting the recipient to attend a lecture by Professor Axel Westman at the Grand Hotel in Stockholm. According to the first version (which Westman himself preferred), the lecture would describe the manufacture of crystalline chorionic gonadotropin. The other version claimed that Westman would account for the 'sensational results' of his hormone research. Neither version mentioned that the actual reason for the event was to launch a new commercial product ('AB Leo to Westman, 27 May 1948'; 'Westman to AB Leo, 31 May 1948' Leo Historical Archives 1948e, 1948d).

Highlighting Westman's presence was a smart move. Today, respected medical professionals regularly attend and even call press conferences to report new research results. But in the 1940s it was a rare event in Sweden, partly because press conferences in general rarely dealt with medicine but mainly because the medical community was particular about maintaining its integrity and its ethical stance. Members of the corps who publically collaborated with private interests in their marketing risked being labelled irresponsible and in the worst case of allying themselves with quacks. Thus it was to be expected that Leo's press meeting aroused plenty of interest and curiosity.

At the conference, Leo's sales reps handed out freshly printed flyers and pamphlets as attendees awaited the main event, Axel Westman's address. It proved to be an academic lecture on female genital illnesses and their treatment, and on the joint effect of sexual and hypophysis hormones on the ovaries ('Professor Westmans anförande', Leo Historical Archives 1948c). Westman also spoke in general terms about how diverse hormone preparations for the treatment of genital illness were produced and the difficulties circumscribing such therapy. He explained that treatment had previously been unreliable and gone amiss since there were no hormonal preparations pure enough or strong enough. He only mentioned the ostensible topic of the evening, pure, crystalline Gonadex, as he was wrapping up, without mentioning his own contribution to the development process, giving the impression that he had been summoned from without to inspect Gonadex with his critical scientific eye in order to give it his seal of approval before the assembled press. According to his lecture notes, Westman stated that

> Swedish pharmaceutical factory Leo has begun shipping a new hormone drug, called 'Gonadex' which satisfies the highest demands. The drug, which is the result of years of effort by the factory's chemical research staff, is of such high-grade purity that it can be manufactured in crystalline form, something that has previously proven impossible. It can be injected directly into the bloodstream without fear of side effects, which makes a high and very precise dosage possible ('Professor Westmans anförande', Leo Historical Archives 1948c).

As in his scholarly articles, Westman was restrained in his assurances of the therapeutic benefits of Gonadex. All he said was that results achieved with Gonadex had so far been 'largely encouraging' and that there was therefore good reason to expect that 'another step forward had been taken in the struggle against functional genital diseases, among which involuntary sterility and abnormal haemorrhaging in young women may gain the greatest benefits' ('Professor Westmans anförande', Leo Historical Archives 1948c). In other words, the ultimate goal was in sight but still somewhat out of reach.

Why did Westman do this? It is reasonable to assume that Westman believed in Gonadex and that he was eager to see the drug implemented in gynaecological practice as soon as possible. He was aware that evidence of the physiological effects of Gonadex was tentative, but he probably felt that the older preparations available were inferior. At the same time, Westman's participation in the press conference should be seen as a component of Leo's well-orchestrated marketing practice. On the strength of his name alone,

Westman attracted journalists and lent Leo's launch of a commercial product scientific legitimacy. Why else had Leo not asked one of its own chemists or engineers to hold the lecture? For it was they, and not Westman, who had succeeded in distilling the hCG on which Gonadex was based.

After the press conference, Gonadex was the topic of the day. A few days later, a women's magazine (*Husmodern*) published a lengthy article entitled, 'A Third of All Sterile Women Can Give Birth with New Hormone Drug'. Shortly thereafter, a popular science magazine (*Teknik för alla*) featured an article on Leo's R&D system, stating that thanks to Gonadex 'women who previously could not be treated for sterility have had their barrenness cured by the new preparation' (Anonymous 1948: 4). Thus, according to news reports, a unique, Swedish cure for sterility had been discovered. And naturally, it was a sensation. One paper (*Morgontidningen*) reported that the presentation of the new drug at the press conference was reminiscent of 'Old Testament prophecy.' This streak of the miraculous was further enhanced by the fact that Gonadex was said to be capable of both increasing and reducing menstrual flow according to need. Moreover, the new drug could also be given to pregnant women in order to prevent miscarriage, to women suffering from climacteric disorders, and even to young boys whose biological development had been arrested at the expected advent of puberty.

Usually, preparations touting so many claims were categorized as 'panaceas' or 'cure-alls' and not unusually associated with quacks and charlatans (Palmblad 1997). But no dissenting voices were raised by the members of the fourth estate. 'In context, the final result is less astonishing but no less impressive.' The context referred to and which contributed to expectations that Gonadex would fulfil its 'lofty promises' can be summarized with the key concepts 'Swedish', 'science-based', and 'industry'; and the link uniting them is Axel Westman. As another paper (*Svenska Morgonbladet*) trumpeted, 'Gonadex is without doubt a major triumph for Swedish research and Swedish industry!' This triumph, illustrated in many of the papers with a portrait of 'inventor' Axel Westman, was of international proportions. 'The new drug is of worldwide significance and a major scientific triumph for Sweden', wrote *Husmodern* (19). It was also pointed out that Gonadex, due to its ability to promote nativity, was of such economic and political value to the nation that the government ought to seriously consider establishing a special national 'sterility clinic' so that it could be maximally exploited.

The unabashedly enthusiastic articles published in the daily press and widely different household readings certainly appear a bit too sanguine in light of Westman's lecture and his modest academic publications on the topic. The enormous emphasis on sterility also appears hasty, to say the least. When Gonadex was registered, it was intended for use on a variety of genital disorders, of which sterility was but one, and one based on vague assumptions, to boot. And yet it was on the cure for sterility the press focused. Why was this?

In the network of relationships that circumscribe the launching of new, untried medicines, the mass media comprise one of the most important nodes. The media contribute to drawing attention to new medical knowledge, thereby raising awareness both within (Phillips et al. 1991) and beyond the medical–academic–industrial establishment, and interpreting its possible meaning for society. It is commonly assumed that the mass media exaggerate, simplify, and distort the results of the life sciences in order to create a 'story' (Ideland 2002). While the case of Gonadex can be understood in this fashion, circumstances were actually more complicated than that. The newspapers did indeed mediate a rather one-sided and uncritical image of the possibilities offered by Gonadex, but this image mirrored the intent of Axel Westman himself, which Leo did its best to inflate at the press conference. The mass media reporting was in other words a result of Leo's marketing practice.

This interpretation takes its cue from Dorothy Nelkin (1995: 11) who argues that the research world 'sells' science to society, e.g. by making extravagant claims in its press information, which creates interest among the public (including journalists) and attracts financiers. There is also empirical evidence for this perception in a document comprised of three typewritten pages and entitled 'Barnlös plötsligt gravid med nytt hormonpreparat' [Childless suddenly pregnant due to new hormone preparation]. As far as can be ascertained, it is a copy of the 'pre-interview' with Axel Westman referred to in the 'Memo re: Publicity for Gonadex' mentioned above, which was distributed before and during the press conference. However, a more fitting term would be 'press release' or, to borrow Nelkin's term, 'a prepackaged source of information' (Nelkin 1995: 119–21). The text features most of the information and word-for-word formulations of sentences that would later appear in the newspaper articles. In order to fully comprehend the mediated image of the values of Gonadex, there is good reason to draw attention to this text.

A well-written press release attracting a large number of journalists is absolutely imperative to the success of any press conference. If said press conference deals with research, then, according to Thomas Gieryn (1999: 187–8), it is essential that the press release meet three criteria: the results to be presented must claim to have huge significance for a large number of people; they should be seen as a new 'discovery'; and finally, they must appear credible, in order for the mass media to believe that they are being given facts, not fed fiction. Leo's press release succeeded in following these guidelines. The significance of Gonadex for the sterility issue and therefore all mankind is already broached in the title—'Childless Suddenly Pregnant'. After that follows a brief introduction featuring assertions that a longed-for scientific breakthrough has finally taken place: 'Sweden first with sensational scientific news', and 'Professor Westman achieves elusive goal after ten years'. The main text opens with the following paragraph, which elucidates the reliability of the results:

> A Swedish hormone preparation of unheard-of efficacy and utility will be available for sale in the next few days after one and a half years of rigorously concealed experimentation. [...] For the man behind the achievement, our internationally renowned and celebrated gynecologist Professor *Axel Westman*, this drug means a major personal triumph and possible invaluable weapon in the struggle against sterility, to which he has been tirelessly committed for so many years. The value world science places on this discovery, which Sweden is the first country to present, is evident in the efforts of the Americans, who have spent months trying to get their hands on the results after some information about the experiments was leaked. Without success, we should add. ('Barnlös plötsligt gravid', Leo Historical Archives)

Note that 'the Americans' was an important signifier in Swedish research and research policy at a time when not only science but also the national self-image was becoming increasingly oriented towards the US (Nilsson 2012: 276 f.; Wennerholm 2004: 143). Invoking the interest expressed by the international community was a sure-fire method of rousing attention in Sweden. Only after this is a more detailed description of the actual significance of Gonadex for sterility brought up. This was probably the information—not mentioned in the academic articles—which really caught the attention of reporters.

> The preparation has in fact been ready for some six months now, and among the approximately thirty cases of e.g. childless women on whom Professor Westman has hitherto tested the drug, *one* in particular has been especially gratifying. For three years, the professor tirelessly but in vain struggled to cure the sterility that had embittered the marriage of a female patient and her husband. She wanted to leave no avenue unexplored, even though all indications were that she would be forced to give up. However, under the influence of the new hormone preparation she became pregnant almost immediately. ('Barnlös plötsligt gravid', Leo Historical Archives)

Much of the rest of the text is reminiscent of Westman's lecture at the press conference. But it is also stated that the Karolinska Hospital would continue to show its gratitude for the contributions of pregnant women to the production of Gonadex, which was 'on the brink of becoming an international commodity'. The impression is that a need for vast quantities of urine is the only thing standing in the way of Gonadex breaking into the world hormone market. But it was not quite that simple.

In the months immediately following its launch, Leo was bombarded with requests from individuals wishing to try the new wonder drug. Gonadex was also put to use in many types of treatments, including sterility treatment, but had a hard time living up to expectations. Physicians, scientists, users, official and unofficial inspectors—everyone had an opinion about Gonadex. Some also had an opinion about Westman. In an article entitled 'Vetenskaplig reklam och praktisk erfarenhet' [Scientific advertising and practical experience], a well-known Stockholm-based physician named Waldemar Gårdlund expressed deep discomfort at the prospect of hormone therapy, including the use of Gonadex. But he was particularly critical of the way in which Gonadex had been portrayed in the material handed out at the press conference, sneering at 'the pretentious and ostentatious manner in which [Westman] pounds the advertising drum for his hormone studies' (Gårdlund 1948: 1746). For Gårdlund, this was a question of ethics, not compatible with the contemporary moral economy of medical science:

> We require certain standards from a representative of science, the highest possible level of trustworthiness, particularly in the case of the holder of a chair in medical studies, inasmuch as he is charged with imparting logical, medical methods, therapy and *medical ethics* to a new generation. He must serve as an inspiration to young people.

That Gårdlund used the term 'Leo-Westman' in his article can be understood as an attempt to expose and also cast aspersions on Westman's dealings with private industry, as if he was selling himself. There were, ultimately, values of trust at stake. Apparently, it was one thing to cooperate with the industry in the lab or in the clinic, but to do so in public was not acceptable—yet.

This public critique was followed by trouble. Four months after the press conference and half a year after seeking a registration, Leo was contacted by the NPL, which informed the firm that Gonadex would probably not be approved as a pharmaceutical specialty. Interestingly, the reason was not that Gonadex lacked the positive effect on patients promised in its advertising (which it did), nor that it proved to have harmful side effects (which it had). The problem was rather the manner in which the strength of Gonadex was indicated on its packaging. During the biological tests conducted at the NPL, it had emerged that the preparation was 'considerably weaker' than its standardized, international counterpart (Rydin to AB Leo, 8 October 1948, Leo Historical Archives 1948b).

It soon turned out that not only the NPL had its doubts about Gonadex; on the first of December 1948, Leo's director revealed to Westman that reports of standardization problems had also been received from two independent instances in Holland and in Denmark. And the registration application was rejected (Rydin to AB Leo, 3 December 1948, Leo Historical Archives 1948a). Although Gonadex later made a comeback, and in fact remained in the market until 1986, it was not much used and was never looked upon as a success. Rather the opposite.

Conclusion

The aim of this chapter has been to shed light on the flexibility of the moral economy of pharmaceutical R&D in the first half of the twentieth century by focusing on three types of exchange relationships between medical science and pharmaceutical industry—consultation, procurement, and mediation—and the values these practices incorporated or gave rise to.

As has been demonstrated, the drug Gonadex can be interpreted as a result of interaction, cooperation, and reciprocity between Leo, Professor Axel Westman and his clinic, and the external hormone lab at the Karolinska Hospital. As a consultant, Westman was definitely involved in both the design and testing of the drug. When Leo chose to invest in a 'line of hormones' during the 1940s, Westman suggested—for both scientific and medical reasons—that human chorionic gonadotropin be rendered in as pure a form as possible, and he conducted the clinical tests of the substance at the Karolinska Hospital's women's clinic. The purification process itself, however, was developed and carried out at Leo's plant in Helsingborg, while standardization was done at the hormone lab.

Westman also participated in the Gonadex project in many other ways. He helped establish the infrastructural arrangements necessary for Leo to gain access to the raw material—urine of pregnant women—from which the hormones could be extracted. Contact had to be made with public institutions and Westman's name opened doors. Moreover, he indirectly participated in the marketing campaign by acting as a mediator between the firm and physicians, journalists, and consumers. If Westman, with his social and symbolic capital, had not agreed to be its poster boy and scientific guarantor, Gonadex would never have attracted the attention it did.

Thus Westman was an invaluable asset for Leo. He may in fact be described as an early Key Opinion Leader (KOL) (see Sergio Sismondo, Chapter 2, this volume). But Westman also profited from collaborating with Leo far beyond the fee he was paid for his consultancy services. Thanks to Leo, Westman gained access to Sweden's first and foremost hormone lab, with the assistants, lab animals, and purified hormones that come with it, which was of benefit to the medical practice and scientific studies he and his colleagues were conducting. Association with that lab, which moved with Westman from Uppsala to Lund and finally to Stockholm, probably also played a decisive role in his securing further research funding from internal and external financiers. Attempting to draw an inviolable line between Westman's consultancy work and his basic research must in these circumstances be considered a fool's errand.

This moral economy of the pharmaceutical company–medical scientist collaboration behind Gonadex must have been hard to grasp for an outsider. Leo boasted of its alliance with Westman. During the launch of Gonadex, Leo explained that it was Westman who was the 'man behind the work'. This is not so odd since the connection with outside expertise heightened the company's reputation, credibility, and competitive edge in the pharmaceutical market (as illustrated in Figure 3.1). In fact, the company was so eager to be associated with Westman that it continued to nurture the relationship long after it

TEAMWORK IN
MEDICINE, CHEMISTRY
AND PHARMACY

The basis for our success is our close and personal relationship with physicians and other scientists. The research work carried out in our modern laboratories is directed by the leading specialists of the day. Their scientific knowledge combined with our technical resources form a perfect team which ensures that our products are in complete accord with the most recent medical advances.

Figure 3.1. Advertisement on Leo's successful collaboration with physicians and other scientists published in August 1948 in the maiden issue of *Acta Endocrinologica*, edited by Axel Westman.

realized that Westman was no longer capable of contributing any research of commercial value to their production efforts.

The image created by the media was that Westman was responsible for the scientific handiwork that was the basis of Leo's hormone drug, which the news media later described as a 'Swedish triumph'. This image—that the company only exploited the results of Westman's pure research—was later perpetuated in a PhD dissertation in economic history on 'The transfer of knowledge from university to company: A study of the importance of university research for the introduction of new products in the Swedish pharmaceutical industry, 1945–1985' (Norgren 1989: 65, 93), which used the connection between Westman and Leo as a case. However, I have not been able to find any conclusive evidence of such a straightforward, linear transfer of knowledge from the hormone lab to the company.

For his part Westman only rarely mentioned in public his relationship with Leo. His message during the 1948 press conference and in the mass media was that Gonadex was the result of the hard work done by Leo's 'chemical research staff'. This was not altogether misleading, because they were indeed the ones who achieved purification, if at his urging. But we must not forget that Westman's argument in favour of extracting hormones not only had commercial motives; he also needed them to develop his research and clinical practice; they had a scientific value, which subsequently would also have a medical and in the end political value. Westman's modesty came as no surprise since in the eyes of some of his contemporary peers in the medical sciences, being officially connected to commercial industry hardly enhanced his reputation, credibility, and competitive capital; quite the contrary. A public relationship of this nature could still be seen as something negative, which is suggested by the vehement criticism levelled immediately after the launch of Gonadex.

Summing up, it can be said that while Leo exaggerated the significance of Westman's science for Gonadex, Westman in turn toned down both his significance for Leo and Leo's significance for his own research. The significance of industry was cloaked, a phenomenon also reflected in Axel Westman's obituary, which makes no mention whatsoever of his thirty years of collaboration with Leo. Hence, the moral economy of Gonadex cannot be traced only from public statements; it is only by taking a close look the actor's everyday practices in context that the co-production of values and R&D can be unmasked and understood.

It should finally be mentioned that the 1948 press conference was not an isolated incident. Rather, it can be seen as an expression of a new, public-relations trend gaining ground in Swedish society at large after the Second World War. Articles on new medical 'breakthroughs' began to be featured so often in the press that the professional medical community felt compelled to call for a more temperate dissemination of information via the mass media. One critic who aired his grievances publically was the physician Leonard Brahme. In an open letter printed in *Svenska Läkartidningen* under the rubric 'Framsteg och publicitet' [Progress and publicity] in 1951, he urged his honoured colleagues to show some restraint. It is worth quoting, as it clearly illuminates an essential tension within the contemporary moral economy of the life sciences:

> Recently, the undersigned was asked by an influential political figure whether it would not be prudent to refrain from announcing 'medical advancements' until they had undergone more thorough critical medical analysis. He continued, 'I am fed up with the many newspaper articles about all the wondrous medical progress being made, which are probably also producing hope in the infirm which cannot always be redeemed. Doctors have a great responsibility with regard to the harm such articles may cause.' I agree entirely with this individual's opinion. Dubiously reliable recent reports rolling off the daily and weekly presses with conveyor-belt regularity about miracle cures are becoming the rule. This is something we must carefully monitor and something which is certainly far from desirable. For obvious reasons I shall refrain from going into detail but a practiced physician knows the kind of troublesomeness these articles can create and the disillusion they often cause. *In the interest of the sufferers, I feel that it is time for the leadership of the Swedish Medical Association to bark a resounding 'Attention!'* We are embarking on the wrong path. We must learn to forgo unnecessary publicity. Should some successful result excite the enthusiasm of the physician in question, said enthusiasm should be tempered by a critical mind. In that way, we shall be spared the many sensationalistic articles in the press and liberate sufferers from unnecessary delusion. (Brahme 1951: 415)

ACKNOWLEDGEMENTS

This chapter draws on my book *Hormones of Life: Endocrinology, the Pharmaceutical Industry, and the Dream of a Remedy for Sterility 1930–1970* (Sagamore Beach, MA: Science History Publications, 2011). Work on that book, as well as the present chapter, was funded by a Pro Futura Scientia Scholarship from Riksbankens Jubileumsfond and a Young Researcher Award from Umeå University. Thanks are due to Sergio Sismondo and the editors of the present volume for critical comments, and to Stephen Fruitman for language editing.

4 The third manuscript

Rules of conduct and the fact–value distinction in mid-twentieth-century biochemistry

Sven Widmalm

> Valuations are necessarily highly subjective but should be influenced by correct information and facts . . . If we can not agree about values we may at least facilitate mutual understanding by clearly stating the facts on which our evaluation of values is founded. (Tiselius 1970: 13)

In mid-September 1969 an illustrious group of scientists, scholars, writers, and businessmen (plus a few students) gathered at Södergarn, a conference centre in one of the loveliest suburbs of Stockholm, for the first ever 'cross-cultural' Nobel Symposium. Thirty papers were presented on 'the place of values in a world of facts'. A galaxy of intellectual and scientific stars was present: Jacques Monod, Joshua Lederberg, E. H. Gombrich, W. H. Auden, Gunnar Myrdal, Arthur Koestler, Glenn Seaborg, Margaret Mead (the only female), and so on.

The opening address was given by the biochemist Arne Tiselius (1902–71), one of the most influential elder statesmen in Swedish science at the time. More than two decades had passed since Tiselius received his Nobel Prize and almost as long since he gave up being an active researcher and instead focused on managing his laboratory, not least various collaborations with industry and with research groups at home and abroad. In the 1950s he emerged as a dominant figure in Swedish research policy and had become more and more active on the international policy scene—in his capacity as chairman of the Nobel Foundation but also in matters related to broader political issues, for example through participation in the Pugwash peace movement. The cross-cultural Nobel Symposium was his brainchild (Kekwick and Pedersen 1974: 417–18). As the quotation above indicates, Tiselius viewed the problem of fact and value dualistically: in science there was 'a clear distinction between false and true' whereas valuations were 'highly subjective' (Tiselius 1970: 13). The idea behind the symposium was to encourage new ways of thinking that could bring science to bear on important issues (environmental, nuclear, etc.) where subjectivity seemed to reign.

Since the 1960s it has been commonplace to assert that science is value laden and that facts without valuation are meaningless (Dupré 2007); or, as Hilary Putnam has it, that 'without values we would not have a world' (Putnam 1982: 11). But within the framework of a largely positivistic mid-twentieth century science, pressured by politicization as well as commercialization, the fact–value distinction helped legitimatize the scientific enterprise not least by emphasizing its autonomy. This was done by ascertaining that scientific facts are impersonal, as well as by an increasing emphasis on peer review, where scientists supposedly demonstrated their ability to evaluate *each other* impartially. Robert K. Merton's early attempt to systematize the normative structure of science stressed norms (universalism, disinterestedness) that idealized impartiality with regard to research results as well as researchers (Merton 1973a).

The importance of understanding the evaluation of peers, as well as the facts they produce, from sociocultural perspectives has recently been emphasized in philosophy and history of science (Daston and Galison 2007; Wray 2007). This paper discusses an instance of such evaluation in mid-1950s biochemistry, in connection with suspected misconduct. The case concerns a controversy over who first published on the purification and structure of MSH (melanocyte-stimulating hormones, or intermedins; a class of proteins that have various physiological effects including stimulation of melamin production, thus affecting pigmentation). There was a dispute concerning the publication of the purification and the amino-acid sequence of so-called β-MSH, and accusations of misconduct were directed against the leading American biochemist Choh Hao Li (1913–87). What Arne Tiselius would call the 'Li affair' took place exactly at the same time as Merton worked on his well-known paper on the importance of priority in science, interpreting this as a symptom of the centrality of the norm of originality (Merton 1957).

The research tradition spearheaded by Merton focused on the scientific norm system and infused norms with social agency. It produced important insights into the workings of scientific culture and we learn even more from the Mertonians if we recognize that they tended to mirror concerns that occupied senior colleagues in academic science (Shapin 2008).[1] The work of Merton, Harriet Zuckerman or Warren O. Hagstrom provides us with a semi-official version of the then current ethos of science, and this is most certainly true of Merton's analysis of the function and dysfunctions of the phenomenon of priority.

[1] Arne Tiselius was for instance instrumental in gaining Harriet Zuckerman access to the scientific elites on which she completed a pioneering sociological study. Zuckerman (1996: 4) wrote that 'Tiselius embodied the Nobel establishment' and credited him for having inspired her to turn her thesis into a book by giving her access to this establishment. Letters from Zuckerman to Tiselius are at the Archive of Arne Tiselius at the Archive of the Biochemical Department, Archives of Uppsala University (ATAUU), F16: 38.

Though attention concerning conflicts and misconduct in relation to priority claims has grown since the end of the last century, historians have not done much analytical work to elucidate them (Claxton 2005a, 2005b).[2] This paper makes use of Merton's insights but focuses not so much on norms as on *rules of conduct*. Norms may be seen as reifications and perhaps internalizations of (some) rules that are recognized implicitly or explicitly within a scientific community; rules are in this sense more fundamental. This has been argued in legal scholarship where Richard Delgado (1991: 936) has claimed that normativity should be studied in the same way as research is studied in the sociology of knowledge, i.e. as an activity rather than as the implementation of a set of guiding principles. This is perfectly logical if research is seen as a form of 'collective action', as described by Elinor Ostrom (2000). She listed eight 'design principles' underlying successful collective action that are all, more or less, applicable to the scientific community. Not least did she emphasize the need for a system of sanctions to hinder repeated rule breaking and free riding (principle 5) and for a system of conflict resolution (principle 6). These were central issues in the Li affair. The actors that I follow interpreted rule breaking in terms of trustworthiness and individual character, corroborating Steven Shapin's (2008) claim that personal 'virtue' is of central importance in late modern science. As the actors saw it, conduct had to be monitored in order for collaboration and competition to function. This may also be understood from the point of view of a moral economy of research. Proper conduct was a matter of adjusting to accepted rules guiding the exchange of symbolic capital within a framework of limited resources.[3]

To some extent it is possible to reconstruct the sequence of events in the Li affair from the voluminous paper trail that was generated by the actors. It is striking that leading members of the biochemical community were willing and able to produce such extensive documentation concerning the exchange of information among peers so that it could form the basis for a quasi-judicial process against one of their own number. The very existence of the paper trail illustrates the importance of the issues under consideration; it gives witness to the scientific community's obsession with proper conduct in information sharing, and also to the applicability in mid-twentieth-century science of Ostrom's sixth design principle, of a 'mechanism for discussing and resolving what constitutes a rule infraction' (Ostrom 2000: 152).

[2] Despite claiming to be in part historical these articles deal primarily with contemporary issues. For an up-to date discussion and recommendation concerning editorial policy, see the latest version of the white paper of the Editorial Policy Committee of the Council of Science Editors (CSE) (Scott-Lichter et al. 2012). One genre that has thrived is the treatment of misconduct as criminal behaviour, often focusing on the few *causes célèbres*, like that of J. H. Schön or the discovery of the structure of DNA. An exception is the detailed discussion in Kevles 1998.

[3] On the moral economy of research, see the Introduction (Chapter 1) to this volume. For a discussion of the moral economy of research in Uppsala biochemistry in the 1950s and 1960s, see Widmalm 2014.

The story that will unfold here is thus a product of the collective construction of evidence by leading members of one side in the conflict, particularly in Cambridge and Uppsala. There is probably a similar paper trail preserved among the Choh Hao Li papers at the University of California, San Francisco.[4] As this collection, which I have not been able to consult, would reflect Li's side of the story it is likely that it would produce a somewhat different narrative from the one I have unearthed.

The following reconstruction of the Li affair will present a coherent narrative from the point of view of Li's accusers, which they themselves would probably have recognized.[5] This narrative does not result in a definite conclusion concerning the issue of misconduct. It was hard to construct a version of the truth that could be collectively recognized as such, and risky to promulgate value judgments on peers. The latter problem was described using the fact–value distinction: the informal process against Li supposedly unearthed relevant facts; valuation was left to the broader community of peers, or possibly a committee of some sort at Berkeley.

The narrative will reveal rules explicated during the investigative process. These were associated with the functionality of the research system as a form of collective action, and with the 'probity' of individual researchers, rather than with any clearly articulated norms. It will be suggested that the monitoring of rules and behaviour may—no matter what the outcome—be seen as an objective in itself, as these processes demonstrated a will to uphold collective action. I should add that this interpretation may look more functionalistic than it is. The following case study will have little to say about how effective the policing of researchers' conduct was from the point of view of the relevant scientific community; the point is rather that scientists acted *as if* they could guard the system's functionality. Such collective action was therefore performative, and I suggest in the concluding section that the audiences for the performance were at least two: the scientists themselves and, implicitly, governments that funded and expected great things from an increasingly bloated system of academic research.

The Li affair

Choh Hao Li was one of the most academically successful American biochemists of his generation (Cole 1996; Zulueta 2009). With a chemistry degree from

[4] According to the description at the Online Archive of California, carton 17 of series 1 (correspondence, 1939–87), sub-series 2 (subject correspondence, 1951–82) contains 'information on the -MSH controversy (1956–57)'. <http://www.oac.cdlib.org/findaid/ark:/13030/tf738nb543/dsc/#c02-1.8.5.2.3> (accessed 26 July 2012).

[5] Jerker Porath has corroborated the broad outlines of my narrative. Interview, 23 August 2012.

Nanking he followed his older brother to the United States in 1935, hoping and eventually succeeding to carry out graduate work at Berkeley. There he became a protégée of the endocrinologist Herbert Evans at the Institute of Experimental Biology. Li focused on chemical and biomedical studies of pituitary hormones; his most important work would be concerned with the human growth hormone. Having become associate professor in 1949 he obtained a Guggenheim fellowship and promptly sailed off to Uppsala in order to work with leading experts there on techniques for separating (thus purifying) large molecules like proteins.

In particular he collaborated with, and befriended, Arne Tiselius, expert in chromatography and electrophoresis. At this time Tiselius held a chair in biochemistry but still cohabited with his mentor The (Theodor) Svedberg at the physical chemistry institute; his own biochemical institute was in the pipeline and would open in 1952. A third generation of separation specialists was emerging at Uppsala at this time, among whom a central character would be Jerker Porath (1921–), co-inventor of gel filtration and other techniques (Widmalm 2014). Porath would visit Li's lab in Berkeley during 1951 and 1952 and would contribute to Li's field through work on the purification of hormones. Another researcher who joined Tiselius's laboratory in the early 1950s was Paul Roos (1928–2006), who would specialize in the study of the growth hormone. Both Porath and Roos, although still PhD candidates, would become involved in the controversy concerning MSH.

On his way to Uppsala Li stopped at the biochemistry department in Cambridge, where Frank Young (1908–88), who had similar research interests, was about to become director. Li stayed for a month hoping to learn techniques for protein analysis being developed by Fred Sanger (1918–2013). He then went to Uppsala where he was apparently so happy that he considered staying for some time, having been offered a visiting professorship funded by the newly established Swedish Medical Research Council (Zulueta 2009: 150–7). At Berkeley, the President Robert G. Sproul, Li's mentor Evans, and in particular Wendell M. Stanley (the latter an old friend of Tiselius) managed to convince Li to return, offering him a professorship in biochemistry. By 1950 Li was a full professor at Berkeley's biochemical department and was establishing Berkeley's Hormone Research Laboratory (HRL). According to R. David Cole, Li's experiences in Cambridge and in Uppsala 'set the research approach of Li's lab for years to come' (Cole 1996: 225). Li's career flourished; scientific accomplishments and awards 'continued to roll by with almost predictable regularity' (Zulueta 2009: 166). Cole, who joined the HRL in 1952, presented the following rose-tinted image of the scientific life there:

> The research staff consisted of David Chung, Peter Condliffe, Jonathan Dixon, Irving Geschwind, Ieuan Harris, George Hess, Anthony Levy, Harold Papkoff, Ning Pon, and Jerker Porath. Through the years many famous researchers visited

> and many worked for a while in the lab. Li made a point of generously sharing the time of the famous ones with all of us; it was a heady experience for a graduate student... Although my recollection might be colored by nostalgia for the sweet simplicity of graduate years I recall a lot of laughter and excited discussion; it was fun and hard, driving work. My impression is that C. H. was successful in maintaining an ambience of joy, excitement, and strong friendship within the HRL throughout its history. The family-like affection of the large numbers who were associated with C. H. in the HRL was evident in enthusiastic reunions on the twentieth and thirtieth anniversaries of the founding of the HRL and on the celebration of C. H.'s sixtieth birthday. Many who attended came from the far ends of the world; the HRL was a markedly cosmopolitan (and democratic) community. The HRL had a personality—in many ways the personality of C. H. himself. (Cole 1996: 226–7)

What happened in 1956 might be seen as a breakdown of scientific community relations, or at least of the *appearance* of well-functioning collective action in science (Jerker Porath has emphasized that competitiveness and extreme demands of productivity were also characteristics of the culture at Li's lab).[6] The controversy that erupted in that year pitted Li against many of those who had been involved with his career and work so far: Sanger, Young, and J. Ieuan Harris (1925–78) in Cambridge; Tiselius, Porath, and Roos in Uppsala; Stanley and others in Berkeley. The driving force behind the controversy was Harris, a junior researcher with a grudge.

Collaboration or competition?

The first document, chronologically speaking, in Tiselius's dossier on the Li affair is a letter from Li to Harris from 29 February 1956. The letter was not only preserved by Tiselius but was, almost a year later, included among the papers concerning the affair sent to the biochemistry department in Berkeley to be used in an institutional investigation of suspected misconduct. It was considered important because it documented Harris's conduct when the race to solve the structure of MSH was about to begin. From a technical point of view two problems were important in the controversy that would ensue: purification, a specialty at Uppsala, and determination of the structure, or the amino-acid sequence, a specialty in Cambridge where Sanger had recently succeeded in determining the structure of insulin, for which Tiselius and others would shortly award him a Nobel Prize.

Li's letter was in reply to an earlier letter by Harris where a suggestion had been made that MSH could be investigated through collaboration between Cambridge, Uppsala, and Berkeley. Li's response was encouraging but non-

[6] Interview with Jerker Porath, 23 August 2012.

committal. There was no sign of discord at this stage. On the contrary Li reminded Harris that he wanted him to send a photo of himself—perhaps aimed for a collection of portraits of collaborators at the Berkeley institution.[7] As far as Harris was concerned, however, such appearances of concord were misleading.

What this letter was supposed to show, then, was that Harris told Li at an early stage, before research work had begun, about the plans for collaboration between Cambridge and Uppsala on the structure of MSH. As collaboration would have had to have been arranged by the laboratory directors at the respective universities, Harris was slightly out of bounds when suggesting that Berkeley join in. A recurring underlying theme in the exchanges that the Li affair gave rise to were the authority, power, and responsibility of laboratory directors: they had the final say in everything concerning collaboration and competition but also bore the brunt of responsibility when things went wrong. In fact Tiselius and Li had already discussed in September 1955 the possibility of collaboration, and Tiselius had reached the conclusion that Uppsala had nothing to gain by such an arrangement as he 'felt that we were ahead of the Li group':

> It seems to me that in cases like this there are only two alternatives: either close collaboration which necessitates joint publication or rather independent work, where information is exchanged only when definite goals are reached.[8]

For Harris to suggest collaboration after Tiselius and Li had made an implicit agreement (Tiselius said the discussion had been rather vague) was a transgression but, he thought, a defensible one:

> You will... appreciate that it really was neither my *right nor responsibility* to write to Li at all about MSH. I did so purely because I was personally concerned about the impending competition, and wanted to warn him that we were in fact going to work on MSH.[9]

It should be noted that both these comments on academic rules of conduct were made after the controversy had begun and should perhaps be seen as a result of it. At the same time Tiselius's views seem to have had support in the scientific community at the time—they were not questioned by colleagues during the Li affair, and they correspond fairly well with views presented in Warren O. Hagstrom's broad investigation of attitudes to teamwork and competition in American science carried out in the early 1960s. Hagstrom stated, as a general truth, that '[p]rofessional scientists whose research is likely to lead to the same discoveries are... invariably either competitors or

[7] Choh Hao Li to Ieuan Harris, February 29, 1956, ATAUU, F16:24, dossier 'MSH–Li'. All subsequent archival material quoted is from this dossier.

[8] Tiselius to Frank Young, October 13, 1956.

[9] Harris to Tiselius, October 5, 1956.

collaborators' (Hagstrom 1975: 69). When competing, he wrote, researchers 'may sacrifice other scientific values to obtain' priority:

> [S]cientists may violate the norms of free communication in science in order to protect their competitive standing, and they may default on their obligation to present only thoroughly verified results in the interest of publishing quickly. (Hagstrom 1975: 100)

It is unclear, however, from Hagstrom's account to whom one had such an 'obligation'. Instead he seems to show that there were differing views regarding quick publication in the form of abstracts or 'letters', which is interesting from the point of view of the Li affair where the dispute evolved around such brief communications. Hagstrom identified two common practices: either one used brief communications in order to ascertain priority and then quickly followed up with a substantial paper giving all the usual technical evidence; or sometimes authors did not follow up a brief communication with more detailed presentations of the evidence behind their claims, thus leaving colleagues to either accept or discard their results. Hence brief communications that were not followed up by substantial papers were liable to create confusion. Hagstrom—whose main concern was to analyse what norms hindered the efficient production of knowledge—discussed these matters in terms of 'rules of behavior', distinguishing them from matters that had to do with 'higher order social norms' (Hagstrom 1975: 93–8, 254). From my perspective, such rules are of central importance as they constitute the fabric of the moral economy of research that underlies collective scientific action.

By February 1956 the stage was set for a race concerning the structure of MSH. Harris and Tiselius had both hinted to Li (they claimed) that they would attempt to solve it. Tiselius seems to have thought that the meeting in September made clear to both parties that each would work on MSH but not collaborate.[10] Uppsala's contribution to the collaboration consisted of the purification of MSH by Porath and the participation in the sequencing work at Cambridge by Roos.

Setting a trap

Roos arrived in Cambridge in April, by which time all exchange of information between the Uppsala/Cambridge group and Li's lab had ceased as both tried to get ahead in ascertaining priority for the structure of MSH. By the end of May the amino-acid sequence had been successfully determined by him and Harris.

[10] Tiselius to Young, October 13, 1956.

During this period Harris had asked Tiselius how to handle communication with Li, and Tiselius had said that news about the eventual success should not be spread until a paper announcing the results had been accepted for publication, thus articulating another rule of conduct for scientific competition that seems to have been generally recognized. By 18 June the Cambridge group had sent a letter to *Nature* and received a note acknowledging receipt but no word (as far as we can tell) about whether the paper would be accepted. Two days later Harris sent a letter to Li announcing that he and Roos '*have been taking a look at the structure of MSH*'. The letter ended with a flippant wish that Li was '*doing good work on everything—except MSH!*'[11] The emphasizing of these words was made by Li who would send a copy of the letter to Tiselius, the point being to show that Harris implied that work on MSH was underway whereas in reality it was finished. He would hear from Sanger about the note from *Nature* on 18 June, therefore assuming (he claimed) the paper had already been accepted. He implied that Harris, by not mentioning this, had deviated from the scientific rules of conduct.[12]

According to himself Harris had not broken the informal rules guiding science communication in a race for priority. But by his own admission his intent had been somewhat devious. He had attempted to set a trap for Li: the letter 'was meant to be a test of his sincerity in the matter'.[13] If 'all had been above board' Li should have replied that Berkeley also was working on the structure. 'Instead he rushed a partial structure into press' and did not reply to Harris's letter until three weeks after his own paper was submitted.

On 30 June Li and co-authors Irving I. Geschwind and Livio Barnafi sent a manuscript with the Berkeley group's own structure of MSH to Marshall Gates, assistant editor of the *Journal of the American Chemical Society* (JACS). The paper was received on 2 July and was refereed by Paul H. Bell, a chemist working in a private company. On 17 July his comments were sent to Li together with a request that the manuscript should be shortened by 160 words.[14] In this manuscript the sequence of MSH was not complete. The fact that the determination of four amino acids (position 14–17) was not yet 100 per cent certain was clearly indicated in the text; furthermore the suspect sequence was marked out by parentheses. Hence Harris's gambit had succeeded: Li's group did rush things, apparently in order to ascertain priority (as they still did not know that the Cambridge structure was being accepted for publication and that the game was in effect already lost).

[11] Harris to Li, June 20, 1956. [12] Li to Tiselius, October 16, 1956.
[13] Harris to Tiselius, November 5, 1956. [14] Marshall Gates to Li, July 17, 1956.

Presenting results

The *Nature* article (a brief 'letter') would appear in print on 14 July and on 10 July Harris sent the manuscript to Li. One of the great matters of contention in the coming controversy would be exactly when this manuscript reached Berkeley. It was sent by airmail and should have arrived in a few days.[15] Li, however, would later claim that he received the manuscript only on 23 July. For a while it seemed as if everything in the evolving controversy hinged on these dates.

On 16 July, before revisions had been made, Li sent the first Berkeley manuscript to Cambridge, i.e. the version that still contained the uncertain (and, as it would turn out, incorrect) sequence.[16] Exactly what happened next was a central question in the controversy and I will go over the details below. There was no controversy however concerning the fact that the revised manuscript was received by Gates at JACS on 24 July, that the (manuscript) paper by Harris and Roos was discussed at a 'Thursday seminar' in Berkeley on 26 July *at the latest*, and that Li sent a manuscript version of the Berkeley paper to Sanger and Young on 2 August. In this version the sequence in positions 14–17 had been changed so as to be identical with the one in Harris and Roos's paper; and the parentheses as well as the sentences indicating uncertainty had been excised. But no reference to the Harris–Roos paper was provided.

Sanger became the whistle blower. On 8 August he wrote to Gates pointing out the lack of a reference to the Cambridge paper.[17] Gates replied that there had been small changes made in Li's paper after it was submitted for the first time and that the order of the uncertain sequence had indeed been changed 'so as to agree with the sequence given in Harris and Roos' paper'.[18] This was worrisome Gates thought: 'From this distance it seems likely that change was made after examining the manuscript of your colleagues. If so, this aid was not acknowledged.' At the same time Gates was unwilling to press Li on the matter as this would mean questioning Li's 'scientific probity'. Besides, as Harris and Roos had published first, their priority was already guaranteed.

With Sanger's reaction the cat of misconduct was however out of the bag and Gates had to confront Li with the matter. He wrote him (with a 'blind copy' to Sanger) suggesting three courses of action. First, Li could revert to the first version of the manuscript, parentheses and all. Second, he could add a reference to Harris and Roos. The third suggestion is more intriguing: 'The

[15] Harris to Tiselius, October 5, 1956.

[16] A general overview of the chronology of events and the different charges against Li et al. are given in Young to E. E. Snell, December 19, 1956.

[17] This letter is not among the papers in Tiselius's dossier, though Gates's reply is.

[18] Gates to Sanger, August 15, 1956.

sequence given in the second version of your manuscript be used but the received date changed to the date at which this second version was received in this office (July 24, 1956)'.[19]

One point of the third suggested course of action must have been that the original received date (2 July) would have been before 14 July, when the *Nature* paper appeared, and that a change of date would make it clear that Cambridge had priority even though they were not quoted. (If Sanger had not reacted and 2 July been given as the date of reception the matter of priority would have become rather muddled.)[20] Another more complicated issue had to do with the question of exactly when the change in the MSH sequence in the Berkeley paper was made. The date 24 July would, to those in the know, indicate that revisions were made before the Cambridge paper arrived in Berkeley. Li chose this alternative and on 5 September the paper was published stating in the first footnote that the '[o]riginal manuscript' was received on 2 July and by the end of the paper: 'Received July 24, 1956' (Geschwind et al. 1956).

It is important to note that the course of action chosen by Li was actually proposed by the journal editor Gates. He suggested that it was unnecessary to quote Harris and Roos, as indeed he had also indicated to Sanger. He also lent authoritative support to Li's version of events, that the last changes in the manuscript were made on 24 July—something that turned out not to be the case. This reminds us of the fact that the relationship between journals and high-profile researchers is symbiotic, the status of each affecting that of the other. But it soon became obvious that Gates and Li had miscalculated. Their views on what may pass as proper conduct were not accepted by other high-profile players.

Publicity

One thing that spurred action from Li's critics was the publicity generated by the Berkeley paper (but not by the Cambridge paper). The *New York Times* (*NYT*) published an account of the Berkeley group's work, attributing to them priority for the discovery of the structure of MSH as well as its purification. This prompted Paul Bell, referee of the Berkeley paper, to contact the science editor of *NYT*, Waldemar Kaempffert, pointing out that Li had not mentioned the Cambridge paper even though Bell knew he had seen it 'before his galley

[19] Gates to Li, August 22, 1956.

[20] This point was made in letters from Young to C. N. H. Long and D. L. Thomson, October 16, 1956.

proofs were returned'.[21] This information came from the disgruntled Harris whose dissatisfaction with Li's behaviour was a driving force behind the tide of criticism that was now mounting. Bell quoted an eyebrow raising accusation from Harris that would soon become widely known: '"We have sent a manuscript to Li and expect to find a similar sequence published from Berkeley in due course."' The implication was, again, that Harris had set a trap for Li; the letter from Bell to Kaempffert was apparently meant to help it close on him. Bell seemingly did all he could to further Harris's cause. He broadened the accusations against Li, hinting at a history of misconduct: 'It would appear that Doctor Li ignores the practice among scientists of giving credit to others when he is aware of such prior work. Please consult your ACHT [adrenocorticotropic hormone] files for another example of his failure to give proper credit.' (The ACTH controversy was hinted at by several actors but never fully described.)

Bell later had to defend his conduct, claiming that he had taken for granted that Kaempffert would treat the letter as confidential. But the newspaper man did not. He promptly reprimanded Bell for being libellous and sent a copy of his letter to Li who sent copies to other actors in the evolving controversy. Kaempffert referred to Li's great standing in the scientific community when dismissing Bell's accusations.[22] To Li, Kaempffert wrote: 'It may be that some of your work on the structure of intermedin was anticipated in England, but that anticipation could hardly have come to your notice through the columns of Nature or Harris' transmission of his communication to Nature in time to be recognized.'[23] The fact that it was Li who spread copies of Bell's letter indicates that he was pretty sure its content would reflect badly on his accusers and not on him. Harris's boss, Young, on the other hand, described the comments by Harris that Bell quoted as 'out of their context and jocular remarks'.[24]

To Tiselius, Bell wrote (cc. to Sanger) that he thought the missing reference to the Cambridge paper might be excused because of the time factor and that he blamed himself for not noting the lack of references in Li's paper to earlier work on the *purification* of MSH. By now this point had surfaced as another serious issue. Three groups had published on the purification of MSH before Li's. One of the earlier publications was co-authored by Jerker Porath at Uppsala University's biochemistry lab. This made Tiselius, in his capacity as director of this institution, doubly involved in the matter. The purification problem was in a sense much more straightforward as there was no doubt that previous work had been accessible well before the Berkeley paper appeared. According to Bell this omission was therefore 'inexcusable'.[25]

[21] Paul Bell to Waldemar Kaempffert, September 17, 1956.
[22] Kaempffert to Bell, September 24, 1956.
[23] Kaempffert to Li, September 24, 1956.
[24] Harris to Tiselius, October 6, 1956.
[25] Bell to Tiselius, October 19, 1956.

Indictment and defence

In mid-September Harris wrote to Tiselius enclosing the article from *NYT* and another from the trade journal *Chemical and Engineering News* (*C&EN*) where it was stated that Li's group had been first to purify and reveal the structure of MSH:

> They have obviously been written from information supplied by Li, and the contents are so unjust and completely contrary to the facts that I feel we should take steps to present the truth concerning our own work on the structure of MSH, and also the previous work... on the isolation of the hormone.[26]

Harris related the sequence of events as seen from his perspective. He pointed out that he had written to Li about the planned collaboration with Roos early in the year in order to make sure that they would not work in direct competition with Berkeley and that the somewhat vague answer was 'the sort of evasive reply which I have come to expect from Li'; neither did Li reply to the letter from 20 June, and now a 'trusted friend' in Berkeley had written to Harris:

> that Li was spreading false and malicious accounts along the lines that Porath and I had 'stolen' the problem from his laboratory and that he knew nothing of our intentions until he had received our manuscript: In so far as these false statements reflect discredit upon both Porath and myself, as far as our scientific reputations in America are concerned, I feel that something should be done to put the true facts on record, if necessary in Berkeley itself.

Harris also mentioned the ACTH affair, where Bell was among the damaged parties. Harris was now attempting to establish that Li's behaviour followed a pattern, an accusation that he would explicate in greater detail later on. This was of central concern also to the senior figures that supported Harris's campaign against Li. The main point was not to establish priorities—that was in effect not a problem, as Li had published *after* his competitors, on both the purification and the structure of MSH. It was rather to protect the moral integrity of the scientific community as such by not allowing one of its leading representatives to get away with misconduct. Tiselius was anxious to emphasize that 'this affair is not just one of those rather sad priority quarrels where one group just happens to come a few weeks ahead of the other'.[27] The 'good reputation' of the Berkeley group was at stake.[28] 'Perhaps', suggested D.L. Thomson at McGill, 'the influence of H.M. Evans is infectious!' indicating a wider culture of dishonesty at Berkeley.[29] The actions now taken were aimed to protect community values, and rules that were supposed to guide not only the

[26] Harris to Tiselius, September 19, 1956.
[27] Tiselius to Young, October 26, 1956.
[28] Tiselius to Young, December 4, 1956.
[29] Thomson to Young, October 23, 1956.

behaviour of individuals but that of institutional collaboration as well as competition.

Harris suggested, after having conferred with Sanger, that a letter signed by Tiselius and Young should be published in *NYT* as well as *C&EN*. Furthermore it was desirable to engage American support '[s]o that this should not appear to be merely anti-American'.[30] Therefore Hugh Long at Yale was asked to sign the letter, not because he had a personal stake in the matter, but because Aaron B. Lerner who worked in his department was among those who had published before Li's group on the purification of MSH. Likewise B. J. Benfey and J. L. Purvis at McGill had published on the purification and ultimately the director of their laboratory, David Thomson, would also sign.

The scheme, then, that was put forth by Harris after consultation with Sanger, was to get a group of laboratory directors whose junior staff had been wronged to sign a letter that would denounce the Berkeley group. Young supported Harris fully, claiming he had behaved with 'scrupulous honesty'. Young now made the important distinction between fact and valuation that he would invoke throughout the controversy: 'The deductions which might be drawn from the facts are obviously arguable, but the facts themselves are, in my opinion, incontrovertible.' He drafted a letter to *C&EN*, which he 'tried to keep as factual as possible'.[31] Tiselius found the matter 'very depressing' but went along with the Cambridge plan, though he thought Li must first get a chance to exculpate himself.[32] He immediately wrote asking him to explain what had happened.[33] The result was two letters, one from Li and one from Geschwind. (The latter was actually first author of the paper, and Li second, but there was never any hint by any of the controversy's participants that anyone but the laboratory director was responsible for the Berkeley group's conduct; it is obvious from the documentation that he would have had last say on all matters pertaining to what was published where and when.)

Geschwind wrote that he, as 'senior author' of the Berkeley paper, must take full responsibility for its content. He gave some background to the MSH work at Berkeley and then a detailed description of how the work had proceeded.[34] Geschwind and Barnafi had done the analytical work and most of it was completed by early June. Importantly Geschwind claimed that the revised Berkeley paper had been submitted on 21–2 July and that the Cambridge paper as well as the relevant issue of *Nature* had arrived later than that.

Geschwind hedged the main issue—exactly when the correct version of the structure of MSH was inserted into the Berkeley paper. He wrote that he and

[30] Harris to Tiselius, September 19, 1956.

[31] Young to Tiselius, October 6, 1956. This was pointed out to the editor of C&NE: 'The note . . . is purely factual and makes no deductions from the facts as they are set out there'. Young to R. Kenyon, December 1, 1956.

[32] Tiselius to Harris, September 27, 1956.

[33] Tiselius to Li, September 27, 1956.

[34] Irving I. Geschwind to Tiselius, October 1, 1956.

co-authors never saw any galley proofs, implying that the final version sent to *JACS* had contained the correct sequence. But he did not *say* this, only that 'speculation regarding the structure' had been removed, since the Berkeley group 'were aware that it was incorrect'.

Li on his part accused Harris of foul play, claiming that the letter from 20 June was the first he had heard about any MSH work in Cambridge and implying that Harris would have known what went on in Berkeley since his visit there a few years back. He enclosed a copy of Bell's letter to Kaempffert at *NYT* to demonstrate that some kind of defamation of him was being stage-managed by Harris and Bell. This material was sent also to Sanger in order to discredit Harris. As for the matter of who knew what when, Li saw the incident as a matter of manuscripts passing each other like ships in the night, making independent publication without mutual referencing natural and excusable: 'It is clear to me that the work was carried out completely independently and not knowing what was going on between the laboratories, I think I was justified in not mentioning the work done by other laboratories at this time.'[35] Li also mentioned that the matter of purification of MSH would be dealt with in two longer manuscripts that had been sent to *JACS*. Hence he implied that it was not to be expected that such information was given in a brief communication establishing priority.

The plot thickens

Tiselius sent copies of this exchange to Harris. This was routine as the complicated negotiations evolved. There are few documents in Tiselius's dossier on the Li affair that have not been copied and sent to more than one person. Hence the paper trail quickly grew although different actors would of course have different versions of the documentation depending on what position in the network they occupied. The thickening paper trail indicates the quasi-judicial character of the process, showing that it amounted to the kind of process against free riding that Ostrom described as typical of collective action.

Li used copying as part of his defensive strategy. He sent Sanger an 'extraordinary collection of letters, copies of letters and manuscripts' that he hoped would exonerate him but which had rather the opposite effect, as they triggered Harris to compose a detailed response aimed at undermining Li's

[35] Li to Tiselius, October 1, 1956. Li presented the same version of events to Kaempffert, September 27, 1956.

credibility which he sent off to Tiselius (offering to provide more copies of various documents in order to butt up his various claims).[36]

Harris implied that Li had not only failed to quote his and Roos's paper but that he had attempted to steal their results. In order to show this he focused, as a Marple or Poirot might have, on dates and postal service. Li claimed to have received the manuscript on 23 July and the printed article one week later. As airmail (it is 'indeed very good', wrote Young menacingly to Li)[37] usually took 3–5 days, Harris thought this highly unlikely. But he had a witness up his sleeve who could testify that Li was lying, namely the postdoc, Dorothea Raacke, who had been at Berkeley the last weeks in July and then had left for Cambridge. She claimed to be absolutely certain that the Cambridge paper had been discussed during the Thursday seminar on *19 July* rather than a week later, as Li and Geschwind claimed. As yet, and 'for obvious reasons', she did not want to come forward publicly with this information, but Harris was sure she was right.

Harris piled detail upon detail to convince Tiselius that Li's version of events was a fabrication—including internal scientific evidence that, as 'Roos can tell you', showed that Li's group could not have solved the thorniest part of the structure using the methods they claimed to have used. Why Li had sent the final version of his paper to Sanger and Young on 2 August, containing the correct structure but no reference to Harris and Roos, while at the same time confessing to having seen their paper already by 23 July, was 'very difficult to explain'.[38]

Harris's conclusion was that he himself had acted correctly in every instance whereas Li did not follow the rules of conduct of 'reputable scientists' but rather acted in a way that was typical 'in political circles'. This way of behaviour threatened not only Li's own reputation but scientific trust in a more general sense: 'Li is getting quite reckless, and is responsible for a great deal of suspicion and mistrust among fellow scientists.' And the latest transgression was not an isolated event: 'His conduct of the ACTH affair was quite scandalous.' This was a matter of concern for a much broader scientific community than the small group directly affected by the MSH incident. Word had spread and scientists in laboratories throughout the US were 'shocked' by what had happened. The conclusion was inescapable: 'If we allow Li to get away with things and to "save face" over his conduct of the MSH work, I shudder to think of what he may do next.' This amounted to a rotten-apple theory of scientific misconduct that again corroborates that actors perceived of the problem of rule breaking as a threat to collective action.

Harris had full support from Young who drafted a letter to be sent to *C&EN*, enrolling both Long at Yale and Thomson at McGill as signatories.[39]

[36] Harris to Tiselius, October 5, 1956.

[37] Young to Li, November 20, 1956.

[38] Harris to Tiselius, October 5, 1956.

[39] Young to Tiselius, 6 October, 1956.

Tiselius had qualms, as he considered Li a personal friend. He nevertheless found that the conclusions drawn by Harris and Young were inescapable and that they were buttressed by what he knew at first hand of the matter.[40] Tiselius agreed that Harris seemed to have acted according to a proper code of scientific conduct whereas Li's behaviour was distinctly suspicious. As Young thought it would improve the chances of getting the letter to C&EN accepted if it was submitted by a Nobel Laureate (whose portrait had appeared on the magazine's cover, to boot), Tiselius agreed to front Li's public denunciators.[41]

Meanwhile Young had confronted Li with the claim that the Cambridge manuscript had arrived in Berkeley on 19 July at the latest.[42] Tiselius was very anxious to know about his reaction as the question of this date now seemed to be 'the crucial one in the whole affair'.[43] According to Young, Raacke and others at Berkeley might be willing to testify about the seminar date if Li stuck to his story about having received the manuscript on 23 July.[44]

The third manuscript

And stick to his story Li did. In a letter signed by him and his two co-authors he flatly denied that he had lied about the date and he claimed there was a local paper trail—calendars etc.—to prove his case. He now mentioned, surprisingly, that the manuscript sent to *JACS* after the reviewer's comments had been addressed was not the final version 'but rather an intermediate manuscript' where the order of the amino acids had been changed into what would turn out to be the correct sequence but where the parentheses were retained to indicate that some uncertainty still remained. A few days later, when the sequence had been ascertained, the parentheses were removed and the manuscript was finalized. There was not space or time to go into chemical details in this brief communication, which was instead done in a longer paper still not published. If the group was intent on 'claiming priority or committing piracy', and if they had had the Cambridge manuscript at hand on 19 July, there would not have been any need for this 'intermediate manuscript with its parentheses and unknown sequence'.[45] To Young, Li wrote: 'If circumstances had allowed we would certainly have added a footnote in connection with their [Harris and Roos] paper. After all, the work of a former associate always reflects the

[40] Tiselius to Young, October 13, 1956.
[41] Young to Tiselius, October 16, 1956.
[42] Young to Li, October 15, 1956.
[43] Tiselius to Young, October 20, 1956.
[44] Young to Tiselius, 22 October, 1956.
[45] Geschwind, Li & Barnafi to Young, October 26, 1956.

training he has received from previous associations.'[46] This seems extraordinarily arrogant as there was no doubt, however events were interpreted, that there *had* been time to add that footnote. And, as we know, Gates claimed that Li had been given the choice to do so.

Harris, getting more and more worked up, told Tiselius that by the admission 'that there was a *third* manuscript' the Berkeley group had 'played into our hands':[47]

> It is no longer important to prove that they had the manuscript by the 19th July, because the really important change in their own manuscript was clearly made after the 23rd July, and was very likely made sometime in August. Yet Li in his letter to Young has the audacity to state, 'if circumstances had allowed we would certainly have added a footnote...!' Perhaps by 'circumstances' he means, that if doing so had not meant that they would have had to share the credit for MSH with others!

This is obviously a different interpretation of the events than the one Li's group intended to convey. They thought that the fact that they had submitted the correct sequence on 21–2 July, albeit in parentheses, proved that it had been arrived at independently as long as it was accepted—which they claimed to be able to prove by means of documentation—that the Cambridge manuscript had arrived on the 23rd. To Harris, on the other hand, the fact that there had been a third manuscript proved instead that they had finalized the paper only after having had its results corroborated by the Cambridge group.

The long letter to Tiselius now transformed into a kind of rant with Harris describing Li as 'desperate', suggesting that he 'feels forced to "invent" the truth'. He blamed Li's 'incredible vanity' and speculated that he was unable to acknowledge Cambridge priority as 'he had already planned the press publicity propaganda to convince the world yet again what a great man he is!' He again brought up the ACTH affair, where Li had produced 'very bad work' showing that he was in fact not much of a researcher at all. But 'as so often happens in America, a man can get to the top by bad work so long as the results appear to be worthy of a column in the "New York Times" or "Time Magazine".' Harris went on and on in this strange letter about various alleged transgressions by Li, painting a picture of incompetence, vanity, and dishonesty. According to Harris, Li was driven by the fear 'that he would "lose face"'; he even claimed that this fear explained 'his whole personality and behaviour as a scientist'. In short, Li's free riding, that made him a threat to scientific collective action, could be explained by his adhering to a different code of conduct from most colleagues, one associated with his Chinese background. The final comment by Harris can only be described as ironic: 'To claim priority for scientific results is bad enough but to defame a persons (*sic*) character in the process is even more serious.'

[46] Li to Young, October 26, 1956. [47] Harris to Tiselius, November 5, 1956.

Editorial matters

Tiselius's reaction to this outpouring was cautious. He did not comment on Harris's character assassination but assured him that he had acted honourably according to Tiselius's own code of conduct—i.e. that one either committed oneself fully to collaboration *or* to competition. As for the information that there had been a third manuscript, Tiselius agreed that this was 'highly interesting' and that the matter should be checked.[48]

By now negotiations concerning the publication of a note in *C&EN* were well under way and Li was informed about what was to be expected. To Long, he expressed dismay at finding him among the signatories and supplied him with the usual bundle of copied correspondence and manuscripts.[49] *C&EN* asked Li to comment on the accusations and he replied in a way that Tiselius described as 'grossly misleading'.[50]

There was some discussion about letting Li off the hook by inviting him to publish a correction in *JACS* himself. Tiselius was in favour of this solution whereas Young was not. He did not think their 'statement of fact' in *C&EN* was polemical and that the 'good reputation' of Li's group was harmed only if they had no good answers to the 'inferences' that could be drawn from the published 'statement of fact'. Besides, a retraction of the note in *C&EN* would constitute a 'moral victory' for Li.[51] Hence Young used the fact–value distinction as a rationale for publishing the charges. The informal court of peers would present the facts; it was up to Li to explain them.

Meanwhile the authorities at Berkeley had got wind of what was underway. Tiselius was contacted by his old friend Stanley who was 'shocked' by the dearth of references in Li's *JACS* paper.[52] Stanley hinted at earlier problems with Li, which had led to their breaking off relations two years previously, and told Tiselius that he had asked the head of the biochemistry department, E. E. Snell, to initiate a formal investigation. Tiselius welcomed this and promised full support from himself, Young, Harris, and Sanger.[53] A few days later Young provided Snell with a first summary of events and an official version of the paper trail (nine documents, mostly letters).[54]

Negotiations concerning the content of the letter to *C&EN* dragged on. Tiselius et al. wanted to publish a fairly detailed description of the traffic of manuscripts and their different versions. The crucial paragraph in the original note discussed the differences between Li's first draft and the final version where the incorrect sequence was changed to a correct one without any information being given about how this was done and with no reference to

[48] Tiselius to Harris, November 9, 1956.
[49] Li to Long, November 7, 1956.
[50] Tiselius to Young, December 4, 1956.
[51] Young to Tiselius, December 8, 1956.
[52] Wendell M. Stanley to Tiselius, December 6, 1956.
[53] Tiselius to Stanley, December 15, 1956.
[54] Young to E. E. Snell, December 19, 1956.

Harris and Roos. This paragraph in effect constituted a call for Li to explain himself or else be considered a plagiarizer.[55] The 'facts' presented here were loaded, to say the least. The magazine, however, attempted to excise the discussion about the Harris and Roos paper completely.[56] We will come back to what was finally published.

The question remained of exactly when the final changes to the Berkeley manuscript were made. In mid-December Young contacted Gates at *JACS*, relating the sequence of events as he perceived it and asking Gates when Li 'was allowed' to excise parentheses and other signs of remaining doubt about the problematic part of the MSH sequence: 'It is expected that an authoritative statement from you of what actually happened, will ultimately nullify the effects of inferences from half-revealed truth'.[57]

Gates was 'sorry to say' that Young seemed to be right in assuming that Li had somehow managed to make changes in his paper after it had been accepted.[58] This, he explained, probably had to do with the work procedure at *JACS* where the last galley proofs were read only by the author and the copy editor, not by the editor. Gates concluded that Li must have made the final changes in these proofs 'without our knowledge' and protested against implications that he had somehow 'allowed' Li to make the changes: 'since if I had been aware that they had been excised [parentheses plus two sentences], I should not have allowed the excision'. According to Gates this 'reflects rather seriously on the scientific probity of Dr Li'. Young thanked Gates heartily for this information and assured him that he never thought Li had been 'allowed' to change the manuscript but had managed to do so surreptitiously.[59] Gates's information was especially damning, as Geschwind had assured Tiselius that there *were* no galley proofs.

This was not the end of this peculiarly complicated investigation however. Gates soon wrote again, admitting that he had been wrong. He had now unearthed a letter from Li, written on 26 July and received four days later, containing instructions for the final changes in the *JACS* paper.[60] Naturally he enclosed a copy. This, then, was the third manuscript: a brief note from Li asking for the removal of parentheses and the excision of two sentences, thereby removing any lingering uncertainty concerning the correctness of the whole sequence. He even provided a page with the paragraph in question rewritten. This was done on 26 July, presumably after the Cambridge paper had been discussed during the Thursday seminar.[61] Gates had also found that

[55] Copy of the original version of the note dated October 26, 1956. This is the date when Tiselius sent the note (by air) to *C&EN*. Tiselius to Young, October 26, 1956.

[56] Kenyon to Tiselius, January 9, 1957. The letter contains a typescript of the final version of the letter by Tiselius, Young, Long, and Thomson.

[57] Young to Gates, December 18, 1956.

[58] Gates to Young, January 4, 1957.

[59] Young to Gates, January 10, 1957.

[60] Gates to Young, January 16, 1957.

[61] Li to Gates, July 26, 1956.

there had indeed been *no* proofs sent to the authors. He thought it still seemed that 'Dr Li has been very reluctant to give proper acknowledgement to the prior work of other (*sic*)'.[62]

So at last the mystery was cleared. It turned out that Gates had been an accomplice to Li's manipulations all along, allowing a change of date of receipt to 24 July, as this would presumably free Li from the suspicion of having changed the structure in accordance with results from Cambridge. The true date of receipt should have been 30 July, when Li's letter—the third manuscript—arrived at Gates's editorial office.

Young sent a new version of the sequence of events to Gates that made the changing of dates glaringly obvious. He thanked him for providing the information but the tone of the letter was distinctly cold. A copy of the letter was sent to Snell, who of course also received a copy of Li's letter from 26 July. Young wrote that he had no desire to comment on the interpretation of what seemed to be the final version of the sequence of events.[63] As the letter in *C&EN* was printed he hoped all publicity regarding the matter was 'finished with'. The letter to Snell ended with an endorsement of sorts of Li:

> In what I hope is my last letter concerning this unhappy affair I should like to express my admiration for Dr Li's scientific accomplishments over a period of many years, and my deep regret should his potential contributions to the advance of knowledge be adversely affected in the future by what has recently happened. In common with many others I have always found Li to be a friendly and congenial colleague... and such personal feelings cannot be subdued however important it may have been to lay bare the facts of recent events.

This must surely be read as a plea not to pursue the matter much further. By now Young would have felt a desire to tone down an affair that threatened to put everyone involved in a bad light. It seemed clear that Li's group had arrived at the correct sequence independently but that they did not remove the last signs of uncertainty from their manuscript until after they had received and discussed the paper by Harris and Roos. This was not theft, but it certainly was not cricket either. Harris had attempted first to trap Li and then to prove that he had stolen his results, and both Young and Tiselius had given him moral support and involved two other laboratory directors in a campaign against Li. As the alleged crime became more of a misdemeanour they must have feared that the affair could become an embarrassment for everyone involved. As for Gates, his shoddy editorial practices would, if they became known, have reflected badly not only on him but on the prestigious journal he helped edit and perhaps on the society that published it. The facts of the case were open to interpretations that might put the biochemical elite in a bad light; in the end it seemed everyone needed to save face.

[62] Gates to Young, January 16, 1957.

[63] Young to Snell, February 6, 1957.

Epilogue and discussion

The letter by Tiselius et al. was published in *C&EN* on 28 January 1957. It did not really amount to a denunciation of Li and his co-authors, more a slap on the wrist (Tiselius et al. 1957). The tone was subdued, the claims reduced to a few factual statements. It was pointed out that the Berkeley group had not cited earlier work on the purification of MSH and that Harris and Roos had published earlier than them on the structure. But the last point—the central issue in the Li affair—was not elaborated. It was not even spelled out that the Berkeley paper had omitted this reference and there was certainly no hint that it was *consciously* omitted. If anything the letter invited the conclusion that publication had been independent. Instead the matter of publicity was stressed. It was said that the false information in the *C&EN* paper was particularly unfortunate as it had received 'extraordinarily widespread publicity'. And then there was an editorial comment regretting what had happened, noting ungrammatically (in order to avoid mentioning specifics, apparently) that the 'California group says it did not claim in its paper that it was the first to do these'. The general impression from the letter and the comment was that *C&EN* had exaggerated, that their distorted view on priorities unfortunately had been reproduced by other media, and that a correction was therefore needed. In the end the matter became a contest of credibility between leading scientists and the media, though the real media culprit—*JACS*—seems to have been too powerful to be dragged publicly into the dispute.

No dramatic consequences seem to have followed from the Li affair on either side of the Atlantic. I have not been able to check what if anything came out of the Berkeley investigation, but it certainly did not put an end to Li's career (though Porath has claimed that his reputation and possibly his later career did suffer to some extent).[64] In 1977 he received an honorary doctorate at Uppsala University (Nevéus 2000: 47, 75). Tiselius's protégés Roos and Porath became successful within their respective fields (separation and growth hormones). Like Tiselius, Young had more or less given up his own research in the mid-1950s (Randle 1990). The creation of MRC's Laboratory of Molecular Biology in 1962, which Young opposed, was something of a blow. It attracted some of the best talent and put Young's department and discipline in the shadow both intellectually and with regard to funding (de Chadarevian 2002: 221–2). Sanger's career is of course legendary as he twice received the Nobel Prize in chemistry, the first time (1958) for work done at Young's laboratory and the second time (1980) for work done at the MRC lab.

As for Harris, easily available biographical information is scant. Like Sanger he moved to the MRC lab as soon as it opened and he remained there until his

[64] Interview 23 August 2012.

accidental death during a camping trip in 1978.[65] He did not rise to the higher echelons of academic leadership. Neither did he drop out of the system but apparently continued doing respectable biochemical work after the Li affair. In 1960 Harris published a brief overview of the chemistry of MSH in which both the Cambridge and Berkeley papers appeared in the list of references. In the text the Berkeley group was given credit for having contributed to the purification of one form of MSH and Harris and Roos for solving the structure (I. J. Harris 1960).

As I have tried to show, the moral economy of research among the biochemical elite in the 1950s relied more on rules of conduct and on individuals' trustworthiness than on what Hagstrom called 'higher order social norms'. This may be understood in the light of the peculiar situation in the natural sciences as a whole after the war: the edifice of government-supported science being erected in this period was founded on what has been described as a 'social contract for science' (Guston and Keniston 1994). Obviously this was not a contract in the formal sense but rather a system of exchange founded on e.g. the scientists' ability to engender trust and to uphold the legitimacy of a rapidly expanding enterprise in the eyes of government—something which in fact corresponds to another 'design principle' (the seventh) identified by Ostrom, namely that collective action is facilitated by sanction from the central power (Ostrom 2000: 152). Despite the fact that such issues did not figure explicitly in the Li affair it may be seen as an example of how a trusting government–science relationship was enacted during a period of government-funded scientific growth in western democracies like the US, the UK, and Sweden. The peer review system, that *was* an issue in the Li affair, embodied this system of trust, not only of government in science but of scientists in their peers. Not only researchers but journal editors were therefore subject to rules of conduct enabling collective action; the formalization of such rules was an ongoing process in the 1950s (Gross et al. 2007: ch. 8). A government-sanctioned system of collective action was hence at stake in the Li affair. In order to defend it the scientists clung to the same idea that Tiselius promoted on the general level of science–society relations, namely the fact–value distinction.

In the Li affair the separation of facts and valuation was however problematic, as facts kept mounting and changing. The factual description of events in the first version of the letter to *C&EN*, and in the first summary of events prepared by Young for the Berkeley investigation, implied that Li had plagiarized results from Cambridge. The factual description of events in the printed version of the letter implicitly put the blame on *C&EN* and other media. Then

[65] For Harris's tenure at LMB, see alumni record at <http://www2.mrc-lmb.cam.ac.uk/about-lmb/archive-and-alumni/alumni/>, accessed 26 August 2014. Harris is listed as 'Dr', not 'Professor'. About Harris's demise, see Lerner 1993: 7–10.

there was the factual description that Young sent to the editor Gates and to Snell in Berkeley, which was aimed at the imminent official investigation in Berkeley as well as the prestigious journal. This description implied that Li as well as the journal had acted shoddily, while stopping short of plagiarism. Unsurprisingly none of these allegedly factual accounts were free from implicit interpretation and valuation.

The discussions about scientific conduct were not cloaked in normative terms, at least not in the Mertonian sense. Conduct was rather associated with values in the sense discussed by Shapin (2008): keeping to them signified personal integrity, good character, or virtue; breaking them was a sign of flawed character or, as Harris hinted in the case of Li, dysfunctional non-western values. In Young's final comments on the Li affair he refused to interpret the facts; instead he chose to vindicate Li as to general professionalism and character. Rule breaking threatened the system by throwing doubt on the virtue of leading researchers, and by inviting conflicts that might have uncontrollable consequences. It came down to choosing between a disease and a dangerous cure. If Li's face could still be saved, this would be a better alternative than an official process against him that might harm the scientific community. That seems to have been Young's message.

The problematic nature of the fact–value distinction was commented on by one participant at the 1969 Nobel Symposium. W. H. Auden regaled the other participants with an ode to Terminus, the Roman 'God of boundaries', that ended with the following stanzas (and atrocious misprint):

> In this world our colossal immodesty
> has plundered and poisoned, it is possible
> You still might save us, who by now have
> learned this: that scientists, to be truthful,
>
> must remind us to take all they say as a
> tall story, that abhorred in the Heav'ns are all
> self-proclaimed poets who, to wow an
> audience, utter some resonant life (*sic*) (Auden 1970: 142).[66]

Poets must be truthful and scientists must abdicate from the pretention to absolute truth; only thus can the fact–value problem be solved. This was, most likely, not what the scientists wanted to hear. The idea that the domain of facts was manageable and the domain of values was messy defined their professional identity. The Li affair implies that the fact–value distinction was at work when the character and credibility of researchers was evaluated by their peers as well as in evaluating broader knowledge claims. It shows that the production of knowledge about character was a social process marked by the need to

[66] The poem 'Ode to Terminus' had previously been published in *New York Review of Books*, 11 July 1968, with slightly different orthography and with the last word being not 'life' but 'lie'.

demonstrate the ability to control fact finding and evaluation, as well as the ability to police the boundary between the two, thereby demonstrating the ability to fulfil science's contractual obligations. In the final analysis the most important aspect of the process against Li, which generated such a rich paper trail, was perhaps not the possibility of reaching a verdict but rather to demonstrate vigilance. The point of the process would then have been the process itself.

ACKNOWLEDGEMENTS

Research for this paper was carried out within the project *Det forskningspolitiska laboratoriet*, financed by the Swedish Research Council (*Vetenskapsrådet*). I am grateful for insightful comments from the reviewers of the paper, Kristine Asdal, and the editors of this book. I am much indebted to Jerker Porath for commenting on an early version of the paper and for open-hearted discussions about its content.

Part II

Markets as Carers for Health

5 A moral economy of transplantation

Competing regimes of value in the allocation of transplant organs

Philip Roscoe

The chronic shortage of transplant organs has long been recognized as a problematic area for medicine and for medical ethics. The considerable attention it has attracted, not only from clinicians but also from academics, the media, and policymakers, has focused for the most part on the supply of organs and issues such as the role of consent and the possibility of using incentives to increase levels of donation. Yet the allocation of organs to patients needing transplants is, perhaps, the more pernicious problem. With long waiting lists for transplants, problems of rationing and allocation will never entirely vanish, even if the supply of organs were to greatly increase. The technicalities of transplant surgery require systems of allocation that take account of the biological characteristics of donor and recipient (Forsythe 2009). For some patients the absence of transplantation will mean certain and almost immediate death, while others can be maintained for longer periods through medication and interventions such as dialysis. Allocation of transplant organs is a technically complex and morally fraught process: it is value-laden through and through.

Normative approaches to the distribution of organs span two poles. Equality of access is exemplified by lottery, while success-based utilitarian approaches emphasize tissue matching and graft survival to achieve maximum survival benefits across the population. Positions between the two must be negotiated on the basis of increasingly complex medical, social, and moral claims (Weir 1995). Disciplinary preoccupations have also become apparent in discussions: bioethicists such as Cherry (2005), Taylor (2005), and Wilkinson (2003), have advocated markets and economic incentives on grounds of autonomy and self-ownership, while medical evidence, objectivity, and transparency, are the central concerns of clinicians (Wiesner et al. 2003). The views of the general public are complex and fail to produce a usable consensus; opinions depend upon age, gender, and social situation (Neuberger et al. 1998; Wilmot and

Ratcliffe 2002). Ironically, individual patients are heard least, and given little option regarding the organ they will receive, although there is an argument that elderly patients should be allowed to choose an available, sub-standard organ rather than queue for a high-quality one (Freeman 2007; Shapiro 2007).

Popular opinion matters to policy, nonetheless. Clinicians are aware of the need to balance public demands and clinical evidence, ensuring a supply of donated organs through the maintenance of trust and positive relations between the medical practice and the general public. As James Neuberger (Neuberger et al. 1998: 979), director of the UK's transplant service writes:

> The processes which will, in effect, result in the denial of a life-saving procedure for many individuals, need to be explicit, objective, just, equitable, transparent and retain public trust and confidence. Therefore, robust processes for both selection and allocation need to be developed and implemented.

Keeping the public happy may be difficult. General principles for allocation among the public include: maximum benefit; the moral worthiness of recipients; preference for younger recipients; and medical urgency (Tong et al. 2010). Prejudices, for example towards those suffering alcohol-related disease, are also clearly shown in public responses (Johri and Ubel 2003). However, the public does value notions of fairness and equality of access over and above measures of outcome, emphasizing just distribution even when the result will be a decrease in lives saved by the organs available (Ubel and Loewenstein 1996).

Though there is no consensus on principles, these disparate approaches do share a conception of moral decisions as prior to, and ontologically separate from, the actual *processes* through which organs are allocated. In the early years of transplantation, it was left to surgeons to enact the requirements of policy and distributive frameworks as best they could (Weir 1995). Where external expertise was brought to bear on problems of allocation it has tended towards practical solutions of the technical complexities: Alvin Roth's work on organ exchange, for which he was awarded the Nobel Prize, is exemplary in this respect (Roth et al. 2005). His 'New England clearing house' is an algorithmic matching device that enables single donations to trigger chains of paired or three-way transplants, helping many to receive the much needed organs. Roth (2007) presents a thoughtful analysis of moral repugnance as a constraint on market design, urging economists to treat repugnance as a genuine dis-utility, not to be washed away by sight of increased welfare gains. Even Roth, however, in concluding that kidney exchange can offer many of the benefits of markets without provoking repugnance, does not fully recognize the complex relationships between notions of value and the material mechanisms that embody them.

Central to this volume, on the other hand, is the recognition that values are neither prior to, nor separate from, their material and organizational settings. As knowledge is enacted in practice (Latour 1987), so valuation is seen as an active and ongoing process embedded in specific practices, tools, and architectures. Value depends upon the epistemic—what is known and what is worth knowing. As the introductory chapter argues, the enactment and stabilization of values are a central part of what there is to explain (see Chapter 1, Dussauge, this volume).

The present chapter explores the articulation of value through the technical apparatus of organ allocation in the UK's National Health Service (NHS). The chapter documents the development of centralized, computer-driven allocation systems for kidneys, and a 'thought experiment' to achieve an optimum result for liver allocation. The chapter is based on empirical work conducted in the first half of 2011. Primary sources have been interviewed and material collected in twelve interviews with members of the transplant community: three consultants (i.e. specialist doctors but not surgeons); two transplant surgeons; five research scientists involved in various aspects of the allocation process; and two nurses involved in retrieval and allocation of organs. Interviewees are detailed in Table 5.1. With one exception interviews were conducted in person; I was also able to visit a laboratory and a research seminar. Additional data on allocation come from the large quantity of publicly available records. The NHS publishes its allocation policies, and the advisory committees publish minutes of the quarterly meetings. The research scientists also made available to me detailed documentation of allocation protocols and material from presentations made to health service committees. The data have been analysed through a process of comparison (Boeije 2002) to produce a plausible (Ahrens and Chapman 2006) narrative account of the development of current standards.

The developments in question are a consequence of the growing recognition of transplantation as a mainstream surgical procedure and a reliable treatment

Table 5.1 Interviewees

Reference	Role summary	Interview length	Recorded by
Consultant 1	Medical specialist	45 mins	Telephone interview, taped
Consultant 2	Intensive care doctor	1 hour	Tape
Consultant 3	Medical specialist	30 minutes	Notes
Nurse 1	Transplant coordinator	2 hours	Tape
Nurse 2	Transplant coordinator	1 hour 15	Tape
Research scientist 1	Laboratory manager	2 hours	Tape, notes, laboratory visit
Research scientist 2	Academic	1.5 hours	Tape, notes, seminar attendance
Research scientist 3	Analyst	1 hour	Notes and presentation materials
Research scientist 4	Analyst	30 minutes	Notes
Research scientist 5	Analyst	1 hour	Notes and presentation materials
Surgeon 1	Transplant surgeon	1 hour	Tape
Surgeon 2	Transplant surgeon	1 hour	Tape

for renal and liver failure. The change in the status of transplantation has greatly increased the demand for transplants, swelled waiting lists, and led to an 'organ donation push' that has recast organs as a public resource, donation as an 'opportunity', and donor bodies as having rights akin to the living.

And yet, for all this scarcity, an organ is of little value without a body. It may not be bought or sold. It has no price. But it can offer someone a renewed life, depending on the existence of a suitable recipient, and its value comes about through use: its steady work in the body of the patient. Even a poor quality kidney might have value, should a recipient be found who would accept it. Thus allocation becomes a crucial moment in the production of value; the value placed on transplant organs by scarcity, by supply and demand, is made visible only by the matching process. Allocation protocols render the interdependence of organ and recipient calculable and malleable; it is through matching processes that the relations of value are discovered.

Notions of value and exchange such as this allude to a moral economy of a kind, less developed or overtly economic than those in cadaveric procurement (Anteby 2010), or cord blood (Brown et al. 2011). Nonetheless, values are more than subjective measures, containing 'normative presumptions about the ways in which markets should be organized' (Reinecke 2010: 564); there exists a reflexive relationship between transcendent values, valuation enacted by material devices, and the professional norms of health-care workers. Transcendental norms are embedded in cultural antecedents and discourses (Boltanski and Thévenot 2006) surrounding the practice of transplantation. There we find the virtues listed by Neuberger et al. (1998): clarity, objectivity, justice, equity, transparency. Shadows of politics are also visible in the pure calculations. Political choices motivate priority to paediatric recipients or those from ethnic minorities, or a move towards a utility model emphasizing younger recipients, to increase survival benefit across the population (Shapiro 2007). Finally, there are the professional ethics of practitioners, manifested as a commitment to care and personal involvement with patients.

This chapter shows how the implementation of allocation devices radically reconfigures each realm of value: the meaning of good clinical practice is transformed as equity and utility are enacted and themselves translated (Latour 2005) through models, simulations, and algorithmic protocols. At the same time, frames of valuation overflow, reclassifying certain behaviours as resistance or unethical practice. The protocols not only dictate who receives, but also *what matters*. Through the empirical site of organ allocation, the chapter extends the contribution of this volume in unravelling the technical, epistemic, and economic embeddedness of value in the life sciences—emphasizing the instability and flux of moral arrangements, a moral economy (Daston 1995) perpetually unfolding and being worked out.

The chapter will first of all consider the earliest aspects of allocation and the development of a discourse of transplant practice at a national level. I will then discuss the development of a formal allocation process for kidneys that took place in the late 1990s and early 2000s, before examining a current 'thought experiment' in the liver transplant sector. The chapter considers notions of resistance to these reforms, and then discusses the enactment of value in the moral economy of transplantation.

Institutionalizing transplantation and transplant organs

Transplantation is a relatively new medical technique, and just as great efforts have been placed into increasing the sophistication and success rates of surgery itself, so much attention has been focused on the problems of queuing and allocation that attend it, particularly since the mid-1990s (Persad et al. 2009; Weir 1995). Nonetheless, the rapid expansion of transplantation in its early years resulted in a historically dependent patchwork of protocols and policies, varying both geographically and across transplant specialization (from organ to organ).

The UK is divided into eight transplantation zones, each centred on one or two large teaching hospitals (for example the John Radcliffe in Oxford, Addenbrookes in Cambridge, or the new Royal Infirmary in Edinburgh). As recently as the mid-1990s, transplantation for all organs operated at a local level. Each centre maintained a list of patients awaiting transplantation, and when a donor became available in a zone, local surgeons were responsible for retrieving organs and allocating them on a discretionary basis according to observed clinical need and suitability of matching. Guidelines did not prescribe any way of assessing clinical need. In the case of liver transplants, for example, matching considerations centred solely on similarity of donor and recipient size and blood group; surgeons and consultants were therefore left with considerable discretion in their allocation.

If, as was frequently the case, there was no suitable recipient, organs would be offered to other centres on a rota basis, centrally managed by the NHS, and monitored by a 'balance of exchange' where centres were expected to finish the year having supplied as many organs into the system as they had received. The balance of exchange was managed locally by the nurse coordinators. This kind of scheme still exists in the case of certain organs and categories of donor, and has some advantages, according to Research Scientist 3: it allows the autonomy of local centres, it is quick to implement and easy to maintain, and is excellent at optimizing activity at a centre level.

Kidneys were the first organ to be regularly transplanted. In the late 1960s and 1970s relatively unsophisticated retrieval techniques and tissue typing meant that all kidneys were dealt with on a local basis. Subsequently, a nationwide system was established to distribute one of each pair of kidneys retrieved, while the other was maintained by the hospital. Inevitably, this led to concerns from peripheral units that patients in the large hospitals, where retrievals were conducted, had a higher chance of receiving a kidney (Surgeon 2). By the mid-1990s it was apparent that there were great differences in waiting times and the accompanying mortality rates from centre to centre. Increasingly, surgeons found this situation unacceptable. In the mid-1990s, Surgeon 1 became chair of the Kidney Advisory Group, the committee responsible for oversight of renal transplantation.

> When I took over as a member of that group and then taking over as the chair of [the advisory group for kidney transplantation], I became particularly interested in allocation of organs, and particularly allocation of kidneys. And I asked for the statisticians [within the NHS] to give us a snapshot of allocation and the effects of allocation policy across the United Kingdom and it was clear to me that there was inequity of access... both in terms of patients getting onto a waiting list, and once they were on waiting list, the allocation of a kidney to those individuals. (Surgeon 1)

These discrepancies (Oniscu et al. 2003) are in many cases still unexplained, and may relate to exogenous factors such as demographics and education of the local population, as well as surgical procedures and the willingness of surgeons to take risks in using less than perfect organs in cases where there may still be a marginal benefit to the transplant recipient (Consultant 1). Nonetheless, they were considered cause for action.

Surgeon 1's statement makes explicit what he considers to be the primary concern of any rationing in allocation mechanisms in organs: equality of access, understood in terms of geography, ethnicity, and social demographics. Behind his statement lies an increasing awareness of transplant organs as a public resource: transplantation had become a standard surgical procedure that could be deployed to tackle increasing levels of renal failure and public health problems. Scientific advances, making the impossible achievable, and then routine, force a consideration of the ethical concerns involved in a new everyday practice. In the US, the Task Force on Organ Transplantation, established in the late 1980s, held that donated organs should be considered public resources, held in trust by surgeons and procurement organizations (Childress 1989). In the UK, the Human Tissue Act (2004) wrote the same claim into statute. European presumed consent goes further still, upholding the state's claim to the bodies of the deceased. Recasting organs as the raw material for an everyday medical procedure has clear ramifications. As a public resource, organs must be shared equitably and in a transparent manner. Consultant 1's most strident objection to the local system is its opacity, in that definitions of clinical need may differ from one centre to another, and too much is left to the judgement of individual surgeons.

The decision to recognize transplant organs as a national resource has ramifications in many related areas. In intensive care, for example, where resources are inevitably rationed, the discourse of transplantation—'the organ donation push' (Consultant 2)—has seen the rights and wishes of would-be donors taken increasingly seriously, even after their death. A would-be donor may be admitted to intensive care, and despite being clearly beyond recovery, may still be cared for:

> if it was that patient's express wish that they would be an organ donor, it is in that person's best interest to provide them with a period of critical care, knowing that they're not going to survive, but knowing that they wish to be a donor, and that there is a secondary benefit, other people who get the organs will benefit. (Consultant 2)

Despite a transformation in the understanding of transplant organs, transplantation as a medical practice exists in a liminal position, half institutionalized, and half gift economy. The provision of an intensive care bed for the maintenance of a cadaver, even for a few hours, may come at the cost of treating another patient. Surgical retrievals, while funded in part by direct payment from a central fund (Surgeon 2) still need to work around the routines of a busy hospital and not to 'piss off' (Nurse 1) the everyday users of surgical spaces. Despite the obligation of hospitals to provide theatres, staff, and anaesthetists, transplantation may sometimes meet with objections. Nurse 1 recalled one anaesthetist's reply to a request for help: 'I didn't do my training to come and anaesthetize a dead person'.[1] The NHS has countered by creating specialized roles such as the CLOD (Clinical Lead on Organ Donation) whose task it is to build institutional support for transplantation; developing transplantation is a job that depends greatly on public relations and relationship building (Nurse 1; Consultant 2). Meanwhile, surgeons are recast as entrepreneurs who are required to submit business plans to departmental heads bidding for resources—theatre, nurses, and anaesthetist—and then claw back costs from peripheral units whose patients are being treated. What counts, in practice, is often down to the energy and political skills of individual practitioners.

Dealing with a national resource forces a discussion of what constitutes fairness, a discussion played out in negotiations between medical practitioners of all levels of seniority. For example, admittance to the liver transplant waiting list is determined in one hospital by a large meeting attended by many of the fifty-strong transplantation group: surgeons, doctors, nurses, social workers, and transplant coordinators. 'Utility' and 'equity' are concepts that exercise these meetings, as are transparency and a constant battle between number-driven medics and the holistic, pastoral view taken by carers and

[1] Organ retrieval is always carried out under general anaesthetic: reasons for this appear to be largely technical, although the possibilities are discussed by Sharp (Sharp 2006).

social workers. The result is a sophisticated and philosophically literate medical workforce: transplantation is, as Surgeon 1 notes, 'ethics in practice'. Utility, equity, and compassion are the concepts that most of all frame subsequent discussions on the allocation of organs: what, exactly, does fairness mean?

Engineering fairness in kidney allocation

Surgeon 1, aided by colleagues from the advisory group, instituted a programme of reform that was intended to systematize the allocation of kidneys in the UK, ensuring equity of access across the nation. The effect of these changes has been to move from discretionary allocation to a situation where 'every kidney has a patient's name on it' (Surgeon 2).

The advisory group settled upon tissue matching as the guiding principle for the new allocation scheme. This arrangement focuses on the utility and efficiency of the transplant, guaranteeing the highest level of success and post-operative function for each kidney. In broad terms the 2006 kidney allocation scheme identifies three levels of antigen (HLA) match, and prioritizes the highest match. However, within a waiting list of 7,000, there are likely to be a number of patients who have a perfect HLA match with the donor, and 25 per cent of transplants are carried out at perfect match levels (Research Scientist 4). Therefore the scheme features a number of other filters for deciding who should receive a kidney. Surgeon 1 explains the rationale:

> Because there are 6,500 or 7,000 people on the waiting list, so at every level [of match] you are going to almost have competition, for want of a better term, so even for a full house match it may well be that there are four or five people for the full house match... one of the most important things is the hierarchy of matching as your prime indicator, so that's the utilitarian thing of the best match being likely to be the best longevity of the organ, that's the prime thing. But then once you have that if there are points systems to then decide where the kidney goes, that allows you to perhaps allow better equity across the UK.

A process based purely on utilitarian HLA matching does not achieve distributional equity through the random distribution of the population. Factors such as ethnicity play a large part in matching, and even with a number of finessing points within the matching equations themselves (equating unusual tissue types with close match standard ones) matching needs to be mitigated by other factors. There is, according to Surgeon 1, enough flexibility within the tissue matching to allow other factors to improve upon equitable distribution, although utility remains '*the prime thing*'. Therefore the score also includes: waiting time; HLA match and age combined; donor-recipient age difference;

location of patient relative to donor; two technical factors relating to HLA matching; and blood group match. Changes made in 2006 emphasized waiting time at the expense of HLA matching, resulting in a higher likelihood of matching for ethnic minorities and increased equity of transplantation.[2]

The inclusion of waiting time is important because it appeals to a natural (or least national!) sense of distributional fairness: the queue (Consultant 3). Waiting time is therefore positioned as the first factor in the list and yet, in practice the second factor is the crucial one (Research Scientist 4). Waiting time is measured in points, with patients gaining one point for every day they spend on the list; a wait of three years equates to 1095 points. The second factor combines HLA match and age: a perfectly matched patient under the age of 30 receives 3,500 points—equivalent to a ten-year stay in hospital; a relatively well matched patient under the age of 30 receives 2,000 points; a perfectly matched eighty-year-old man receives just 500 points. The exact point distributions and formulae are shown in Figure 5.1. The weighting of this calculation means that subsequent levels of the points system become little more than tie-breakers, although location may also play an important role, with patients receiving 900 points if they are at the same centre as a donor.[3]

The shape of these curves was determined through an iterative process of statistical modelling and discussion by the steering group (Research Scientist 3). Some factors, such as the weighting towards transplants in the same centre, may be for what Surgeon 1 described as 'legitimate scientific' reasons, such as the reduction of 'cold ischaemic time' (i.e. the time that organs are out of a body). At the same time, it may be possible to discern some traces of previous systems, which gave priority to locality, fossilized in the points systems. The reasoning behind points systems may be fundamentally pragmatic, too: '[If you had a young kidney] your instinct says you would want to give that to a young person . . . We know that the older kidneys are not going to . . . last for so long' (Surgeon 2).

In this comment we can distinguish both a medical pragmatism and a concern for individual patient coexisting, perhaps, with the rational utility of the points system. And despite the stated intention to increase equity of access, it does appear that the points system emphasizes certain goods, such as youth. Surgeon 1 describes the prioritization of young transplant patients thus:

[2] NHSBT Transplant Activity Report 2009/2010: 21.

[3] The exact formula is: $3{,}500/(1+ (age/55)^5)$ for level I mismatch patients; $2{,}000/(1+ (age/55)^5)$ for level II mismatch patients; $500/(1+ (age/55)^5)$ for level III mismatch patients.

Graph is drawn from formulae. [*Source*: <http://www.odt.nhs.uk/pdf/kidney_allocation_policy.pdf>, accessed 21 October 2014].

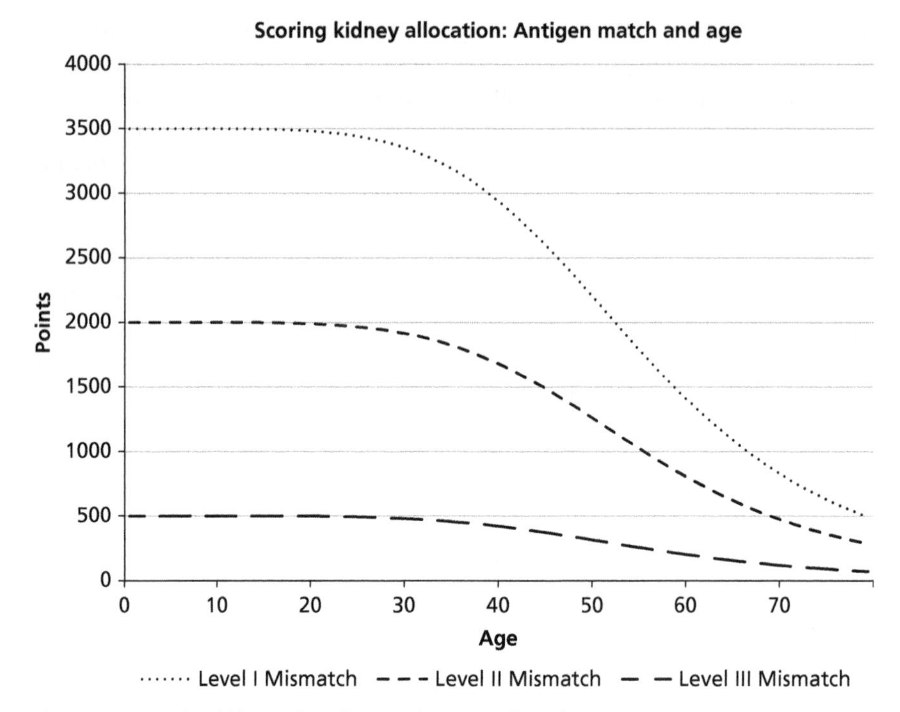

Figure 5.1. Scoring kidney allocation: Antigen match and age

> Let's say you're 18, the average lifespan of a kidney means that it is likely that you will need more than one kidney transplant. If we put a poorly matched kidney into you when you're 18, then you will have antibody levels in the future, and so therefore it's going to be harder to give you a second kidney later on down the line. So more or less, the better match should be for the younger patient. Completely different from the patient who is aged 60. The patient who is aged 60 hates being on dialysis, on the main, and would really like to have their life back, and accepts that if they get a poorly matched kidney it may not last them for a huge amount of time, but it may last until their death, from other causes. So therefore you can afford to get away with a poorer match in the more elderly patient. So . . . the match points count much more when you are young, but drop-off and count much less when you are old.'

The sharply decreasing curves of the points system prioritize younger patients over older patients, even at a relatively low level of matching (Research Scientist 4). The implication here is that first principles, be they queuing or distributional equity, are upheld so long as they do not clash with other objectives that may be legitimately described as 'scientific': the avoidance of sensitization in subsequent transplants, or the reduction of cold ischaemic time. Longer-term strategic decisions, such as the need to remove an accumulation of difficult-to-match patients from the waiting list, were also incorporated into the points system. On the other hand, despite the appeal to scientific reasoning, the intentions of surgeons do not seem so different from the themes articulated by the non-medical public: the notion of a 'fair innings', where a young patient is prioritized for the sake of fairness in life lived, and an older patient should be prepared to accept a compromise (Tong et al. 2010; Wilmot and Ratcliffe 2002). In these accounts professional care and utility are not mutually exclusive, but are expressed in different ways. In fact, it is the work of the allocation system to reframe how practitioners understand the duty of care, as becomes apparent in the following discussion.

The kidney allocation system, while the first of its kind, has met with criticism from specialists in other disciplines. Consultant 1 and Consultant 3 suggested that the Kidney Advisory Group had not adequately established its objectives before beginning its research. The decision to prioritize young people appears arbitrary to some; the transplant community recognizes a systematic problem in teenagers refusing to take immunosuppressive medication which makes them overweight and spotty, consequently suffering graft failure and returning to dialysis (Nurse 1; Consultant 3). They also criticize the kidney allocation system for covering the distribution of organs without covering access to the waiting list. On the other hand, restricting access to the waiting list is a form of rationing and would require an additional process of screening, which may not be popular:

> So you could say, well we'll have an age limit, or will devise *yet another* [emphasis] scoring system and have points, and if you don't get the required number of points for whatever reason be it a bit of age, be it a bit of comorbidity, or we just don't like the look of your face. (Surgeon 2)

The consultants and scientists who worked on the kidney programme deserve credit for being the pioneers of the points-based allocation system that is now being introduced into other specialties, such as pancreas transplantation. It also seems to make life easier for the surgeons. Instead of a moral choice and responsibility that might place a genuinely unacceptable burden upon physicians—the ability to dispense life and death, to arbitrate between the persistence of suffering and providing treatment—they have recourse to a system that can be justified as scientific and egalitarian. The responsibility of the surgeon recedes to a (weighty enough) technical role with no further accountability:

> Once [patients are] on the list, it is equitable... It's an agreed system which to a whole group of people involved in doing it seemed fairest way of allocating... It actually makes it a lot easier for us to say the machine does it [determines the allocation]. And then nobody can actually come back to you and say 'Why did you give it to that one and not that one?' (Surgeon 2)

The points system draws our attention to the relationship between the epistemic—what we know—and practices of valuation. In the case of kidney allocation, the whole architecture of tissue matching is based on a process of tissue typing run from a small number of dedicated national laboratories. Within these laboratories three generations of tissue and DNA testing technology and the technological expertise of the laboratory technicians combine to produce a simple three digit code for every patient, donor, or recipient (Research Scientist 1). This code drives the matching engine. The statistician's model, once agreed, is converted into a software engine that runs the allocation calculations. Matching points are co-dependent between donor and recipient, and therefore the matching run must take place for every new donor, producing a score for every recipient on the waiting list. Final scores are derived by a simple process of adding up the underlying points, giving a clear and transparent system, published online for the public to see. What is presented to the nurses operating the central duty room is a single list of potential donors, ranked by their score, whose coordinating nurses need to be contacted so that offers can be made. The concentrated work that has gone on in establishing the shape of various distributive curves within the points system is hidden from sight, part of the nest of calculations that make scientific work possible (Latour 1999a).

The points system is also productive of values: normative categories, such as equity and transparency, cease to be static and are enacted in the calculation of matching lists. The 'ends' desired by steering groups are folded into the 'means' of algorithms, the technology itself becoming a manner of moral exploration and translation (Latour 2002). Notions of care and clinical good practice are subsumed and transformed by kidney allocation protocols; they in turn diffuse through and subvert the scientific rigidity of the protocols. The liver allocation experiment, on the other hand, tolerates no such impurity.

A thought experiment in allocation: Livers

> Some people have a great deal of difficulty in seeing that what we're trying to do here is a thought experiment, which doesn't necessarily mean anything will change. (Consultant 1)

At the current time, the transplant community is in the process of developing a new universal liver transplant allocation scheme to replace the existing region by region allocation. Liver allocation is particularly difficult as patients who do not receive a transplant will certainly die, while choice of recipient is less constrained by technical issues; the only important factors are blood group and a rough match in size between donor and recipient. Livers are therefore allocated on the existing centre by centre method on the basis of clinical need. (There is also a super-urgent category where patients suffering from acute liver failure, i.e. who are likely to die within a few days, will receive any suitable liver that becomes available nationally). In response to a perceived need for transparency and clinical rigour (Neuberger and Thorburn 2005) the transplantation community is considering moving to a quantitative score, 'based entirely on objective measures' (Neuberger and Thorburn 2005: 586). Consultant 1, who is active on the liver advisory group, described the current situation:

> The surgeons and the physicians together decide to whom [a liver] should go. Well currently, we have about 60 patients on the transplant list, and then you can take the 60 patients and divide them up by blood group by size, but even then you still come down to: 'Here is a donor, and I now have 14 or 15 people left who am I going to do?' And the surgeon and physician decide, we're going to do it in Smith or Jones. That is opaque . . . I can give you evidence to show that it currently is done to some degree on need, but the definition of need can be different in different centres . . . how you define need in Edinburgh may be different to how you define it in Cambridge.

Among the liver community, objectivity and transparency are considered as important as justice and fairness in the maintenance of public trust, and thereby of donation levels (Neuberger et al. 1998). In June 2009 the liver advisory group agreed to develop a new, national scheme, based entirely on clinical evidence. At the heart of the new scheme is the agreement that worthiness of recipients—potentially defined in many different ways—should be understood in terms of *outcomes*. It abolishes centre-based organization in favour of a national allocation protocol that treats donated livers as a national resource and aims to maximize the national benefit, calculated on the basis of the *total years of life saved at a population level*. Consultant 1 explains:

> Very few people have looked at the whole process of allocating health resources, not on the basis of an individual's need or an individual's utility but on the basis of the population life years. And that we think, because transplantation is a public good, is the way we should judge transplantation. Not on the basis of how it helps an individual but how it helps the whole population. So what you do is you add up the number of life years of all the people who died before they had a transplant, all the people who had a transplant and then died, and all the people who had a transplant and are still alive then you come to the total population life years, and you see which allocation process maximises everybody's life years.

As is the case with the introduction of the kidney and pancreas allocation protocols, the liver scheme focuses on delivering empirically grounded scoring systems that can then be used in the allocation process. The advisory group has developed 'UKELD' (United Kingdom end-stage liver disease score), an enhanced version of the USA's model for end-stage liver disease (MELD) (Wiesner et al. 2003), which gives an indication of the seriousness of the patient's liver disease. UKELD has already been implemented by the steering group as a means of screening entry to the transplant waiting list and of determining urgent treatment. The Northern Liver Alliance (comprising

Scotland and Northern England) has suggested a MELD score of 25 as a priority band necessitating more immediate treatment.

MELD and UKELD offer points of comparison for simulation exercises and the advisory group is therefore conducting a 'thought experiment' on allocation policy. The experiment will test three differing allocation schemes (Research Scientist 5), following a model set out in Schaubel et al. (2009): a need-based scheme; a survival-based, or best outcome, scheme; and a 'transplant benefit' scheme representing the net gain per patient, or estimated survival with a transplant, less estimated survival on the waiting list. The factors that underlie these models are both clinical (e.g. bilirubin and sodium in the bloodstream) and nonclinical (e.g. age, body mass index, and ethnicity). The absence of tissue matching means that the scores are not co-dependent with the donor. At the time of interview, Research Scientist 5 was in the process of running a simulation using data collected between August 2010 and January 2011 by the transplant service. The experiment set out to calculate the population life years that are saved by each allocation scheme.

Consultant 1 suggested that the use of population life years represents an 'innovative and useful' way of assessing the value of medical interventions. Figure 5.2 illustrates the point. Basing allocation on survival measures alone sees mortality shifted pre- or post-transplant: a best outcome approach, offering organs to the healthiest and youngest candidates on the list produces huge gains in post-operative survival, but results in high waiting list mortality; a needs-based, or sickest-first, approach improves waiting list mortality at the cost of lower port-operative survival. As the first graph shows, it is difficult to compare the regimes. The population life years approach, on the other hand, visualizes total life expectancy on both sides of the operation, across the whole patient group. An allocation regime can then be chosen on the basis of the greatest contribution to population life expectancy.

The introduction of survival measures as the most appropriate means of determining allocation, and the use of calculative regimes driven by clinical scoring methods, offer another way of articulating values. Where kidney reforms have centred on the pursuit of transparency and equity, thereby reframing clinical practice, the liver experiment understands transparency in terms of the objective rigour of clinical evidence. Population life years are perhaps the inevitable conclusion of the initial recognition that transplant organs are a public resource, with clinicians putting aside all factors beyond population benefits. Once again, through measures of transplant utility, notions of good clinical practice, of care, of equity and transparency itself, are radically transformed.

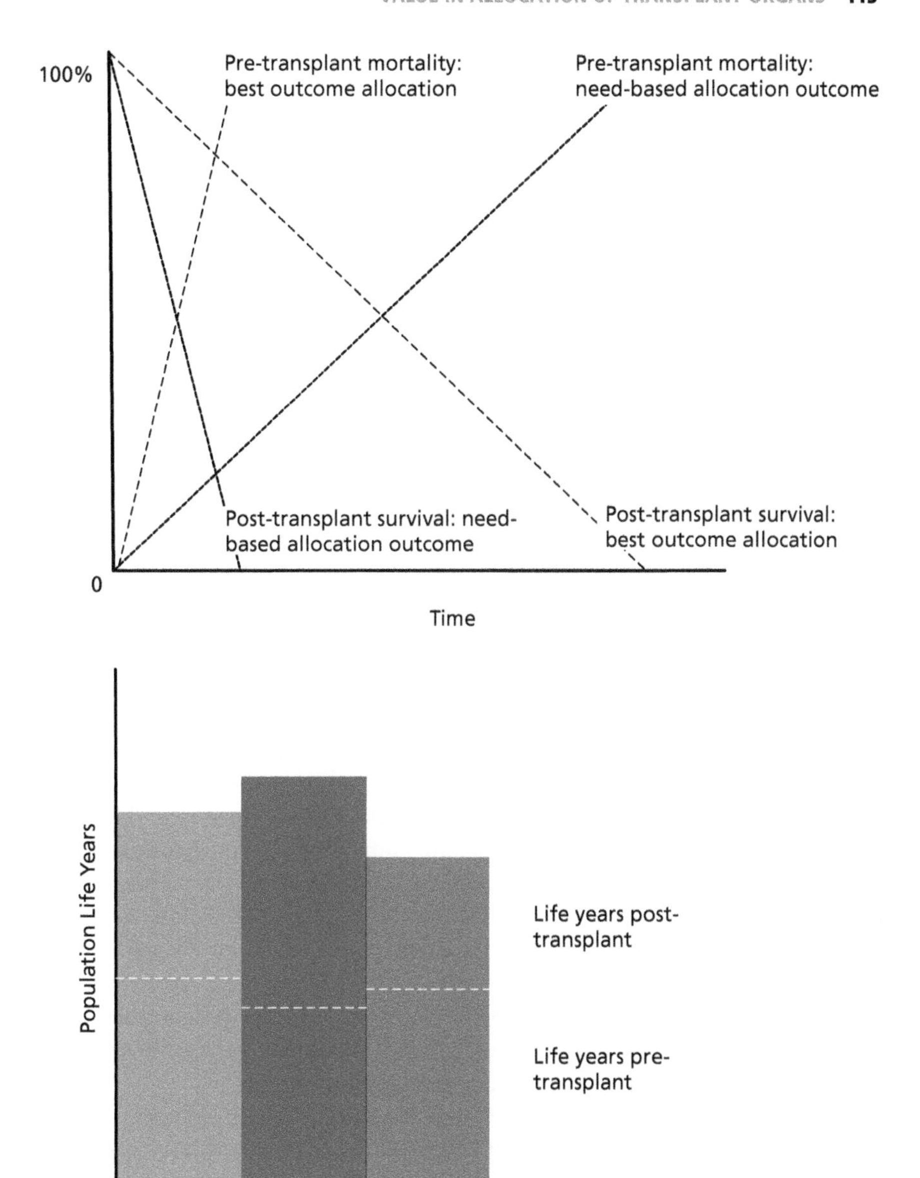

The top graph shows the problems associated with survival measures in determining allocative regimes: mortality rates shift pre/post-transplant and resist comparison. The population life years approach, shown below, renders transplant benefits immediately commensurable.

Figure 5.2. Competing liver valuations

Ghosts in the machine: Compassion and complexity

The new allocative regimes attempt to simplify and close down the complexities of organ allocation, leaving in place a single, transparent, and objective calculative framework. Yet interviews suggested that the closure is only partial, with ghosts of personal relationships and other professional values such as compassion haunting the allocation process. A social worker might, for example, object to the clinical decision to list a patient or conduct a transplant, on the grounds that the patient lacks adequate social support mechanisms. Carers build personal relationships with their patients, and develop a sense of local obligation and ownership that resists the national distribution of organs and transplantation:

> You can't help but believe that it's wrong. I can see from a utilitarian point of view that it's possibly right, but you're sitting there... [we] have a donor in [our region], and we will get Joe transplanted, because poor Joe has really struggled, and then an organisation remote says, actually big brother says that this liver goes to somebody else. Okay, that's great for somebody else, they may be sicker, they may require it. But you're not ours...' (Nurse 2)

Point scores and algorithms render such considerations exogenous, irrelevant to a clinical decision, and the implementation of a national ranking system reconfigures the meaning of good clinical work. While the nurse cherishes personal relationships with her patients, a commitment to the methodology of scoring includes the recognition that the organ may make a similar or greater difference to someone else (Surgeon 2). Personal attachment to patients becomes problematic, and a hindrance to the pursuit of transparent, just outcomes. But it remains difficult for carers to relinquish all control. Standardization has meant giving up the possibility of helping individual patients who are struggling with medication or dialysis where under the previous, local system surgeons could use their discretion to help such people. Similarly, the category of 'compassionate listing', where centres would send out faxes to ask for help in locating an organ for a particular patient, usually a child, has disappeared.

Perhaps, then, it is no surprise that practitioners talk of how the new systems might be manipulated. For example, given predetermined scores to access a waiting list, or to qualify for a 'top band' transplant, physicians could—hypothetically—manipulate the bodies of their patients to qualify for a higher score: an infection or a change in diuretic might be enough to lift a MELD score to a qualifying level.

> All of these are entirely illegal and not looked upon in the profession as kosher, but nonetheless... it probably goes on to a certain extent, although to be fair nobody does it deliberately, because the patients themselves are precarious.' (Nurse 2)

Under the revised allocative regime, caring for patients means surrendering them to the centralized process, not bending rules and risking patient health to nudge individuals up the score chart. Interviewees considered it unlikely that anyone would be so reckless as to act in this way; 'gaming' the system like this seems to exist only as an idea, a shadow of resistance to the totalizing algorithms.

Interviewees also speculated about surgeons exercising their sole remaining area of discretion to procure organs for patients who are in need and not favoured by allocation protocols. Surgeons are allowed to refuse an organ. When the central duty room offers, surgeons are expected to assess the health of the potential recipient and his or her suitability for transplant, and accept the organ only if the patient is well enough for surgery. Yet patients are out of the hospital and it may not be easy to make such an assessment; the surgeon may accept the kidney in good faith only to find that the patient is too ill to receive a transplant. In this case, the surgeon must notify the duty room and another recipient will be found, but if the kidney has already set out on its journey it will still be delivered, and can be allocated at the surgeon's discretion. Once again, evaluations based on data from central allocation can read that story differently: discretionary allocation to a fall-back patient has become a moment of subversion, of resistance to the algorithmic process. A pragmatic, professional judgement on the ability of a patient to undergo a transplant has been reclassified as another way of outsmarting the system, something worthy of comment:

> When you look at the national figures there are certainly some units where there are an awful lot of kidneys that do not end up in the person they were first offered for. (Surgeon 2)

These speculations are, perhaps, only that. They are the result of colliding worlds of value, the conflicting regimes of critical care and organ supply, of national equity and personal care, of centralization and systematization versus expertise. As allocation regimes struggle for closure, placing clinical objectivity at the centre of allocation regimes, so values that previously stood as central to clinical practice may become increasingly marginalized. Nevertheless, clinical judgement and personal compassion remain evident within the transplant community, and closure remains precarious and incomplete.

Discussion: The moral economy of transplantation

Contests over value encircle the practice of transplantation. In the literature, we find appeals to autonomy, ownership, justice, and equity; to the desert of recipients; to scientific rationality and effectiveness. These disparate arguments are united by the notion that value is stable and separate from clinical practice, and that it can therefore be used as an organizing framework for transplantation. In this chapter I have set out to demonstrate that values and practice exist in a complex relationship; as normative discourses feed into the organization of allocation protocols, so the protocols reconfigure the meaning of those norms. In the centre of this web of value sits the organ itself, understood in terms of its compatibility with recipients, or its potential contribution towards population health. The real value of an organ arises through its use in the body of another—the longer, better life that it can offer the recipient—and so becomes apparent in relation to the embodied characteristics of those who might receive it. Compatibility and survival benefit structure the practice of transplantation and yet are invisible without database work and algorithmic sorting. The whole architecture of transplantation depends upon an epistemology that is entirely relative, made concrete by allocation mechanisms.

I have explored the unfolding moral economies of organ transplantation since the 1990s by identifying the changing practices for allocation of transplant organs. Several key themes emerge as practitioners struggle to synthesize moral claims and clinical considerations: a national resource, equity, and most of all, utility. Redefining organs as a national resource changed the way that allocation is understood in the community, forcing clinicians to abandon local discretionary allocation in favour of more systematic and transparent approaches. The transplant community tends to classify approaches as either utility, need, or equity-based, or a 'fudge' or 'hodgepodge' of the three, but utilitarian approaches run much deeper in the community than the comments of practitioners might suggest.

Utility is the organizing principle that dominates the kidney allocation scheme, through the choice of HLA matching as the key measure, and through factors such as location or age. The reduction of cold ischaemic time, a key measure in improving transplant success, is a utilitarian measure; deserving individuals may miss out, but outcomes across the population are improved. Waiting time, a proxy for equity, is given prominence in descriptions of the scheme, but appears in practice to have a light weighting in terms of points awarded. In the case of the liver group's thought experiment, all three scores (need, utility, and transplant benefit) are rendered into population life years, itself a utilitarian measure. Population life years are a vision of general utility, directing the organ where it will contribute most. The population life year calculation is a classic utilitarian approach, directing resources to maximize aggregate benefits, irrespective of its effects on individuals (Ryan 1987). Indeed, the criticism levelled by one nurse is that doctors, when assessing whether to accept individuals onto the waiting list, are simply too 'clinical', ignoring personal circumstances that might make transplantation profoundly inappropriate.

This criticism—that clinicians are over-clinical—opens up a discussion of the role of professional care and judgement in a practice organized by allocation protocols. We can see that the understanding of care has been radically reconfigured by the implementation of algorithms. From the predicate that organs are a national resource, we arrive soon enough at the allocation by survival benefit proposed under the liver thought experiment. With transplantation framed in this way, the good clinician is committed to the needs of the population, rather than individual patients. The organ, with the potential to do good anywhere, must be directed where it can do the most good, and that is achieved by a rigorous compliance to the evaluatory framework demanded by the algorithm. Good practice becomes good measurement, reporting, and a dispassionate handling of patients in one's immediate care. Yet the process of closure is far from complete, and ghosts of older, more subjective realms of valuation—compassion, personal attachment, and strategic gaming—can be glimpsed in the narratives of practitioners.

This chapter highlights the materially and technically embedded nature of value in the transplant community. The debates on valuation, political objectives, and clinical practice that take place as steering groups discuss allocation protocols have been finalized and shut into the 'black boxes' of allocation protocols. From there, they shape the world as their designers intended. Allocation algorithms are 'performative' (MacKenzie 2006): there has been a striking similarity between the predicted distribution of transplants and the outcome of the system during its first few years of operation (Research Scientist 4). The algorithms also bring about a more general change in the way that practice should be done: specifying how to calculate means redefining what counts, what matters. New ways of calculating and visualizing illness give rise to new understandings and principles for action (Mol 1999): knowing *what counts* is a vital step in deciding how to act.

In a similar way, scores and rankings act as cognitive framings that organize the decisions of participants, and shape the world accordingly. Scores for liver health make visible a granular ranking of sickness, and a likelihood of mortality that can be manipulated and parsed by medical scientists. Categories of illness, such as the Northern Liver Alliance's priority band, expose patients to a further politics (Bowker and Star 1999) of rationing and need. Valuing, in the case of transplantation, is inseparable from knowing: value depends upon the epistemic—what is known and what is worth knowing. How participants come to understand and make sense of what counts—in a very particular context—has been the subject of this chapter.

Finally, as this chapter has demonstrated, the stabilization of value can only ever be partial and temporary. The push to clinical objectivity leads practitioners to assert the other values, such as compassion and individual expertise, that they have accumulated over many years of medical work—'all that matters to [the patients] is that we know that they are sick' (Nurse 2). Frames of valuation persistently overflow and must once again be reframed (Callon 1998). The moral economy of transplantation, as this chapter has shown, is unsettled, fluid, and elusive. It remains a work in progress, and will remain an important site for studies of medical ethics and organizational values for many years.

6 Critical composition of public values

On the enactment and disarticulation of what counts in health-care markets

Teun Zuiderent-Jerak, Kor Grit, and Tom van der Grinten

Defining or composing public values–The case of health-care Markets

Within the emerging field of valuation studies, the question on how to care for values that are classified as 'public' is a key concern. This concern stems from a differentiation within the policy sciences between 'societal' and 'public' values: societal values are those that are desirable for society as a whole, whereas those values become 'public' if governments needs to intervene to ensure them (van der Grinten 2006). Public values thereby become not merely an important discursive achievement that allows governments to distinguish the values they need to ensure and the ones they can leave alone; they also present politicians with the pertinent problem what their 'intervention' might consist of.

In recent years, such intervention has increasingly taken the form of *defining* what values are public. This would suffice since, according to policy theory and practice, though government has a mandate to *classify* values as 'public', the actual work of *ensuring* public values can be delegated to other parties (WRR 2000). This is how market arrangements entered the scene of ensuring public values: they are seen as an efficient way to do so, since they require minimal state intervention. Positioning markets as a solution to the problem of caring for public values matched developments such as the rise of new public management (Pollit and Bouckaert 2000). A clear definition of public values by governments was however seen as vital for making this strategy work, as only then can market actors be expected to implement them properly.

The idea of deploying markets to ensure public values is far from undisputed. One of the most heated debates in this regard is on the role of markets when ensuring public values in health care. Every so often, the introduction of market mechanisms in health-care regulation leads to warnings that health-care markets are an oxymoron (Godlee 2012; Palm 2005). But regardless of such critique, health economists' persistent claim that regulated competition can be used to implement public values, has been attractive to policymakers at least since the 1980s (Ashmore et al. 1989). Ever since, they have focused many of their efforts on ascertaining public values through the development of 'regulated' or 'managed competition' in health care (Enthoven 1988; Enthoven and van de Ven 2007).

In spite of their differences, the critics and the proponents of health-care markets share the assumption that, once defined, values remain static in the implementation process. However, as scholars from social studies of markets have shown, this is far from the case. Repeated studies have indicated that the development of market instruments actively shape the very values they were supposed to implement (Callon et al. 2007; MacKenzie and Millo 2003; MacKenzie et al. 2007; Sjögren and Helgesson 2007; Zuiderent-Jerak and van der Grinten 2009). Therefore, various authors have proposed a shift from the *unambiguous definition* of public values that can be ensured through delegation to market actors to a focus on the *composition* of public values in governance arrangements that are deployed to ensure and shape them in practice (Callon 1987; Latour 2007; Zuiderent-Jerak and van der Grinten 2009). As Bruno Latour has pointed out in his *Attempt at a 'Compositionist Manifesto'*, the word composition 'underlines that things have to be put together (Latin *componere*) while retaining their heterogeneity' (2010: 473–4). Public values are not therefore to be defined outside of the practices they are supposed to govern, but are assembled within them; they are immanent rather than transcendent. According to Latour, composition is therefore the opposite of critique. The main problem he sees with critique is that it is 'predicated on the discovery of a true world of realities lying behind a veil of appearances' (Latour 2004: 274–5). Composition, according to Latour, rejects such claims to deeper empirical truths as much as it rejects postmodernist rejections of empiricism:

> [C]ompositionists want immanence *and* truth together . . . [N]othing is beyond dispute. And yet, closure has to be achieved. But it is achieved only by the slow process of composition and compromise, not by the revelation of the world of beyond (Latour 2004: 478, emphasis in the original).

Consequently, the problem that compositionists have with markets for domains such as health care is not that marketization violates the complexity of practice by reducing it to quantified outcomes, as is often proposed by economic anthropologists (e.g. Miller 2002) and some authors within science and technology studies (STS) (Mol 2008). Rather, the issue is that within the present policy practice, the instruments for shaping these values in action are highly limited. This problem emerges since, as Bruno Latour and Vincent Lépinay have argued, both economists and economic anthropologists tend '*not [to] sufficiently quantify* all of the values to which they have access' (Latour and Lépinay 2009, emphasis in the original). The privileged status that economics grant to financial measurement of values leaves other values insufficiently quantified so that they cannot be brought into the equation of what counts.

The reasons for the preference for using *financial* market devices are, to some extent, obvious since, as Gabriel Tarde pointed out, they have one crucial advantage over many other possible quantifications: 'wealth is something much simpler and more easily measured; for it comprises infinite degrees and very few different types' (Tarde 1902, as cited in Latour and Lépinay 2009: 14). To counteract this tendency, Latour and Lépinay (2009) propose that economics focuses on the inter-comparison of a much broader spectrum of values, for which it needs to develop a much wider range of 'valuemeters'.

In the context of debates on health-care markets, this advice however seems quite similar to long-standing health economic initiatives to develop instruments for example to measure Quality Adjusted Life Years (Sjögren and Helgesson 2007), which means that the suggestion by Latour and Lépinay can be explored empirically in this field. In this chapter we therefore analyse the potential of extending the range of valuemeters in markets, to see whether the presence of such market devices indeed leads to a wider-ranging composition of values that extends value practices beyond the realm of financial valuation alone.

To do so we turn to an analysis of the introduction of 'diagnosis-treatment combinations' (DTCs) in Dutch health care. DTCs are a funding scheme that singularizes treatment processes into packaged health-care products consisting of diagnosis and treatment and for which one price is paid to hospitals by insurance companies. It provides an alternative to activity-based funding, where hospitals receive payment for each activity they perform. We explore the introduction of this new funding scheme in order to address three questions. First, what is the intended result of this Dutch market for hospital care according to various policy actors and market builders? Second, what work has to be done by various actors to make markets function? A focus on the visible and invisible work (Star and Strauss 1999) of making markets, allows for a symmetrical analysis of how unintended, as well as intended, effects come about in practice. Such work is often left out of the analysis of markets by jumping to their effects, which are then quickly classified as 'normal' effects of well-functioning markets on the one hand and 'market failures' as phenomena to be repaired on the other.

Our third and central question is: what values are enacted in the resulting health-care market practices? This question allows us to explore market devices as elements in the process of the composition of values, and thereby empirically assess the value of the composition-through-devices approach, proposed by some scholars within STS, for the study of value practices that forms the theme of this book. The study consisted of longitudinal research on the consequences of the introduction of DTCs in Dutch health care (cf. Zuiderent-Jerak (TZJ) 2009).[1]

Market devices for hospital care

Since the 1970s, public expense in health care has been on the rise and by the 1980s, the introduction of market mechanisms was already being proposed within the Dutch health-care system as a possible solution to turn the tables on increasing costs (van Egmond and Zuiderent-Jerak 2010). Though initially not being met with much enthusiasm, continuous efforts by policymakers to develop market infrastructures prepared the ground for market mechanism-based changes in the governance of hospital care (Helderman et al. 2005).

The notion that the health-care consumer should have individual buying or co-paying power and thereby steer quality (Berg et al. 2006) has been heavily criticized for being practically unworkable and theoretically flawed (Jost 2007); Dutch policymakers therefore chose not to put this notion at centre stage in the construction of the Dutch hospital market. Rather, according to the 2006 Healthcare Insurance Act, the construction of the Dutch market for hospital care assumed that insurance companies, acting as proxies for individual citizens, could buy good quality care at a reasonable cost on their behalf. As health-care insurance companies are not automatically expected to want only the best for their clients, citizens are positioned as a countervailing power (Light 2000) by being given the option to choose their insurer. Insurance companies have to accept citizens as their customers and health insurance is compulsory for all citizens to avoid 'free riders' in the system. As not all citizens have equal health risks, an extensive risk adjustment instrument compensates insurers for inequalities in health risks in their populations (cf. van Egmond and Zuiderent-Jerak 2010). A nationally defined basic package specifies the care that all insurers must provide. It leaves

[1] Between 2005 and 2007, TZJ conducted ethnographic research in a large teaching hospital in the Netherlands on initiatives that were supposed to bring together quality improvement and a strong position for the hospital in the Dutch market for hospital care. In 2010, TZJ returned to this hospital for a series of follow-up interviews with a specialist nurse, the innovation manager of the hospital, a medical specialist who also chairs one of the specialisms in the hospital, and a division manager. He further conducted interviews with a purchaser for the largest insurer in this hospital's catchment area, with the development manager and an economic expert at the Dutch Healthcare Authority, a regulatory body, and with the expert at the Dutch Association of Insurers responsible for developing a *Diagnosis–treatment combinations purchasing guide* for insurance companies.

other forms of optional non-basic care (such as dental care, etc.) to be insured via voluntary insurance schemes.

In this regulatory arrangement, the role of insurance companies is to negotiate with care providers on the quality and cost of the care they wish to deliver. Such negotiations would ideally lead to 'selective contracts', with insurance companies no longer reimbursing care at all hospitals, but rather contracting only those care providers who provide the best quality at the lowest cost. This is supposed to provide an incentive for other care providers to raise their bar in terms of quality and efficiency to be able to become preferred provider for insurers. Insurance companies are expected to have 'buying power', as they represent large numbers of citizens and thereby many potential clients for hospitals. They are expected to apply this power to stimulate quality improvement and cost reductions.

Prior to 2005, DTCs were developed so that the negotiations between insurers and hospitals could revolve around care trajectories—say, all activities related to a total hip replacement—that were packaged in one product, instead of reimbursing separate activities hospitals carry out in the care for a patient—e.g. the separate payment for diagnostics, anaesthesia, surgery, inpatient days, etc. carried out when replacing a hip.[2] The Dutch Ministry of Health decided that DTCs would be divided into two groups: an A segment, with pricing fixed on a national level; and a B segment, with prices that could be freely negotiated between insurers and hospitals. The A segment mainly comprises complex treatments such as less frequently occurring oncological care or emergency care, whereas the B segment contains care that is less complex and occurs more frequently, such as hernia repair or cataract care. Though both segments can in principle lead to negotiations between insurers and hospitals (the A segment on quality and volume, and the B segment also on price), the fact that prices are fixed for the A segment has as a consequence that this segment is usually not referred to as a health-care market by policymakers, hospital managers, and care professionals. In contrast, the B segment, with negotiable prices, is discussed in market terms. Policy discussions on the development of health-care markets focus on extending the B segment to make prices negotiable for more types of care. Thereby, marketization policy equates 'markets' with 'money', which is exactly what Latour and Lépinay (2009) criticized. The fact that many actors follow the definition of the market as a *financial* instrument brings us to our next question: how and why does this market work?

[2] Since January 2012, the Ministry has introduced some changes to this system, by moving from DTCs to *Diagnosis-Treatment Combinations On the road to Transparency* (DOTs). Besides pointing to the elusive quest for clarity, these changes mainly involve a simplification of the DTC structure and a reduction of the available products. These changes do not impact on the analysis in this chapter and therefore are not discussed in detail.

How care providers and insurance companies make markets work

In the wake of the developing market arrangements, the Dutch Ministry of Health started a large scale improvement programme called 'Better Faster' which would 'prepare the hospital sector for the new care system' (Ministry of Health Welfare and Sport 2005). Better Faster supported hospitals in improving patient safety and logistics through a series of national breakthrough collaboratives and other quality improvement programmes. Hospitals started analysing their care processes in terms of waiting time, throughput time, length of stay, number of interventions in the process, and number of visits to the outpatient clinic. In effect, the hospitals analysed how their care processes could be organized differently and how quality improvement could lead to gains not only in terms of organizational efficiency and patient experience but also in terms of the profits that a hospital made on care trajectories or, to put it in quality improvement jargon, what the 'cost of poor quality' was. For this, improvement teams in hospitals developed business cases that compared current with desired trajectories to calculate the financial implications of their quality improvement efforts.

However, these business cases required hospital financial departments to know the cost of, for example, individual interventions, outpatient clinic visits, and the cost of staff. This was not the case. For most hospitals it was a huge task to produce costing data and they tended to prioritize calculating the costs of interventions in the B segment over calculating costs that were only relevant to the A segment. Once available and integrated in these business cases, the costs sparked interesting discussions between hospital management, doctors, and quality managers. In many cases quality improvement and cost reduction seemed feasible by reducing length of stay or by omitting redundant interventions that at times were the result of poor coordination between professionals (Pronk 2006; Zuiderent-Jerak 2009).

Improving care processes and assessing financial consequences were not the only tasks hospitals started to carry out. They also developed 'dashboards' for internal steering, to ensure that once a care product had been sold at a low price, it would also be delivered accordingly, rather than falling back to the previous (more expensive) situation which was no longer reimbursed, and which would lead to the hospital facing financial loss. For colon cancer patients, overviews of the number of visits between colonoscopy (diagnosis) and surgery (treatment), for example, were readily available and were contrasted with the norm for such care set by the improvement team for this care trajectory (see Figure 6.1).

Even with all of this being achieved, hospitals still had to face the substantial challenge of getting care insurers interested in quality improvement. This

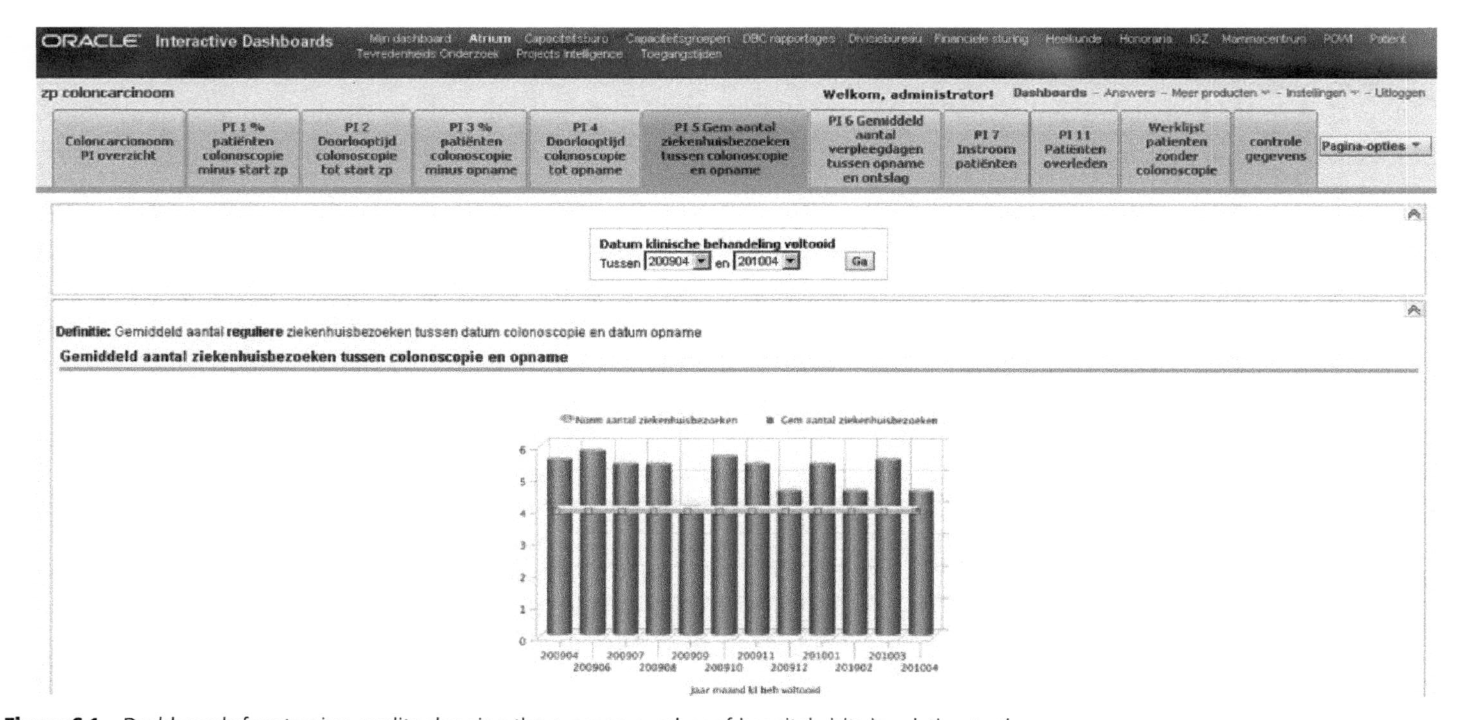

Figure 6.1. Dashboards for steering quality showing the average number of hospital visits in relation to the norm

turned out to be particularly hard work with insurers who did not always display the expected level of interest in quality matters during negotiations. As the quality manager of the hospital put it: 'It's up to us to bring everything to do with quality to the negotiating table, and they're hardly interested'. Since insurers did not seem to have an a priori focus on quality, the hospital went to some lengths to position quality as a relevant issue, trying to pitch their quality achievements. At times this led to fairly archetypical forms of commodification. The quality manager continues:

> We have to put quality on the agenda. They're not asking for it, so we have to present it. In a few areas, and these are increasing, we know we have something extra to offer, a discerning product, so we put on a big song and dance—always spontaneous, not very structured—but quite a show really, just to make sure the insurer notices. Part of the show involves producing brochures on our discerning care product. We have brochures for our departments of obstetrics, gynaecology and paediatrics, nice brochures filled with graphs, protocols and the details needed by various target groups in the insurance company. We show them off, we say, here, take a look at this, this is what our care looks like. That's all part of our repertoire.

Despite producing such sales brochures, it remains hard for a hospital to sell quality in the terms they would like. In the debate on health-care markets, negotiations are sometimes more focused on financial aspects than on medical quality. Both health economists and policymakers often explain this phenomenon away by pointing to prevailing 'information asymmetry': if information about quality is not readily available to all parties, the negotiations will focus on the information that *is* available, and that tends to be of a financial nature. The solution for this problem is generally not seen as a problem of markets as such, but is scheduled into a promissory future: as long as quality gets defined in terms of quality indicators, these can be brought into the assessment. The notion of information asymmetry supposes that once transportable performance indicators *are* available, they *will* be taken into account in quality and price negotiations. However, in this case the problem seems to be that even readily available quality information only becomes part of the equation in very particular instances. And those instances tend to be where cost, again, is a central factor. As the purchaser of the insurance company put it:

> If you want to be a preferred provider, then your price has to be below average. That doesn't mean that as soon as someone else goes ten euro cheaper, they would get moved to pole position . . . We assume that quality and affordable care can go together. That means that as soon as you [the hospital] do something right but it turns out to be more expensive, we would be less interested than if it happened to be less expensive.

So even *when* valuemeters, in the form of indicators for throughput time measured in dashboards and presented in brochures *are* readily available, quality becomes relevant only when it saves on costs. If quality always came at a lower cost, this would pose no conflict, but in that case health care could do with simpler techniques for measuring best quality at lowest cost. As this was the initial assumption for many players in the hospital care market, the market in this hospital ironically worked better in terms of negotiating for quality and price in the early years based on relatively poor information (Zuiderent-Jerak 2009) than it seems to be working now, after a longer period of sometimes frustrating experiences, but based on better information on quality and cost. The dashboards on quality and cost parameters per care trajectory that were only managerial dreams in 2007 had actually materialized by 2010, and yet in some cases it proved harder to bring quality and cost together in annual negotiations.

This problematizes both the notion of information asymmetry and of value-meters, and brings to the fore the importance of market belief in making market instruments work. On the other hand it shows the importance of sustained analysis of how markets develop over time. Market practices change, which can have dramatic consequences for how capable market practices are for coming to a composition of values which is not dominated by price.

When an insurer defines quality in terms of cost reduction, hospitals seem to have two possible strategies to continue improving care that may come at a higher cost, both of which the hospital is pursuing: creative bookkeeping (not necessarily in the usual pejorative sense of the term); and playing the patient card. Creative bookkeeping has of course become associated with scandals, greed, and the misappropriation of funds. Leaving such normative judgements aside, hospitals have imaginative bookkeeping strategies, creatively ensuring that the costs incurred for delivering additional quality are borne by insurance companies. Such creativity sometimes leads to adding up certain items in order to then be able to charge a different DTC. As a division manager explained:

> Outpatients with oesophagus carcinoma get a PET-CT scan. If deemed necessary after this scan assessment, they also get an endo-echo test on the same day. Because of their illness they are quite frail and so we admit them on a day care basis, giving them a bed to recover in, in between these two big, heavy diagnostic tests. Because of our regional specialization and given the relatively long distances many patients have to travel, day care treatment is all the more important. It's how we reduce the number of hospital visits *and* manage to create a means of recovering some of our additional costs.

Bookkeeping DTCs so that various interventions get turned into a short admission, is a strategy that health economists refer to as 'upcoding' (Steinbusch et al. 2007), which is defined as 'the practice of miscoding and misclassifying patient data to receive higher reimbursements for services provided' (Lorence and Richards 2002: 423). What may be seen as a pragmatic solution within the present definition of quality insurance companies embrace is seen by some health economists as a 'hospital-acquired disease' (Simborg 1981). One of the main problems with upcoding is that there is no countervailing power that prevents creative bookkeeping from turning into simple money-grabbing. For an efficiently organized hospital it may be a way of making ends meet, but it is easily seen as endangering the public value of affordable care.

An alternative that does not involve upcoding is to play the patient card. In the model of the Dutch hospital care market, the countervailing power of insurance companies are individual patients who are expected to 'vote with their feet' and change insurer if not satisfied with the way the 'third-party' insurance company is ensuring their interests (Schut 2009: 70). This encourages hospitals to target patients and their representatives more directly so that insurance companies are willing to broaden their definition of quality. This is exactly what the hospital studied here is doing. To frame the importance of quality improvement that is so hard to sell to insurers directly, this hospital has also chosen to address citizens and other parties by signing a 'contract with society', which includes announcements on highly specific care agreements per diagnosis in local newspapers and in the hospital's quality journal. As the quality manager explains:

> It started with our anniversary in 2004, 100 years of [this hospital], that was our first contract with society. We used the jubilee year to spend many Saturday mornings talking with many patient groups in our auditorium and asking general things like, what do we want from each other? The care guarantees [of what patients can expect per diagnosis] are actually a specification of what started then. Now our contract with society gets adjusted annually and has become far more specific: What do we deliver to our Parkinson's patients? What can [a patient] count on? When is something not good enough? And what penalty card can you hand in where? We've got the support of a management system on our side: are we still delivering what we agreed to deliver?

However, a problem with this strategy is that it assumes that insurers can sell better quality at a higher cost to individual clients, while in practice they face reputation problems: insurance companies have a hard time purchasing quality that comes at a higher price, as it is difficult for them to convince insured parties that this prices goes into better quality care, rather than into higher profits. This is however not the only problem that insurers face: it is also not easy for them to actually negotiate with hospitals on quality care.

A HEALTH-CARE PURCHASING GUIDE AS A VALUATION DEVICE

One of the problems that insurers face when having to negotiate on quality with hospitals is that they could—in theory—negotiate on a very large number of DTCs, as initially around 30,000 such products were defined. This would require insurers to have an enormous and highly qualified staff with detailed knowledge of every instance of care delivered at hospitals—a daunting task

Figure 6.2. Purchasing guide for contracting care by insurance companies

that is obviously unfeasible. So one of their core activities in ensuring they can at least negotiate on some trajectories is to limit the number of DTCs on which they negotiate. As the purchaser of a large insurance company put it:

> We cannot review all DTCs down to the last digit. So we drew up a list of priorities that is based mainly on revenue and volume, let's say a top 20 or a top 15. And we also looked at what we find important, like breast cancer and diabetes. So those aspects were also taken into account. These actually are also large volumes, so that was a good match.

But if insurance companies were delegated the task of ensuring public values, the question becomes: what about the other DTCs? For those combinations that also allow for negotiation on price (the B segment), the Dutch Association of Insurers, the sector organization representing the providers of care insurance in the Netherlands, publishes an annual *DTC purchasing guide* (Figure 6.2), subtitled *Quality as a compass when purchasing care.*

The Dutch Association of Insurers compiles and publishes this commercially available guide for negotiations between health-care insurers and secondary care providers with various other actors, including scientific medical associations, and structures the profiles around available clinical practice guidelines. Furthermore, performance indicators developed on a national level are directly integrated and the opinions of patient associations on several diseases are taken into account.

The purchasing guide was developed as an immediate response to the new role insurers play when negotiating with hospitals on the content and price of care. This purchasing guide frames to a large extent what negotiations focus on. It is an interesting market device for several reasons. First, it seems indispensable in reducing the enormous work of health-care insurance companies to doable proportions. Second, though this framing is crucial for a pragmatically functioning hospital care market, the idea that public values are delegated to insurance companies, who act on behalf of their insured, seems in need of revision. With its hybrid forum of doctors, patients, and guidelines, this hot potato of composing quality in relation to cost seems to have landed on the Dutch Association of Insurers' plate rather than with individual insurers. This raises questions about the form and place where public values are indeed shaped and what is taken into this de facto national framing of quality of care. Because of this, the framing of the guide warrants further scrutiny.

One way the purchasing guide deals with the overwhelming number of theoretically negotiable DTCs is to cluster them. As the respondent of the Association of Insurers explained:

> Let's take a simple example: cataract. There are now three DTCs: one in outpatient clinics, one as day treatment and one with admission. You can say, all right, you can assign three prices to that, or you can say—and that always has been our primary aim—all well and good: let's put them all in one basket and make a combined profile. So we reduce these three to one. There are three different codes, from outpatient, day treatment and admission, but as far as we're concerned, you can put one price to this [cluster]. That's one way we made it doable.

Besides clustering, another way of simplifying negotiations on large numbers of DTCs is to specify which ones can be excluded. The same respondent explains:

> Wherever it says 'no' [*niet-onderhandelbaar*: non-negotiable] we said: 'we don't make a profile for that'. This actually means two things. Either it's nonsense: this DTC shouldn't even be listed, meaning: price equals zero. Or it is so rare: we're not going to negotiate, you give your price and I'll see if I think: Ouch, let go of my arm'... So that's how we've approached it. We looked at how many DTCs there are in the B segment—something like 10,000—and we've been able to reduce these to let's say 200. Then again, our focus has always been on high volume DTCs.

In this quest for doability, DTCs are increasingly selected and combined in clusters. A relatively small number of 'baskets' thus covers a large percentage of delivered care, especially in certain types of treatment such as eye care. Thereby, the purchasing guide assumes that a smaller number of DTCs can be used in negotiations, all the while still pursuing quality on a larger scale:

> At some point I can start to discuss eye care. Someone who's good in cataract procedures, wouldn't that person be good in glaucoma procedures as well? . . . Once I know the core points, where I can say, well that's organized well, then the rest will follow.

This assumption is highly understandable in the light of the creation of a doable health-care market, but quality improvement researchers have observed that improvement to one stream of care is often at the expense of other patient groups. This phenomenon, generally referred to as 'carve out' (Silvester et al. 2004), has been particularly noted in eye care where the dominance of cataract often leads to well-organized treatment pathways that are completely isolated from other forms of eye care.

These problems do however not stop this negotiation infrastructure from at times being highly consequential. As the associations' respondent told us, the purchasing guide is not only an instrument for insurers, but also for hospital directors:

> The board of directors at [a large hospital in the western conurbation of the country] took this guide to their doctors, saying: 'Well, look here!' And within a week admission length was cut by two days . . . [The guide gave the directors] something on paper that isn't theirs but has objective credibility, they didn't make it up. It's the professional organizations saying what they think things should look like on average. So they took the guide to their specialists and said: 'Looks like you're deviating from the norm. You *can*, but then we'd like to know why.' Apparently there was no valid reason, so this guide produced results: wonderful!

In this sense the purchasing guide sets a de facto norm, even though it was supposed to be a 'mere' negotiation aid. Apparently, the threat of future negotiation combined with a standard created partly by doctors themselves can produce results—and possibly—quality gains without the need for actual negotiation.

As we have shown in this section, the hospital has had to do much work to create valuemeters that shape public values in the market for hospital care. The challenge for insurers seems no less daunting, and the results for the relation between markets and public values seem equally ambiguous; this despite the valuemeters the sector organization of health-care insurance companies developed in the form of the purchasing guide. Having analysed some of the work various actors carry out in the market for hospital care, let us now return to the question of which values are produced.

The composition of public values in the Dutch market for hospital care

One aspect of the health-care market that has come explicitly to the fore is that all public values, not merely affordability, tend to get framed in the light of financial devices. Quality improves through the market devices we encountered, but generally only the quality that comes at a lower cost. Interestingly, all actors, whether based in hospitals, insurance companies, or agencies like the Dutch Association of Insurers, grant an ontologically privileged status to the price mechanism to ensure public values. One of the clearest indicators for this is that for those aspects of care not included in the DTC B segment, there is no purchasing guide. Insurers indicate that quality is only quality if it includes efficiency gains and hospitals point out that quality not associated with a financial advantage cannot be sold.

As a consequence of this situation it is harder to sell good quality than to divest expensive care—a policy that respondents referred to as 'managing bleeders and feeders'. This shows that, whereas the aim of the market policy was to get health-care organizations to compete for the favour of the insurance companies on the basis of differential quality, in fact they were becoming more similar due to the market devices that were developed to meet this aim.

Negotiations on quality do seem to have started, which most certainly is a major gain for hospitals, patients, insurers, and policymakers alike, in comparison to earlier times where discussions between insurers and hospitals only focused on volume and money. But according to the respondents, through the central positioning of price mechanisms, these negotiations shape the definition of quality as positively related to cost reduction, despite the wide range of valuemeters being developed by many actors to countervail this definition of the public value of quality of care.

Conclusions: Critical composition of value practices

As we hope to have shown, the market practices under study have complex relations to public values. What 'the market' is and what 'public values' are, is never clear in any fixed or static sense. Our approach to studying this process of the composition of public values through market practices has allowed us to analyse the *work* that many actors need to do to produce such effects. It also opens up the study of markets as 'political issues' (Barry and Slater 2002: 287) as it allows us to analyse how markets and public values shape each other. If market devices shape public values in specific ways, rather than merely implementing predefined values, the development of market practices is a highly relevant empirical domain. Government agents as well as market researchers will want to understand how this occurs, with what consequences, and possible alternatives to remedy bad value-composition.

We feel there are two reasons why this should not lead to a re-politicization of the development of market devices through increased democratic control and debate over their development processes. First, as we indicated in the introduction, political debates on the role of markets in health care are generally caught in the question as to whether health care really is or is not a market, making it highly unlikely that such re-politicization would be productive. Second, bringing markets back into the political forum implies that the composition of public values could somehow be brought back under democratic control through discursive action. We would rather propose to accept that public values are shaped in practice and that therefore the relationship between policy aims and consequences can never quite be captured through the logic of implementation. For this the process is too unpredictable and the consequences too unforeseen. Such dynamic unpredictability is better explored through an experimental role for policymakers. They would see market devices not as an operationalization of policy aims, but as experimental practices in which the aim is a good composition of public values. Without such a shift, health economists and policymakers may always have the promise of a better market future but never a prospect for achieving it (Latour 2010: 486). Abandoning the idea that a market can implement public values is a key requirement for composition, for how could one 'assemble anything properly *while not looking at it*! . . . It is impossible to compose without attending firmly to the task at hand' (Latour 2010: 487, emphasis in the original). And such 'attending firmly' is precisely undone by the logic of implementation.

Our analysis shows that market devices can shape specific public values in rather unexpected ways. Though the policy aim in the Dutch hospital market was to increase choice for both clients and insurance companies and thereby provide incentives for quality improvement through diversification of hospitals, ironically, DTCs are producing the very opposite result of a national standard for quality of care. These market devices therefore seem to undo the very aspects that, according to the policy aims, they were supposed to strengthen. Market devices have often been regarded as shaping the setting in line with the policy aims and assumptions under which they are expected to operate—generally captured under the heading of 'performative' market mechanisms (MacKenzie 2004; MacKenzie et al. 2007). However, the market devices we studied made the practices within which they were supposed to operate less favourable to the policy aims—a phenomenon that has been called 'counterperformativity' (MacKenzie 2007).

Furthermore, the market devices studied here lead to specific compositions of the public value of quality that is shaped predominantly in financial terms. Quality is easily defined as directly related to cost savings, which leads to substantial gains in the affordability of care and reduction of length of stay in most hospitals. This is a substantial improvement, given that each unnecessary day in a hospital is both a societal cost and a risk for patients. Yet this specific definition of quality sits uncomfortably with notions of quality that do *not* involve cost savings. Given the extensive efforts to develop valuemeters for such non-financial definitions of quality, the fact that quality has come to be defined in terms of cost-saving is not merely a consequence of a lack of quality information and/or 'information asymmetry'.

Even where information on quality *is* readily available during negotiations, price seems to dominantly shape what counts as quality. Information may be available but still not have any consequence if financial aspects are so much easier to calculate and other quality aspects are both more elusive and less convenient to some actors. This result is not unique to this market arrangement: financial aspects were even more confining in times of fixed budgets that tended to be consumed towards the end of the year and could lead to the closing down of operating theatres for some surgical procedures until the start of the new financial year. It is hard to imagine a more dominant link between quality and its definition by financial aspects.

The point therefore is not that these new market devices have made an issue of price or money. Nor do we want to pose it as problematic that public values are shaped in the practice of operationalizing them in market devices: means always change and translate the aims they are supposed to ensure (Latour 1999b). Interestingly in this study however, we noticed that the redefinition of public values in terms of price is not in the absence of, but despite the wide availability of a range of valuemeters that precisely aim at the broader notion of 'metrology' that Latour and Lépinay (2009) propose. The strategy of preventing such narrow definitions of public values through the development of a wider range of valuemeters therefore seems not only to neatly match initiatives by health economists and other actors who have been developing such meters for decades; it also seems unworkable in practice.

This crucially shifts the focus from the *development* of valuemeters to a study of valuemeters *at work* and points to tentative ways of exploring experimental grounds for the task of composing rather than implementing public values.[3] Such explorations would need to be pursued not merely

[3] In this light, it would e.g. be interesting to experiment with market developments that do not ascribe a privileged status to financial devices and price mechanism, for which experimenters could draw upon the existing economic literature about non-price competition (Gaynor and Vogt 1999; Hammer 1999; Pope 1989). Rather than developing more valuemeters that have to *compete* with the measurement of price, the DTCs with fixed prices (the DTC A segment) may be precisely one of the most promising domains for exploring competition on other public values such as quality. Experiences within the British NHS with the 2006 reform of fixing prices and giving patients the choice of at least five hospitals seem promising in this regard (Gaynor et al. 2010).

through developing valuemeters that would make actors' value judgements 'visible and readable' (Latour and Lépinay 2009: 16), but also by *disarticulating* certain values to allow other values to be more powerfully articulated. Latour's definition of composition, of having to put things together while retaining their heterogeneity, also implies that, when such heterogeneity is lost through the dominance of one value defining the others, a somewhat more antagonistic process of excluding certain aspects—in this case price—may be warranted. If composition is to substantially change existing practices of market development, rather than mirroring a health economic promise of a future with symmetrical treatment of heterogeneous values once the correct information is available, scholarly critiques of the workings of existing market arrangements may not be opposed to composition, as Latour proposed, but may well be dearly needed for the very process of composing public values. When such critiques are not aimed at debunking 'the market' in order to unveil sociological complexity, but are geared towards differentiating between poor and better composition of a heterogeneity of public values—in other words, if critique shifts from being predicated on sociological realism to finding a firm footing in immanence—composition becomes *critical* composition. Wouldn't that be a prospect for social studies of value practices?

ACKNOWLEDGEMENTS

We would like to thank our respondents of this study for their generous participation. We are particularly thankful to the quality manager and other respondents at the large teaching hospital for their continued contributions to research. Previous versions of his chapter have benefited much from feedback by workshop participants and by Daniel Neyland, Erik Schut, Gerard de Vries and the editors. The research was part of and funded by the Market, State and Society program of the Netherlands Scientific Council for Government Policy.

7 The mosquito multiple

Malaria and market-based initiatives

Daniel Neyland and Elena Simakova

This chapter offers an analysis of the challenges posed by market interventions in malaria. Attempts to introduce a market for malaria vaccines are held centre stage. Through this focus, the chapter engages with a central theme of this collection: the coordination of different forms of value. Coordination work, we will suggest, is focused around the notion of market value. We will explore the creation of a $3b value for a malaria vaccine market and ask: where did this value emerge from, how, and with what consequence? We will suggest that market values are assessed in terms of their general properties, linking otherwise disconnected entities, and raising questions of the ontology of newly connected people and things. However, we will also argue that market values are understood through their particularities, their troublemaking disruptions, their role in generating disconnection and dissonance. Hence we will explore the $3b market value as a focal point for the coordination, articulation, and disruption of generals and particulars. The chapter will begin by engaging with the general and the particular, before analysing the specific details of market interventions in malaria.

The general and the particular

Scholars within science and technology studies (STS) have recently turned attention towards matters of organization and business (for an introduction, see Woolgar et al. 2009). Within these moves, markets have formed a clear focal point (see for example, Callon 1998; Callon et al. 2007; MacKenzie et al. 2007). These studies of markets have been used to suggest that markets can be conceived as heterogeneous assemblages, through which values, externalities, connections, prices, and consumption (among many other activities) are accomplished and coordinated. This draws on a history of STS research which suggests that, for example, technologies can be usefully and provocatively conceived as assemblages (see for example, Latour 2005), that scientific knowledge can be understood as constructed (for example, Latour and Woolgar 1979), and that ontology can be addressed as multiple (Mol 2002).

Allied to these moves into markets have been attempts to engage with marketing (Araujo et al. 2010; Cochoy 2009; Simakova 2012), accounts and accounting (Hopwood 2009; Law 1996; Skaerbaek 2009), forms of value and valuation (Beunza et al. 2006; Muniesa 2012; Sjögren and Helgesson 2007). This treatment of value and valuation upholds many of the same analytic principles as the STS move to markets. Hence value can be thought through in terms of constructive processes, assemblages and, importantly for our chapter, as a focal point for both the tense, disruptive, and provocative untangling and coordination[1] of ontological multiplicity. Following the latter line of argument, assemblages can be understood as featuring unstable associations of ontologically unstable entities (with the mutual instability providing further bases for ongoing uncertainty). Constituting assemblages involves work that draws together and coordinates (e.g. through commensuration) incompatible renditions ('versions') of the 'same' thing with distinct values (as e.g. in Sjögren and Helgesson 2007). Assembly work thus can be understood as drawing together and coordinating the means to render distinct things of the same value (for example, stocks in distinct companies could share the same price) or provide a basis for assessing the equivalence of distinct things through a singular line of value from which comparisons can be drawn (for example, in carbon trading mechanisms; MacKenzie 2009a).

However, STS research has not held exclusive sway on these matters of value. From economic sociology (broadly construed[2]) authors such as Boltanski and Thevenot (2006) draw on vocabularies resonant with the STS turn to markets and the economic (for more on this, see Guggenheim and Potthast 2011). Hence we find a pursuit of people and things and associated analyses of symmetry; we find a focus on qualification; investigations of associations; and we find a focus on the construction of value as worlds of worth. Boltanski and Thevenot (2006) also place great emphasis on distinctions of worlds, regimes, and forms of value.

We could take these distinctions as one focal point for value-laden coordination work. However, we also find many distinctions between the approaches. For example, could the work of Boltanksi and Thevenot be said to provide an alternate take on multiplicity in comparison to Mol (2002); a varied grammar of description and analysis? If Mol's take on multiplicity can

[1] By coordination we understand not only decision making, strategizing, and evaluation of market initiatives accomplished by participants, but also the role materiality—in the sense of market devices (as in Callon et al. 2007)—plays in establishing and contesting ontological multiplicity and value matters in market assemblages.

[2] For a more detailed discussion of the conversation between economic sociology and STS see collection *Living in a Material World: economic sociology meets science and technology studies* edited by Pinch and Swedberg (2008).

be summarized as a concern with distributed praxis through which ontologies are accomplished, Boltanski and Thevenot's work could perhaps be understood as providing an ontology of values through their grammars of worth.

Two salient points emerge from the work of Bolatnski and Thevenot (2006) for this chapter. First, these 'worlds of worth' are not singular nor do they necessarily involve the assemblage of entities into comfortable agglomerations: much of Boltanski and Thevenot's work is dedicated to an investigation of particular problematics that arise through these forms of worth. Second, for these authors a central feature of the move to count, value, justify, and place worth on some person or thing, is to move between the particular and the general. However, there is no guarantee of smooth passage between particular and general. For example, particular problematic instances have to be explained in such terms that accommodations can be made by general principles or, more frequently, problematic particulars have to be cast out from generals in order to pay recognition to the higher value of generals, maintaining the latter's aggregate integrity. For our focus on value-laden coordination work, the general and the particular might be taken as a focal point for tension.

More precisely, this suggests that if we are to further explore STS ideas on values in market-based interventions made into the field of malaria, we could usefully explore the coordination practices which accomplish values and valuation and moves between the general and the particular. In order to understand values this chapter will ask: how do participants in market-based interventions move from the particular to the general and vice versa and how do they act to protect market principles (generals) from troublemakers (ill-fitting particulars)?

To address this question, we will engage with the work of Lee (1999) on the general and the particular. Lee (1999) investigates policies relating to children and explores the complex position of children in social order. They are understood in universal terms at times, as specifically different or unique on occasions, they are given responsibility in specific circumstances, and there are those individuals and institutions that in other circumstances are called upon to speak on behalf of and take responsibility for children. Lee suggests that policy interventions, such as Article 12 of the UN Convention on the Rights of Children, engage many of these issues simultaneously. It is both potentially toothless and profound, and importantly for our chapter, it attempts to operate in general and particular ways. Lee argues:

> It offers the most general provision for children that is possible. In order to provide for all children, however, its provisions must be seen to be capable of covering the case of each particular child. Abstract principles are all well and good, but unless they are seen to be applicable they can scarcely amount to a policy directive. The authors of Article 12, then, had to find a way for the general and particular to meet each other on the page. (1999: 457)

It is often suggested that such policy interventions talk on behalf of particular populations (in this chapter sick and potentially sick populations of malaria sufferers), that they distribute forms of responsibility and accountability (Neyland 2012). We propose to engage with the ways in which such policy interventions, and in particular through their focus on forms of value and valuation, draw together on the same page notions of the general and particular. Hence we will explore how attempts to introduce markets, for example, invoke motifs of the general (that firms will pursue profit, will pursue higher returns, that these are universal conditions which underpin the actions of firms and their pursuit of value, that value is characterized by economic gain). At the same time we will analyse the ways in which atttempts to introduce markets are also drawn into the details of the particular (relating to specific pharmaceutical firms, the particular details of research and development projects, the complexities of intervening in particular diseases in particular countries or regions).

This suggests that value-laden coordination work can be envisaged through policy interventions and invocations of value. These might be thought of as bringing into being general ontological statements regarding the nature of value, the nature of business, the nature of disease, while also being engaged with the particularities of specific disease problems. However, we will also look to go beyond this singular directionality, moving from the general to the particular. We will also investigate the work done in and around particular policy interventions to attempt to prevent the particular from shifting to the general. We will argue in the empirical material presented in the following sections, that a discursive specificity move can be seen to be made to prevent particulars becoming generals in certain situations.[3]

Market-based interventions in malaria

Those working to intervene in malaria are said to be engaging with a complex field. First, it has become a familiar argument that only 10 per cent of global medical research is devoted to conditions that account for 90 per cent of the world's disease burden (DNDi 2006; POST 2005). In this sense, malaria is a disease outside of conventions for understanding economic value; it is a disease without a profitable market. Second, treatment usage is said to be ineffective, interventions in some geographical areas non-existent, and knowledge of

[3] The research for this chapter draws on interviews carried out over a two-year period with thirty-four organizations involved in various kinds of interventions in malaria. These organizations included governmental agencies at national and international level, a large pharmaceutical firm, several philanthropic organizations, university scientists, and economists with an interest in malaria. Interviews were analysed to investigate the ways in which participants discursively accomplished positions for various malaria entities. This methodology drew inspiration from ethnomethodology (Goodwin 1994; Lynch 1998).

diseases limited (POST 2005; WHO 2008). Hence malaria is positioned as one in a group of diseases characterized by uncertain medical 'expertise', treatment, or intervention. Third, the scale of malaria is understood as both significant (in the size of the population of sufferers, the cost, and geographical spread, Bayer 2006; Gates Malaria Partnership 2006; GSK 2005; National Geographic 2007; VOA 2005) and uncertain (for example, with estimates of death rates shifting between 1 million and 2 million per year, National Geographic 2007). Fourth, scientific accounts of malaria also shift between certainty[4] and uncertainty[5] (National Geographic 2007; VOA 2005).

Among these uncertainties, vaccines for malaria have been heralded as a way forward. Although interventions in sub-Saharan Africa have included educational initiatives (for example based on promoting the use of bed nets); malaria management drives (through, for example, attempts to reduce mosquito populations); and malaria treatment (through the provision of medicines to people who have developed malaria), vaccine development programmes (coordinated by, for example, the Malaria Vaccine Initiative, a public private partnership involving Gates Foundation funding, GSK[6] pharmaceuticals and university researchers among others) currently attract the most research funding (Global Health Program 2011). The story of vaccines narrated by vaccinologists is compelling; the single dose, delivered swiftly, which provides lifelong protection and may even eradicate malaria (by cutting the lifecycle of the disease). Hence uncertainties might be reduced regarding the delivery of treatment (if a vaccine was developed it might be deliverable in a single dose), scale (it might work for everyone), and science (the cycles of malaria might be cut).

Although compelling, if not seductive, this narration has introduced two key questions into the field of malaria. How might the first uncertainty (detailed above) of malaria's position outside the conventions for markets, profits, and the accrual of economic value be addressed? Furthermore, to what extent can the vaccinologists' narration of vaccine certainty be accomplished?

Malaria and market devices

One means of explaining the ongoing absence of a malaria vaccine has been offered by Harvard economists (Glennerster et al. 2006). They suggest that the market for malaria is economically unattractive and needs to be made more

[4] With a focus on the cyclical nature of mosquito bites, sporozites entering and multiplying in the blood, being picked up by other mosquitoes, and entering other bodies leading to fever, vomiting, coma, and sometimes death.

[5] In understanding the relationships between blood cells, immunity, brains, parasites, mosquitoes, and their feeding habits.

[6] GlaxoSmithKline.

alluring. In order to achieve that, various market devices have been proposed and evaluated as part of coordination work to achieve market value. For example Glennerster et al. (2006) suggest that a malaria vaccine market could be constructed through Advanced Market Commitments (AMCs) or Advanced Purchase Commitments (APCs).[7] They argue that these would act as 'pull factors' to entice pharmaceutical firms into developing vaccines for otherwise less attractive (i.e. less lucrative) diseases. They suggest: 'One proposal to incentivize private sector R&D investments in products for diseases concentrated in poor countries is for sponsors (rich-country governments, private foundations, or international organizations such as the World Bank) to undertake "advance purchase commitments" for desired products, such as vaccine... If no vaccine is developed, no donor funds would be spent' (2006: 67). They argue that this approach is cost effective, involving an outlay of $15 per life year saved.[8]

The first AMC/APC has been recently launched for pneumococcal vaccines.[9] Discussions continue shaping a malaria market equivalent. This would involve European member states, the World Bank, and the Gates Foundation, under the management of the Global Alliance for Vaccines and Immunisation (GAVI). According to Glennerster et al. (2006) the details of the AMC/APC would operate as follows: a group of credible sponsors would provide a legal contract, setting out the total potential market for a vaccine. The value which has been placed on such a market is $3b. The sponsors would then underwrite a price (around $15 per dose). This price would be guaranteed for a certain number of doses (up to a total cost of $3b). Countries that would be eligible would also be established at this stage. After this fixed price, the developer (who will have covered their costs by this point) must then guarantee to sell doses at a cheaper price (say $1 per dose). Sponsors would pay more than recipient countries of the initial $15 dose (say $14 and $1 respectively). Subsequent vaccines would also be eligible for guaranteed price; if a better product comes to market, recipients could switch to another product. The proposal suggests an independent adjudication committee oversees the agreement.

Various principles, which we will term 'generals' and 'particulars', emerge through the AMC/APC proposition. We will begin with the generals. First,

[7] In some cases APCs have a slightly different emphasis in comparison with AMCs. The former are focused on putting in place an agreement to purchase a near-market product; the latter involve producing an agreement to purchase a theoretical future product once available, effectively (attempting) to stimulate general market competition to produce such a product. However, on occasions in the literature, the terms APCs and AMCs seem to be used interchangeably.

[8] Economists suggest that developing-country populations are usually considered good value for anything up to $100 per life year. Glennerster et al. (2006) argue that: 'the US cost-effectiveness threshold is estimated to be as high as $50,000 to $100,000 per life-year saved' (2006: 74).

[9] See: <http://www.gavialliance.org/funding/pneumococcal-amc/>, accessed 1 April 2011.

there is a general principle that private sector firms will be incentivized by a commitment to receive financial return and that a value (in this case $3b) can stimulate action. Second, there is the prioritizing of a vaccine as a general good. Third, cost effectiveness is drawn to centre stage as a principle means for adjudging value. Each of these generals is predicated upon ontological certainty regarding the nature of private sector firms, the benefits of a vaccine, and what constitutes cost effectiveness. Certainty resides in the ability of generals to talk on behalf of their populations, thereby implicating the importance of audiences (imaginary, real, specific) for the practices of articulating value, or valuation. However, as the following interviewee suggests, incentives for private sector firms may be more complex than this initial proposal suggests.

> Interviewee 9 (UK government agency)
>
> My own personal opinion is that people haven't thought through enough who the AMC is aimed at. I mean I think on paper the AMC is a really interesting idea but we need to have a more detailed analysis of who will actually respond to developing potential vaccine candidates for neglected diseases and very often that's smaller bio-tech companies rather than big pharma so the incentives of an AMC are in the wrong time-scales for bio-techs and my view . . . the timescales of AMCs is missing the most innovative part of the industry. I've talked to bio-techs and they're trying to take things forward more quickly than big pharma so we need a more nuanced approach more than just we'll promise to buy all this in 20 years time. We need intermediate milestones which will give the right incentives to the right companies at the right time in development.

For this interviewee the general and the particular cannot sit side by side on the same page and indeed the particulars (of biotechs) question the generals (of a singular model of firms each operating to the same timescale). The ontological certainties of general propositions (that private sector firms operate to a singular temporality) appear to be placed under scrutiny by specific particulars (a broader range of private sector firms, with distinct timeframes) that by necessity must be cast out from the unified aggregate category of the general in order for that category to make sense. The smooth ontology narrated for the private sector within the AMC/APC proposal slips away through this interviewee's discussion of the particulars of biotechs. The passage from particular back to general is cut; the general no longer talks on behalf of its assumed population of particulars. In a similar manner, the following interviewee response can be understood as questioning the general principle that private sector firms in this field were profit seeking:

> Interviewee 14 (Neglected disease consultant)
>
> [O]ne firm actually said to us if there was a profit in this we would leave, because what we're getting out of it now is the good guy benefit, and if we're making money we don't get that and it's not worth it, we just go back to blood pressure. So I think it was very, very clear that they wanted to partner, it wasn't about profits and the incentives were set wrong.

> ... it's funny, because when the economists reviewed our work they said it's, the funniest thing is that governments appear to be, there's a lot of not-for-profit activity being done by companies in the drug field and governments are now trying to monetise it in a sense, and move them to doing it for-profit what they now do not-for-profit, and it's really hard to see why you would do that.

The general principle of profit seeking in the private sector is disrupted by this interviewee through the particular features of pharmaceutical activity in the field of malaria. Rather than entering that field for profit, the interviewee suggests firms are attempting to disavow their status as ruthless profit seekers by working for free. However, with the emergence of AMCs/APCs they are now placed under pressure to contractually agree to receive income—a percentage of the $3b market value. According to this interviewee the value of involvement for large pharmaceutical firms is not profit, but reputation and corporate responsibility, an attempt to generate ethical rather than financial value. The general principle of value through profit in this instance appears directly opposed to the particular principles sought of ethical value through non-profit. The smooth aggregate ontology of private sector firms built into the AMC/APC proposal is disrupted once again; not only are private sector firms in this field divisible into distinct particulars (large pharmaceutical firms and smaller biotechs), they are also divisible by the field into which they are entering (seeking profits through blood pressure and ethical standing through malaria). The particulars in these excerpts could be said to generate dissonance in the smooth aggregate generals of the AMC/APC agreement. These particulars directly question the ability of generals to talk on behalf of their populations and go some way to casting these particulars out from generals.

The AMC/APC proposal can aptly be described as a market device producing generals and particulars. Through the scheme, further generals emerge; that a market can be valued, that contracts can bind values, that dosage prices can be used to calculate values, and that values can be governed through a committee. These generals also sit on the same page as particulars; a specific market value ($3b), a dosage price ($15), and a division of cost (between donors and recipient governments). Within the AMC/APC model the relationship between these generals and particulars appears unproblematic. The twin logics of value invoked of economy and health appear comfortably intertwined. However, this comfort is based once again on ontological certainty. Interviewees were swift to point out tension between the ontological certainty of generals and problems with these specific particulars. Although interviewees did not suggest that placing a value on a market was problematic as a general principle, they did seek to raise questions of the particular valuation of $3b. The following interviewee suggests that the incentives supposedly made available by a $3b evaluation are misleadingly straightforward:

Interviewee 10 (UK based senior vaccine scientist)

[T]he idea behind the Gordon Brown initiatives[10] is that if you put a big enough pot of gold at the end of the rainbow that suddenly private companies will invest in these diseases... that's an untested idea. The initial problem with it is... by and large vaccines are made by very very very big companies. 80% of the world vaccines are sold by six companies. These companies are interested in products that earn at least a billion dollars a year, not a third of a billion dollars a year, and the malaria vaccine and TB vaccine might hit peak sale of half a billion a year. But by and large they are under the threshold. They're not blockbusters.

This interviewee uses the term 'blockbuster' as a means to generate a distinction between compelling and not so compelling financial evaluations (for more on compelling market relations, see Simakova and Neyland 2008). For this interviewee malaria vaccines would remain on the non-compelling side of this division even with an AMC/APC in place. In the preceding excerpt the ontology of valuation (that a monetary figure could be placed on a market and this will compel action) was disrupted through calling upon further valuations as comparative metrics to prove the absence of allure in $3b. Further questions were asked by interviewees of the specific consequences stemming from a $3b valuation of a malaria market:

Interviewee 9 (UK government agency)

If you've got a guaranteed market for something which is good enough but not very good what's the incentive for making something really good?

For this interviewee the smooth certainty of the AMC/APC principles was drawn into question through future uncertainties; what would happen if one 'good enough but not very good' vaccine was tied into a contract and received most of the funding? Would others compete to try and take that contract away; who would decide on which vaccine was good enough to receive a contract; would disputes be handled by the AMC/APC arbitration committee; would they be lobbied by firms seeking funding for their vaccines? Responsibility for ontological certainty remained an unanswered question.

Vaccine certainty

Further questions were raised by interviewees regarding the compelling certainty of vaccines narrated by vaccinologists. At the centre of attempts to build a $3b market value for malaria vaccines was certainty that a vaccine would be

[10] Although the AMC/APC model was devised by US economists, a strong advocate for this model was Gordon Brown initially in his role as Chancellor of the Exchequer and then in his subsequent role as UK Prime Minister.

produced that would work. Uncertainty among interviewees was mirrored in the broader malaria literature with discussion over twenty years of the immanence of a vaccine (see Turnbull 1989) and suggestions made that a vaccine against a parasitic disease has never been successfully accomplished (VOA 2005). The compelling certainty of malaria vaccines (that vaccines could be developed and delivered in a single dose, providing lifelong protection) may still prove elusive (Neyland 2012). However, much of the proposed market-based intervention into malaria is based on this problematic but seductive certainty; the AMC/APC model appears to depend on the notion that a successful malaria vaccine could be developed in response to the financial stimulus of the $3b model. Many of the interviewees in this research swiftly moved to question the status of the current leading malaria vaccine candidate and its future efficacy:

> Interviewee 2 (Senior neglected disease scientist)
>
> I see malaria vaccine as taking money away from areas that we know work to something which is entirely speculative. I'm just a bit cautious about that... It's a really serious issue... [The current leading malaria vaccine candidate] will never achieve anything at all in Africa.

> Interviewee 3 (US government agency)
>
> We also know, based on the... experience [of the current leading malaria vaccine candidate], that here you have a novel synthetic vaccine that gives you at least partial immunity and so, you know, if you take all of these things together, I think that the overwhelming impression is that technically it should be feasible to come up with a malaria vaccine.

In these excerpts we can note that the seductive general certainty of vaccines—that they can be delivered in a single dose and provide lifelong protection—both underpins the value in the AMC/APC model, and is characterized by an uncertainty that varies among interviewees. Is the current leading vaccine candidate effective enough? Is it effective for the general population, or for those groups particularly susceptible to malaria (such as children under five or pregnant women)? For how long would it be effective if parasites were able to continue to mutate in the human body in response to contact with the vaccinated body? Cost effectiveness in the AMC/APC model is also predicated on the seductive narration of vaccine certainty. Cost effectiveness is pushed to the fore as a general basis for adjudging value in the AMC/APC model, but is simultaneously undermined by the ongoing uncertainty of vaccine development. If it turns out, for example, as many predict (VOA 2005), that parasites will evolve in the human body in response to vaccines, rendering long-term judgements of the efficacy of a vaccine impossible, this ongoing ontological uncertainty would invalidate values produced through cost effectiveness judgements. These initial generals appear to derive their aggregate integrity and ability to talk on behalf of populations (such as the private sector) and values (such as cost effectiveness) from a problematic ontological certainty premised on the notion that the aggregates (such as private sector firms, vaccines, and cost effectiveness) on behalf of which they speak are coherent and singular.

Further adding to these uncertainties, the following interviewee raised questions of the notion of scientific innovation designed into the AMC/APC model. When asked why she thought advocates of the AMC/APC thought a market-based intervention would stimulate scientific discovery, she answered in the following manner:

> Interviewee 14 (Neglected disease consultant)
>
> I think largely because it was designed in the US and they have a strong pro-market preference, so all their incentives start with we need to make a market, because that will stimulate people. But in practice, I always say to them are there a lot of people in America have colds, of course there are, there's a big market for a common cold but there's no treatment for it. Why is that? It's because for some things it's not the market, it's the problem is we don't know how to do it.

For this interviewee, the AMC/APC model was premised on a notion of innovation that featured a straightforward relationship between financial incentive and new scientific discovery, which was here rendered questionable from a practical perspective. In this excerpt, her suggestion was that innovation would not simply emerge from greater finance nor could finance determine innovation. Unlike some of the previous interview excerpts, in this instance it appeared that two generals collided. It was not the particular features of malaria that undermined the general principle of the AMC/APC model. Instead it was the general principles of this model which clashed with a further general model of innovation. The disjuncture was thus realized through the drawing together of incompatible generals.

In sum, interviewees raised questions of the ontological certainties of general principles[11] designed into the AMC/APC and also discussed the problematic particular features of malaria intervention. At times the general was drawn into question for its apparent aggregate but oversimplified singularity (for example, in assuming that private sector firms had a singular and unifying set of characteristics), suggesting that generals did not speak on behalf of their populations. At other moments particulars were used to demonstrate a problem with the general (for example, cost effectiveness as a general mechanism of adjudging value would be undermined by ongoing changes in the nature of the malaria parasite and its relationship to a vaccine). On other occasions, the general was maintained as a feasible idea undermined by the particulars of malaria (for example, placing a value on a market was not

[11] This is reminiscent of the performance of 'ontological gerrymandering' in attempts to solve social problems discussed by Woolgar and Pawluch (1985).

critiqued as a general idea, however the specific valuation of $3b dollars was deemed problematic). In the last excerpt, two general issues drawn together into the model were noted as incompatible.

Economists working in the field of malaria appear to have adopted a slightly different stance. For them the general principles of market economics appear to require protection from the particularities of what they want to suggest are the startling oddities of the case of malaria. For economists the move is to try and protect general principles from particular features by preventing any scaling up or promotion of particulars to generals. For example, Farlow (2005, 2006; Farlow et al. 2005;) suggests that the AMC/APC 'model for these vaccines is unworkable, inefficient, and inequitable towards the wide range of potential developers and suppliers of such vaccines' (Farlow et al. 2005: 2). Farlow (2005) also argues that there seems to be a 'set of literature that severely downplays the problematic side of APCs for early-stage vaccines, and that instead paints a picture of a "simple," "straightforward," and "powerful" new tool' (Farlow 2005: 2). Furthermore:

> The case for APCs for early-stage vaccines was not helped by the early decision to trivialize the science of... malaria vaccine development to one that is 'linear,' fixed, simple and static, when for early stage vaccines it is instead highly complex, and dependent on feedback loops, collaboration, and comparison of results and sharing of information. (Farlow 2005: 4)[12]

In addition, Light (2010) suggests that initial *in vivo* market experiments with available vaccines for other diseases common to developing countries have proven problematic, particularly in keeping to initially agreed prices. And Sonderholm (2010) suggests the malaria vaccine AMC overlooks the complexities of ethical issues resonant with distinct populations where the disease is prevalent.

Through this literature we can note many of the features previously highlighted by interviewees, such as a difficulty in relating finance to innovation and a challenge raised by continuing uncertainty in vaccine development. We can also note some new locations for dissonance in pricing instability and ethics. However, for economists such as Farlow (2005), there is a further move to protect the general from the particular by rendering a division between economists' understanding of the Market (as general) and the AMC/APC understanding of the market (as troubling particular). In this sense, the AMC/APC principles are downscaled from generals to particulars. For Farlow, problematic particulars comprise a number of issues with AMC/APC market modelling: why would $3b be the correct amount to stimulate interest among

[12] Farlow (2005) argues that later-stage vaccines require different considerations: 'For currently existing and near-market vaccines, purchase commitments are all about creating stability of demand, incentives to invest in production capacity, the tailoring of an already existing product to new users, the creation of low product prices, and access to vaccines' (2005: 3).

private firms; what steps would be introduced to prevent the first vaccine available taking most of the available fund, limiting any future innovation; how could minimum standards of quality or effectiveness for vaccines be developed; what would AMCs/APCs mean for non-eligible countries; what would be the cost of introducing and running this intervention; what would stop firms lobbying the oversight committees to get their vaccine prioritized; how would intellectual property issues be resolved?

For economists such as Farlow, critiquing the AMC/APC model involves elaborating and maintaining protectionist principles such that the Market as ultimate general retains its ascendancy. Hence the critiques of the APC/AMC neatly delineate the focal points for critique in order to protect the dignity and integrity of the Market. It is this particular model (e.g. assuming one finite pot of money will ensure continuously improved upon products), these particular terms of value (e.g. the $3b to be offered), and these particular assumptions about the members of the model (e.g. that science is relatively straightforward or that pharmaceutical firms are primarily involved with malaria for financial reasons) that are drawn out for critique. It is these particulars that are utilized to argue that the model does not and cannot be represented by the general (the Market). The particulars must be cast out and in their banishment we are given the opportunity to witness their danger; troublemaking particulars which would cause unwarranted trouble for meta-economic orders of value.

Hence we can understand the work of economists such as Farlow as an attempt to disaggregate the coordination work of the AMC/APC model. In place of a series of coordinated efforts to constitute values in the field of malaria vaccination, the economist's practice here is oriented towards bringing particulars to the fore in an attempt to undermine the value of the proposed generals. Value practices here are oriented towards defending the value of economists' general models of the Market, which are deemed endangered by the crass particulars of this AMC/APC model. The AMC/APC model is positioned as undermining the valued Market modelling practices of the economist. Although practices might be understood as constituting values, and alternative practices might be understood as undermining or questioning values, we can also note here efforts to protect long held and valued economic practices for talking on behalf of the Market.

Conclusion

This chapter has featured an analysis of attempts to generate a market for malaria vaccines. The chapter has dealt mostly with particulars, including the specificities of malaria, parasites, and AMC/APC proposals. To summarize our value-laden coordination work, we have decided to end with some general comments about particulars and some particular comments about generals. First, it seems to us that the ontological certainty of statements calling for market-based interventions is of broader prevalence than the field of malaria vaccines. Engaging with the tension generated between ontologically bold propositions (generals) and the details of managing interventions (particulars) appears of broader interest and perhaps provides a route to be further explored. Attempts to situate generals and particulars on the same page seemed to us in our study of malaria vaccines a central feature of the coordination work of value-laden practices. Second, our analysis of malaria suggests that values and practices of valuation could be usefully further explored as moments where particular forms of trouble are generated for generals. The modes of policing of such troubling particulars also provide further routes through which to raise questions of value. In our study the coordination work that went into the production of the AMC/APC model was continually placed under the spotlight by suggestions that the particulars of malaria entities undermined the generals of the model.

Third, however, the reverse also appears salient. It appears that our questioning of particulars can be inverted. In this sense, we can investigate the trouble caused by generals failing to speak on behalf of the particulars they are held to address. Generals without a constituency appear to provoke trouble of their own which perhaps could be the subject of further analyses. In our study of malaria the general principle of firms seeking profit was at times without a constituency (as firms instead sought to evade profiting from malaria at all costs in order to build a surplus of good publicity in case of bad publicity ahead). Fourth, our analysis of markets and malaria has suggested that scalar shifts between particular and general (and in critiques, between general and particular) are possible and problematic. Moves to defend against the scaling up of particulars and protectionist drives to hold particulars in their place and to deny them a perceived higher value through the scalar accomplishment of being promoted to generals are perhaps deserving of further research. At the end of the chapter, we suggested economists themselves were seeking to operate as the appropriate body to police the trouble that may be caused to economists' practices of talking on behalf of the Market by the apparently ill-conceived upscaling of AMC/APC particulars into Market generals.

Fifth, it seems to us that the nature of what constitutes generals and particulars could be further explored. Our analysis suggests that in testing circumstances—such as interventions in diseases such as malaria—generals must be found[13] to have sufficient ontological singularity to retain coherence

[13] We are using the term 'must be found' here as a shorthand for all that work which goes into constructing generals, e.g. advancing and defending the Market as general through e.g. mobilizing economics reasoning and forms of counting, representation, reporting, legitimating, and advancing particular modes of critique.

and sufficient ontological fluidity to be utilized/transformed for a range of particular instances. Furthermore, particulars require focused ontological singularity which can be recognized as a coherent transformation of the general (the kind of transformation which can still be recognizably transformed back into the general) and which is not characterized by such a degree of ontological transformation that to consider it an instance of the application of a general would be to undermine the dignity, integrity, and aggregate utility of the general. Our example of markets and malaria appears to fail, on occasions and through the discussions of specific participants, both the general and particular test.

These five points draw our attention to some significant market-based failures, in coordination work, in establishing singular value, in bringing a market for a malaria vaccine into being. However, treating the idea of 'failure' as a negative and undesirable result of economic initiatives appears a limited assumption, particularly when considering that failure and experimentation are frequently an integral part of economic deliberations (Neyland and Simakova 2012). Furthermore, the critiques of economic models by economists (in our case of the AMC/APC model) appear to depend upon a modification of Latour and Woolgar's (1979) splitting and inversion model of scientific facts; the 'Market' as a general concept is to be split from these particular market models and their particular details of valuation, and hence preserved in a separate space. Critique of the particular model is enabled, while the Market is preserved and reinforced through justifying the economic effectiveness of various market devices. In attempts to protect market value singularity through modelling and market devices, coordination work appears to be equivalent to establishing and maintaining a singular (monetary) value (e.g. $3bn) and also a singular entity 'market for malaria' in order to make the Market work as a general.

Such a split is deemed crucial to economic assumptions and calculations that do not permit the back-and-forth movement between particulars and generals to stand as a basis for analysis and proposed action. And, for the economists, their critique should not be read as undermining the Market, but inversely it ought to be seen as part of the very process of defending and extolling the Market's virtues, thus triumphing worth and value as the ultimate outcome of market work. Hence in place of any assumption that market modelling and coordination work could be achieved more effectively, based on better or more accurate policy analyses—in some way avoiding failure—what we conclude instead is that the tribulations of market models highlight forcefully the continuing importance placed by economists (and other participants) on recurrent work to propose, challenge, and critique the 'correctness' of terms, values, and assumptions directed towards stimulating a viable Market.

Part III

Valuing Human and Non-human Bodies

8 Genetic value

The moral economies of cloning in the zoo

Carrie Friese

One of the surprising aspects of endeavours to clone endangered animals is that these projects have been met with a general tenor of public support as opposed to abhorrence. While Dolly the Sheep raised anxieties about the spectre of human cloning, the first cloned endangered animal was met with public proclamations that even the most reprehensible of technologies could find a good use.[1] While eating cloned animals has been responded to with disgust, reproducing endangered animals through cloning has been understood as a public good associated with environmental conservation.[2] The public approval of cloning endangered animals has, however, stood in stark contrast to its reception within zoos. Within the zoological park, cloning endangered animals has been marked by deep contestation, as the value of somatic cell nuclear transfer has been debated rather than assumed. How are we to understand these debates over cloning within zoos?

While conducting research regarding the debates over cloning endangered animals in zoological parks, many of the zoo people I interviewed would use the term 'genetic value' to discuss and debate somatic cell nuclear transfer and its application to reproducing endangered animals.[3] Cloning projects that

[1] Public opinion polls in the United States, where the majority of zoo-based cloning projects involving endangered animals has taken place, reported that cloning endangered animals was the most acceptable kind of cloning. This was a statistic that study participants commonly referred to during our discussions. See Fox News/Opinion Dynamics Poll (2002).

[2] For a discussion of the public concerns over consuming cloned agricultural animals, see Gaskell (2000); Priest (2000).

[3] This chapter is based on interviews, participant observation, and an analysis of documents. I interviewed 21 people who were actively involved in and 'implicated by' (Clarke and Montini 1993) endeavours to clone endangered animals in zoos. This included: (1) reproductive scientists and geneticists working at research centres affiliated with zoos, universities, and biotechnology; (2) members of and advisors to Taxonomic Advisory Groups (TAGs) and SSPs that manage captive populations of endangered animals through the American Zoo and Aquarium Association; and (3) a field conservationist who works on habitat and species preservation in situ. I also visited zoos and affiliated research centres to observe how cloning and related techniques were being used and to see how cloned endangered animals were displayed. This included: (1) the San Diego Zoological Park and the Zoological Society of San Diego's research centre, Conservation and Research for Endangered Species (CRES); (2) the Audubon Center for Research of Endangered Species (ACRES); and (3) the Zoological Society of London. I also attended professional conferences, including: the 2006 annual

resulted in 'genetically valuable' endangered animals for display and breeding within zoos received far more support than those projects resulting in 'genetically surplus' animals. Specifically, cloned animals that would help the managers of national and international zoo animal breeding programmes—typically referred to as Species Survival Plans (SSPs)—were favoured, as these clones could help in the attempt to create genetically diverse captive populations that did not rely upon capturing wild animals for the zoo. Genetic value thus served as an idiom through which people articulated, engaged, and debated the values of zoo-based, life science research. What do these debates tell us about the values of zoos, and the value of life science research within this institutional context?

This chapter explores the idiom of 'genetic value' in zoological parks, asking what kinds of practices it enables. I consider this mode of valuing wild and endangered animals as a 'moral economy' (Daston 1995; Thompson 1971) that shapes how life science research is conducted and evaluated in zoos. I do this by contrasting two different kinds of endangered animal cloning projects that have occurred in conjunction with zoological parks.[4] On the one hand, experiments in cloning endangered animals that have resulted in 'genetically surplus' animals have been understood, as least by people working in zoos, as embodying the moral economy of biotechnology. Production, profit, and novelty have been conceived as the three moments in this moral economy. In contrast, experiments in cloning endangered animals that have resulted in 'genetically valuable' animals have been understood as embodying the moral economy of zoos. Reproduction, management, and responsibility are the three moments in this moral economy.[5] By looking at the relationship between these

meetings of the International Embryo Transfer Society and the 2006 annual meetings of the Felid TAG that is organized through the American Aquarium and Zoo Association, which is responsible for managing endangered cat populations in US zoos. Throughout the research, I read journal articles, position statements, websites, legislation, book chapters, and newspaper articles pertaining to endangered animal cloning and zoo animal breeding more generally.

[4] See also Friese (2009, 2010, 2013b) for an expanded exploration of these debates on cloning endangered animals in zoos. This work did not, however, use the concept of moral economy. The workshop on the Moral Economy of the Life Sciences helped me see the salience of this concept for understanding debates about cloning. This chapter thus explores material presented elsewhere through an alternative analytic lens.

[5] In her article developing moral economy as a concept for exploring how value systems shape scientific knowledge production, Lorraine Daston (1995) states: 'Trust, civility, and curiosity were thus three moments of the moral economy of seventeenth-century empiricism. It was a moral economy that set evidentiary standards, stipulated the forms of facticity, selected certain objects as worthy of inquiry, and accelerated the rate of that inquiry.' I am directly drawing here upon Daston's statement to articulate the ways in which cloning endangered animals occurs within a space wherein two different moral economies are in tension with one another. In differentiating production and reproduction, I am not positing a fundamental difference rooted in nature/culture. Both production and reproduction are socially organized processes. Rather, I am articulating the ways in which zoos pursue their breeding practices in a manner that is not rooted in the commodity form. I do recognize that breeding animals is a way for zoos to bring in visitors, thus generating capital. But I do not think that these animals can be understood as commodities in the traditional sense.

moral economies and scientific practice, I explore how different values are enacted through techno-scientific development. If moral economy denotes a set of values that regulate exchange, this chapter describes the tensions between two different moral economies: one based on genetic value and another based on capitalization.

Conceptually, the paper builds upon the Introduction, Dussuage et al., this volume, in looking to two different conceptualizations of moral economy in order to understand genetic value as a value practice. Lorraine Daston (1995) has conceptualized moral economy vis-à-vis scientific knowledge production in her argument that shared values, which are both related to broader social life and yet particular to science itself, shape knowledge production. Meanwhile Thompson (1971) has conceptualized moral economy vis-à-vis the marketplace, arguing that shared values shape the kinds of actions deemed legitimate and illegitimate within exchange in order to reproduce shared values as well as resist emergent value systems. Using both conceptualizations of moral economy, this chapter extends the literature on 'bio-economies' and 'biocapital' to include a focus on how the capitalization of life is being resisted by creating alternative economies in bodies and bodily parts.[6] I contend that 'genetic value' is a moral economy through which the incursion of 'biocapital' into the life science practices of zoos is resisted. In making these claims, I address the different ways in which animals are valued. Specifically, I argue that animal bodies are not always 'rendered' (Shukin 2009) as biocapital within economic processes, but are instead valued in varying ways. Drawing upon Stefan Helmreich's (2008) argument that we need to look beyond capitalism to include other arrangements of value, I contend that cloning endangered animals both reproduces and resists the capitalization of animal life. As the editors of this book emphasize, looking to value practices allows us to see how economic and other forms of valuation are related in the everyday practices of science itself.

The moral economies of cloning in zoos

An interspecies modification of the somatic cell nuclear transfer process has been used to birth animals of five different endangered species as part of zoos. In November 2000 the first cloned, endangered animal—an endangered

[6] I use bio-economies and biocapital to denote a range of literatures that explore how the biosciences and biomedicine are increasingly seen as ways to grow national economies, the biological is reconfigured according to and acted upon as capital, and markets in biological parts are socially organized. See for example Cooper (2007, 2008); Franklin (2007); Franklin and Lock (2003); Haraway (2008); Rose (2007); Sunder Rajan (2003, 2006); Thompson (2005); Waldby (2000, 2002); Waldby and Mitchell (2006). For a critical review and synthesis of this literature, see Helmreich (2008); Larsen (2007).

bovine species known as gaur—was born in Iowa. This animal was created through a short-term collaboration between the human embryonic stem cell company Advanced Cell Technology, the agricultural biotechnology company Trans Ova Genetics, and the Zoological Society of San Diego.[7] However, the cloned gaur died shortly after birth. Following this, the same organizations once again conducted another cloning experiment, which resulted in the birth of a cloned banteng (another endangered bovine species).[8] This clone is currently on display at the San Diego Zoo. In addition, three litters of African wildcat clones have been born and two unrelated cloned cats reproduced a litter of healthy kittens at ACRES in New Orleans, Louisiana.[9] African wildcats are not endangered, however. While I was conducting this research, scientists were using the technical processes developed with the African wildcats to clone an endangered sand cat. One live birth resulted in 2008, but the cloned sand cat died about 60 days after birth.[10] Finally, there is a current research programme in cloning amphibians at the Zoological Society of London. These researchers are using common frogs in their interspecies nuclear transfer research, with the idea that endangered frogs may be cloned in the future if the experiments work.

The cloned gaur, African wildcats, and sand cat were created in a context where innovation was paramount, and in this way are understood as embodying the values of 'biocapitalism' (Sunder Rajan 2003, 2006). Here the focus has been on proving that an interspecies modification of somatic cell nuclear transfer works, as part of innovation. These projects resulted in a significant amount of media attention, given the fascination with both 'world's first' stories regarding technology as well as endangered animals.[11] As such, the speed of innovation was of central importance in each of these projects.[12] Many of the people I spoke with thought that publicity of this kind works to bring in further funding for technological innovation within the zoological

[7] Sunder Rajan (2003: 88) has argued that strategic cooperative relations are typically formed between different companies and different kinds of institutions in biocapitalism. His focus is specifically upon collaborations between biotechnology companies and publically funded universities, particularly in the United States (see also Sunder Rajan 2006). Endangered animal cloning projects originally developed in this context, where strategic and short-term relations were formed between not-for-profit zoos and for-profit biotechnology companies. The purpose of these relations was linked to another component of biocapitalism, as defined by Sunder Rajan. Biotechnology companies like Advanced Cell Technology were trying to develop positive public relations for controversial biotechnologies (e.g., cloning and human embryonic stem cell therapies) through the symbolic capital associated with endangered species preservation.

[8] See Lanza et al. (2000a, 2000b).

[9] See Gomez, Jenkins, Giraldo et al. (2003) and Gomez, Pope, Giraldo et al. (2004).

[10] See Gomez, Pope, Kutner, et al. (2008).

[11] See also Friese (2009, 2013a).

[12] On the importance of speed in biotechnology innovation that occurs as part of biocapitalism, see Fortun (1999); Sunder Rajan (2003, 2006).

park, which is consistent with the ways in which technological forecasting generates capital for contemporary bioscience and biomedicine.[13]

With the focus on the speed of innovation, these cloning projects have been hyped as part of 'species preservation'. However, the actual experimentation has occurred at distance from the value practices of zoos and, in turn, species preservation. A scientist involved in the African wildcat and sand cat cloning projects, for example, told me that his laboratory was not as involved in the SSPs as were other laboratories, for which they were criticized. He admitted that the animals they work with, and thus reproduce, may not be the most genetically valuable animals. However, his laboratory's position was that making genetically valuable animals was not the most important factor right now. As such, they use the cells and animals most readily available in their experiments, rather than the cells and animals considered valuable by zoos and, specifically, SSPs. As such, the laboratory has prioritized developing reproductive technologies as quickly as possible, often being the first to successfully use a given technique with a particular species of endangered wildlife. They have in turn marginalized reproducing zoo animal populations, and do not prioritize this type of work.

This scientist went on to tell me that the technologies his laboratory is developing will undoubtedly be important for future species preservation work, which is consistent with the prospective approach to much of bioscience and biomedicine today (Brown et al. 2000; Rose 2007). However, he felt that enacting cloning experiments today according to the values of zoos would inappropriately slow down the technology development process.

The three moments in the moral economy shaping these cloned animals has thus been understood as consistent with the value practices of biotechnology. Production, profit, and novelty have been of central concern. The goal was to produce a novel animal through innovative techniques, which could in turn generate new kinds of capital for the zoo. Specifically, public enthusiasm for biotechnologies like cloning could bring new kinds of funders and funding to the zoological park.[14]

These value practices are not necessarily shared by other people working in zoos today, however. As such, a number of people would use interviews with me as an opportunity to articulate their criticisms of these kinds of cloning projects and the moral economies that are therein enacted. Many people told me that interspecies nuclear transfer is expensive, inefficient, and invasive, and so developing this set of techniques within zoos is a poor use of limited resources. People also worried about the publicity surrounding cloned animals: the public was being provided stories about a technological fix when in fact cloning had limited applicability to ex situ preservation. Indeed, many

[13] See Brown (2003, 2009): Brown et al. (2000): Rose (2007): Sunder Rajan 2006).

[14] See also Friese (2013b).

people wondered what the benefit of these cloning projects was from a preservation perspective. One zoo scientist told me that gaur do not need cloning; they are a 'dime a dozen' in zoos because they reproduce naturally within captivity. An SSP manager for cats told me that African wildcats are not endangered, and the TAG for felines recommends that zoos do not keep this species, in order to make more space for endangered cats. As far as he was concerned, the cloned cats were actually a problem for zoos.

While articulating these criticisms of cloning, many people drew on an alternative set of value practices that they routinely referenced through the idiom of 'genetic value'. This idiom is exemplified by the following statement:

> If you're going to produce an animal, spend the time, the money, and actually do what you need to do to these animals—which is not always non-invasive—it behooves you to make sure that the offspring you produce are not only healthy, but are genetically valuable.
>
> Zoo scientist, interview

This scientist, like other zoo scientists I spoke with, does not think that the reproductive sciences should be organized by the value practices of industry when endangered animal bodies are concerned; rather, reproductive sciences in zoos must be organized according to the value practices of the park. In this context, animals produced through techno-scientific means cannot simply prove the validity of a technique with the birth of an endangered animal. In addition, the resulting animal must be healthy and genetically valuable, in order to compensate for the time of scientists and the bodily risks experienced by endangered wildlife that are a necessary consequence of the biotechnology development process.

But what does genetic value mean? What practices does this idiom refer to? We have already begun to see that the idiom of 'genetic value' refers to and is reliant upon institutionalized practices and procedures, specifically SSPs.[15] These are small organizations within the American Zoo and Aquarium Association that use selective breeding techniques, specifically kinship charts and stud books, to manage captive populations of endangered animals across zoos. Similar organizations operate in zoos across the world, which are increasingly being systematized through a global database of zoo animals along with zoo animal sperm and embryos. The goal of these projects is to ensure that small populations of endangered animals remain genetically diverse as long as possible. This is part of an attempt to create 'self-sustaining populations' within zoos. Zoos today want to have safely kept, reserve populations of endangered animals held in captivity, which are reproduced through selective breeding as opposed to wild animal collection programmes.

[15] See also Braverman (2011, 2012) on SSPs.

'Genetic value' thus articulates the shift from collecting wild animals for display in zoos to reproducing captive endangered animals as part of species preservation, which since about 1980 has marked the reworked institutional identity of zoological parks.[16] The modern zoological park was rooted in the history of extracting material resources from colonized spaces. Hunting, collecting, and transporting were the primary practices underlying this valuation system. As these collection practices were increasingly criticized by people both outside and within zoological parks, breeding captive animals has become the focus. This has helped to legitimize the zoo, making it a moral and valuable institution within its own historical context.

Zoos understand the value of assisted reproductive technologies in this context, and alongside the parallel development of 'frozen zoos'. Here somatic cells, sperm, and embryos from endangered animals are preserved using cryopreservation. The first frozen zoo was started by Kurt Benirschke in 1965, which was institutionalized at the Zoological Society of San Diego in the 1970s. These collections started as a kind of service to future scientists, who would only be able to do genetic analyses of species that had gone extinct through such a resource. As such, somatic cells were the only kind of cells collected. However, frozen zoos quickly expanded to include a reproductive function. When reproductive scientist Barbara Durrant joined the Zoological Society of San Diego, she began to include sperm to the collection on the basis that these cells could be used to help reproduce endangered animals for future zoos using assisted reproductive technologies, including artificial insemination and, potentially, cloning. Today, there is an international inventory of endangered animals living in captivity, which includes frozen sperm within its database.

The cloned banteng is the one example of an experiment in interspecies nuclear transfer that was conducted according to the breeding protocols of contemporary zoological parks.[17] The cell used to create this clone came from a banteng who died at the San Diego Zoo in the 1970s before reproducing. This banteng was considered 'genetically valuable' because he did not have extensive kin relations within the captive population. Fibroblast cells from this banteng had been cryopreserved in the San Diego Zoo's Frozen Zoo™. The SSP manager told me in an interview that the cloned banteng, along with cryopreserved sperm from a deceased banteng, are the two most valuable animals in the captive population. What this means is that their reproduction will create rather than diminish the total genetic diversity in the population.

In this context, many people working in zoos would evaluate cloning projects according to the extent to which this research resulted in 'genetically

[16] See also Friese (2009, 2010); Hanson (2002); Rothfels (2002).

[17] See Ryder and Benirschke (1997) on the logic behind this cloning project.

valuable' individuals. We see the idiom of genetic value used to evaluate cloning projects in the following statements:

> The gaur wasn't the best [animal to clone] because it wasn't that genetically valuable... So the banteng was much better.
>
> Zoo scientist, interview

> [There are] two potential big values of cloning. One is getting samples from deceased individuals that are valuable to the population. And again, depending on how genetically valuable that animal is, it might be a good idea to try and clone it.
>
> Zoo scientist and Species Survival Plan Manager, interview

There was thus more support for reproductive science research and technology development that was conducted in accordance with the values of zoological parks, through the logic and practices of genetic value.

Indeed, one of the key critics of cloning endangered animals is currently pursuing a cloning project that is directly shaped by the managed breeding protocols of zoological parks. Bill Holt was an early critic of endangered animal cloning, as he was concerned that excessive hype from the biotechnology industry masked how difficult it has been to transfer assisted reproductive technologies from domestic to endangered animals. In this context, he felt that cloning endangered mammals was inappropriate. But he did start to think that cloning other species such as frogs might be a feasible research project for the zoo, on the basis that amphibian cloning has been occurring since the 1950s.[18] In this context, Holt's postdoctoral researcher, Rhiannon Lloyd, is asking questions about mitochondrial and nuclear DNA in cellular development, using interspecies nuclear transfer with two common species of frog. If the technique proves to be feasible, Holt and Lloyd intend to use the technique as part of amphibian preservation projects, which is pressing, given the staggering rate of amphibian endangerment and extinction today.[19]

Genetic value can thus be understood as a 'moral economy' (Daston 1995). Lorraine Daston (1995: 4) has defined a moral economy as a 'balanced system of emotional forces, with equilibrium points and constraints'. These emotional forces denote a mental state that is collective rather than individual. Daston thus uses moral economy to refer to culturally-mediated and historically specific values, which inspire and shape scientific knowledge production. Daston's conceptualization illuminates the ways in which the practices of zoological parks—ranging from life science research to population management—are

[18] See Holt, Pickard and Prather (2004).

[19] As a taxa, amphibians are generally considered to be more threatened than either mammals or birds (Beebee and Griffiths 2005; Stuart et al. 2004). An Amphibian Action Plan (Gascon et al. 2007) was devised in response, which brings multiple components of conservation together including habitat surveillance, captive breeding, and cryopreservation.

expected to be organized according to the moral economy of the park, or genetic value.

Reproduction is here the key moment in the moral economy of zoos, through which the park can exhibit its responsibility by contributing to—rather than hindering—endangered animal preservation practices. Rather than produce a novel animal, the goal is to reproduce a captive population through managed breeding. The focus here is not on capital accumulation, but rather legitimizing the zoo as an institution. Specifically, the zoo in part expresses its standing as a legitimate and responsible institution by breeding its captive animals as opposed to collecting animals from the wild. Emotional debates thus arise when the scientific practices of zoological parks contradict the moral economy of the zoo. Making animals which are 'genetically surplus' to SSPs is an affront to the selective breeding practices of contemporary zoos. The debates over cloning are thus not about cloning per se. These debates are instead rooted in questions about which moral economies should shape technological research and development in zoos.

Genetic value versus genetic capital

People working in zoos commonly pointed out that selective breeding is not organized according to the needs of profit in the park, which differs significantly from agricultural applications. Indeed, selectively breeding plants and animals is an area where there has been sustained interest in producing capital-intensive bodies by exploiting biogenetic relations. British agriculturalist Robert Bakewell (1725–95) is most famously known for his role in elaborating this system of valuing animals in agriculture. Historian Harriet Ritvo (1995) has analysed these breeding practices. She argues that Bakewell created 'genetic capital' by transferring the value of an animal from the location and environment in which the animal dwelled to the biogenetic genealogy from which the animal came. Specifically, agriculturalists of Bakewell's time generally located the value of animals in the health and size of herds and flocks, wherein individual animals were considered more or less interchangeable. Bakewell instead looked to the reproductive potential of specific animals that displayed exceptional and highly desired traits as the source of value. Studbooks were created in this context both to organize agricultural production through selective breeding as well as to prove that genetic capital had in fact been created. These methods allowed for the creation of a global market in domesticated animal bodies rooted in the commodity form (see Derry 2003; Franklin 2007).

The selective breeding practices associated with domestication have long been held in opposition to the collection practices associated with the wild animals held in zoos. Zoological economies have traditionally been based in hunting and collecting. Here, wild animal collectors would kill adult animals in a pack and capture their young (Hanson 2002; Rothfels 2002). These young animals would then be transported great distances before arriving at the zoo. However, these collection practices in part contributed to the endangerment of some species (Hanson 2002). In response, animal importation practices began in the 1970s to be increasingly overseen. Zoos today thus focus on breeding their populations of 'wild' animals. All but one zoo worker I spoke with commented that they hoped never again to have to collect an animal from the wild for display in a zoo.

Genetic value thus articulates the changing collection practices in zoos, which are meant to not only 'moralize scientists' (Daston 1995) but also the zoo itself. As such, genetic value represents a new kind of economy in wild animal bodies. Rather than collect animals from the wild, the goal instead is to make animals in the park through genetic and reproductive science. Rather than deplete small populations of in situ animals, the goal is instead to create captive populations that can stand as a back-up or reserve while serving an educational function regarding the importance of conservation. And rather than being a consumptive institution associated with colonialism, the goal instead is to remake the zoo as a non-consumptive institution that contributes to global conservation goals. Genetic value is thus a discourse through which the zoo asserts itself as an ethical institution, wherein science does not raise moral problems but instead provides ethical solutions. Genetic value is meant to remediate the previous political economies of the zoological park based in colonial extraction with a non-consumptive mode of auto-reproduction that is based in species preservation.

Genetic value thus differs from genetic capital, a point that study participants frequently made by emphasizing that selective breeding in zoos is not about creating profit but is instead more akin to a gift economy. Responsible care of endangered animal populations held in captivity is the organizing principle. One researcher articulated the gift economy upon which SSPs are organized as follows:

> In the conservation world we tend to manage these populations—for example, you may have fifty different institutions in the U.S. that all have White Rhinos. But they're managed as one population. They are moved around to different zoos to maximize breeding. So we don't tend to think of these animals so much as our animals and that's because it doesn't usually translate into a commercial profit.

However, three life scientists working in zoos with whom I spoke commented that cloning has fissured, or least brought into question, this gift economy due to concerns about commercialization and profit, particularly among biotechnology companies but also among other zoos. In this context, the question of who owns animals resulting from cell lines cryopreserved in frozen zoos was raised. The zoo scientist cited above continued:

> You could ask who owns a cloned animal. Is it the company that produced it? Is it the person that donated the sample? Or is it the individual who owned the animal from which the sample came? And as a result of that people have gotten a lot more careful about what we call biomaterial transfer agreements. Because we have institutions asking other institutions for samples all the time, whether it's a skin biopsy or a serum sample or a milk sample or a faecal sample and all the rest of it. And what's happening is institutions are starting to tighten up by sending out biomaterial transfer agreements that stipulate what you can and cannot do with the sample... As we start to see the publicity surrounding things like cloned animals or we become aware that there is potential with these new technologies to take some of these genetic resources and exploit them for commercial purposes, institutions are starting to become more aware that they need to protect themselves. I don't see this as being restrictive. I think as long as people are assured that they're not going to be effectively ripped off by somebody else they are still going to want to share samples knowing that this research is ultimately going to benefit—hopefully benefit the animals in captivity.

Zoos are willing to share animals in breeding programmes along with bodily parts for research if deemed beneficial to zoos and zoo animals. However, they have become concerned that cryopreserved cell lines could be used in such a way that will not be centrally organized by the needs of either zoological parks or animal populations living in captivity, but rather with a commercial interest. They are, in other words, concerned that the capitalization of 'life itself' would intrude into the gift economies of zoological parks. Genetic value thus requires a gift economy, one to which biocapitalism was understood as a threat.

Based on my conversations with people working in zoos, exchanges in endangered animal bodies and bodily parts were a key site through which such concerns were articulated, debated, and enacted. The moral economies of the park were calibrated by and in turn calibrated the market in endangered animal bodies and bodily parts. For example, one scientist told me a story of how zoos would refuse to donate cells to researchers for the purpose of cloning if they disagreed with this research trajectory. Here research materials needed to do cloning are withheld as a form of protesting cloning research, which was believed to be organized by the moral economy of industry rather than the zoo. Another scientist, who had been involved in cloning animals without 'genetic value', told me that his zoo's request for endangered animals was denied by the SSP. He suspected his cloning work was the basis for the denial. Here animals were withheld as a kind of punishment for failing to comply with the animal breeding practices of the park. Genetic value is reinforced in and through the economies in endangered animals as well as their bodily parts, specifically cryopreserved cells, sperm, and embryos that stand as proto-individuals within the kinship charts and studbooks commonly used by SSPs.

In delineating 'moral economy' as a concept, E. P. Thompson (1971) argued that the food riots in eighteenth-century England cannot be explained through economic reductions (e.g., unemployment, hunger, and distress). Rather, the food riots were a response to the changing social organization of rights and customs. Specifically, the paternalist model rooted in provision, upon which the poor depended to ensure the price of bread in times of dearth, worked to legitimize the actions of the crowd. The food riots thus expressed the moral economy of the poor. The incursion of a 'free' market rationality was resisted by imposing prices through crowd action within the marketplace (Thompson 1971: 117). 'Moral economies' thus denotes a set of historically specific expectations regarding the morality of social relations, which enable and legitimize certain forms of action and protest when incurred. I want to draw on Thompson's delineation of moral economy to contend that zoos reproduce genetic value as a moral economy in and through the market of endangered animal bodies and bodily parts. By exchanging or not exchanging cell lines, by exchanging or not exchanging zoo animals, genetic value is reinforced. In the process, zoos legitimate and delegitimize certain kinds of social relations within parks and among other social actors, including scientific practices. These exchanges serve as a means to criticize and delimit the incursion of a moral economy associated with biological industries from entering the zoological park.

Genetic value as a bio-economy

In his classificatory analysis of the different 'species of biocapital' populating the social studies of bioscience and biomedicine literature, Stefan Helmreich (2008: 474) notes that we need to ask not only what happens when biology is capitalized, but also whether capital is today the only modality through which the economic and biological encounter one another. Looking to anthropological scholarship on organ transplantation (e.g., Cohen 2005; Lock 2002; Sharp 2006), Helmreich suggests that markets in bodily parts are not all overdetermined by capitalization and the commodity form. Indeed, SSPs exchange animals and animal gametes but do not understand this as related to the commodity form. Zoo science is conditioned by the moral economy of the park, through genetic value, in order to emphasize the responsible reproduction of animals that are not—as many zoo workers would comment—'genetically surplus'.

Indeed, the trope of 'genetic surplus' was frequently used to criticize the initial research projects in cloning endangered animals. One zoo scientist noted that the initial cloning experiments, not unlike much of the development of reproductive technologies, produced undesirable forms of surplus that zoological parks then needed to accommodate. She stated:

> Unfortunately, when it became very popular to do artificial reproduction in zoos, a lot of animals were produced that were actually genetically surplus. What was the point of that? Just to prove that you can do that? I mean it's interesting from a scientific standpoint, but the bottom line is a lot of individuals were produced that were surplus to the SSPs, which is surplus to the captive population. So these animals took up space in zoos without having any desirable genetic input into the population. So there was very little thought given to the eventual value of those animals produced.
>
> Zoo scientist, interview

Surplus is, of course, central to Marx's ([1867] 1990) articulation of capital, and has been used to critique the excesses and waste evident in contemporary forms of biocapitalism (Cooper 2008; Sunder Rajan 2006; Thacker 2005). Cloned endangered animals, which are not genetically valuable, embody this surplus. These animals take up limited space in zoos; they require significant resources to make them live across their life course. But they don't 'give back' to the population in a meaningful way, according to the zoo's value practices.[20] Indeed, from the zoo's perspective, these resources could be going to the life of another animal that would assist in sustaining the genetic diversity of an endangered population. In this context, zoos must accommodate the surpluses and excesses of biocapital. And thus the zoo becomes another site to resist and critique the creation of these forms of waste.

Protesting biocapital in the zoo is not particularly surprising. The zoo has handled its own institutional crisis by redefining the park as being involved in species preservation and, thus, environmental conservation. As a movement, conservationists have emphasized the way in which the commodity form and surplus value have created havoc with the materiality of the planet. Species such as the gaur and the banteng are, for example, going extinct because their ecosystems—previously home to a slew of different species—are increasingly being turned into coffee plantations as part of the global food trade. In a review of Stephen Gardiner's recent book on climate change, Malcolm Bull (2012) notes that contemporary environmental politics, through global warming, have been rooted in creating a new morality through which our contemporary actions are amplified by reference to their consequences. The future is not reckoned with to generate capitalization in the present here, as seen with the generation of biocapital. Rather, the future is reckoned with as a connection

[20] Matthew Chrulew (2011) has noted that the facilitation of life through SSPs has not meant that death is disavowed in the zoo, which is—I would add—consistent with Foucault's (2003) delineation of biopower. He notes that some zoos kill animals that are surplus to the SSP. This kind of life, which is reliant upon another animal's death, is consistent with Shukin's (2009) analysis of animal capital. However I did not encounter culling in US zoos; news reporting similarly indicates that culling genetically surplus animals is not supported by US zoological parks (Kaufman 2012).

through which present day actions are assessed by way of their consequences, particularly for future generations. Debates over the moral economies of cloning in zoos must articulate the value practices of science with the value practices of environmentalism within the current institutional context of the zoo (Friese 2013b).

However, Nicole Shukin (2009) has argued that conservationist projects, often presented as alternatives to and critiques of capitalism, are frequently complicit in the hegemony of capital. This is because saving nature has been linked with saving capital. She uses the animal rendering industry as an example in delineating this argument conceptually, wherein animal fats are recycled so as to produce capital. Shukin also uses the shift from hunting wildlife with a gun to hunting wildlife with a camera in delineating this argument through a material example, wherein the film used to preserve wild animals relies upon celluloid made from animal products.[21] In the context of Shukin's critique, it is crucial to point out that zoos do generate capital through display. Cloning and other technologies allow for the display of some endangered animals to persist not only by producing animals but also by legitimating the conditions in which they live.

However, I would argue it is inappropriate to collapse genetic value and biocapital on this basis. These are two different labels for moral economies, which enable different kinds of life science research to be practised within zoological parks. These scientific practices enact different values and are thus valued differently. While cloning endangered animals may be complicit with the reproduction of capital in a general sense, it resists the logics and practices of biocapital both by delineating exchanges in animals and animal bodily parts as well as in loosely regulating the forms of experimentation that can legitimately take place within zoos. Animal life can be made in 'biopolitical times' in very different ways, and is not always rendered as capital.

Conclusion: Moral economies and value practices of cloning

This chapter has explored the practices of genetic value in zoological parks as a 'moral economy' (Daston 1995; Thompson 1971). Genetic value has shaped how cloning endangered animals has been both conducted and evaluated.

[21] Shukin (2009) importantly seeks to recoup and make visible the animal matter that is erased from Marxist understandings of how capital is reproduced. It is worth noting that the domestic animal matter underlying cloning endangered animals—as egg cell donors and as surrogates—has been rendered visible in zoos as a way of critiquing the incorporation of scientific practices associated with biotechnology into zoological parks. See Friese (2010) for a discussion.

I have argued that there have been two different kinds of endangered animal cloning projects that have occurred in conjunction with zoological parks, one shaped by the values of biocapital and the other informed by 'genetic value'. Through tracing the contestations over these moral economies, I have argued that in zoos the capitalization of life is being resisted by pursuing and enforcing alternative economies in bodies and bodily parts. The everyday practices involved in cloning animals engage different values, and the value practices of the zoo—embodied by the idiom of genetic value—are reinforced. Evaluating scientific practices on the basis of this moral economy works to regulate scientific knowledge production practices within the zoo.

In making this argument, I have built upon Lorraine Daston's (1995) delineation of moral economy as a process through which scientific knowledge practices and exchange relations are informally regulated. Unlike Mertonian norms, moral economies are here understood as historically and culturally created and thus subject to change and contestation over time (Daston 1995: 6). This aspect of moral economy in science is interlinked with Thompson's (1971) use of moral economy in the market. Changing values are legitimately contested by intervening in and disrupting exchange. As an analytic, moral economy thus allowed me to see how the zoo has appropriated new technologies (e.g., breeding) in the context of institutional change, and how corresponding values are used to regulate scientific work (e.g., cloning). The concept also allows us to see how scientific practices involved in experimentation enact values that are related to economic and other domains, and are thus are sites for reproducing and changing the values of science and zoos.

9 Enacting values from the sea

On innovation devices, value practices, and the co-modifications of markets and bodies in aquaculture

Kristin Asdal

At the table

I knew very well it was the season for it. But I had never imagined, or I had simply forgotten the taste, the texture, and the look of it, the amazing whiteness of the meat, the perfect and aesthetic ways in which the meat folded itself out in delicate layers. The whole thing sprang to my eyes, my senses.

Not surprisingly, this whiteness is something for which the cod is famous, praised, and much valued (Kurlansky 1997). No wonder I had no regrets over the price I had had to pay for it. The codfish I was offered was terribly expensive, more than kr 250 per kilo, if I remember correctly. But now, at my dinner table, I could not care less. I was only happy I had bought enough, a sufficient quantity to include my neighbours at the table.[1]

In recent years the *Gadus morhua*, a fish more widely known as cod, has not only presented itself to the consumer at the dinner table, it has become the object of extensive and intensive innovation strategies. Hence, it has turned public because of a number of innovation documents and strategies with the stated objective of turning the precious Atlantic cod into a farmed species.

In the Norwegian context salmon is by far the most important farmed species. But efforts to farm cod are in themselves nothing new and have a long history, also in Norwegian fisheries. During the last decade these efforts were intensified and heavily supported by governmental initiatives. In 2001 there was great optimism for the farmed version of the *Gadus morhua* and the government declared that they shared in this optimism: sustained long-term

[1] I am grateful for very helpful comments on earlier versions of this chapter from Svein Ole Borgen and Bernt Aarset in the 'Go Food project', as well as from Carrie Friese and the editors of this volume. I would also like to thank my colleagues in the project 'Newcomers to the Farm'—John Law, Marianne Lien, and Gro Birgit Ween, as well as Stefan Erbs and Bård Hobæk for their assistance.

efforts would now be invested in developing the cod into a farmed species (Aftenposten 2012).[2] This marked the beginning of a long-term innovation strategy, and this strategy and related innovation documents serve as the point of departure, or site, for this chapter which will explore value practices in the life sciences—hence in my case the enactment of values from the sea and the remaking of the *Gadus morhua* into a farmed entity. In doing this, the assumption is that the practices and sciences 'in the field' are not the only privileged sites from which attempts are made to enable such potential future values—so also are the ongoing series of policy documents and strategy plans which value and evaluate such efforts. These documents are technologies of politics (Asdal 2004, 2008) or, as I will elaborate on them here, innovation devices for valuing and timing values. This implies that rather than simply representing and being a source of an extra-textual reality with the 'real' practices taking place elsewhere, the documents can be understood as lively value agents.

Analysing these documents as innovation devices for valuation and timing will be an integral part of this chapter. But there is also another way in which these documents serve as windows of opportunity for exploring value practices in the life sciences. In order to enable a viable commodity, both markets and biological entities need to be worked on, to be modified. The innovation documents are sites for exploring this *double entendre* and for seeing such processes of modification as inextricably linked: they are sites for exploring value practices in the life sciences as what I suggest calling practices of co-modification.

In laying the grounds for the above approach I will first discuss a set of selected contributions from science studies and economic sociology as well as academic efforts to grasp the mechanisms of the bioeconomy. Next I will research the innovation documents in more detail. In elaborating on these I will posit and employ the suggested conceptual and analytical approaches for studying value practices in the life sciences. This means that rather than going in depth on *one* selected concept, I will suggest what one might call a conceptual toolbox which will be explored in more depth and in other contexts. Taken together, the concepts and approaches I propose are:

- documents as innovation devices and value agents;
- the notion of co-modification. This builds on processes of turning some things into commodities, hence commodification, and making something

[2] Prime Minister Jens Stoltenberg in Tromsø, August 2001 (Aftenposten 2012): '*Det rår stor optimisme knyttet til torskeoppdrett. Regjeringen deler optimismen og vil nå ta et krafttak for å utvikle torsk som oppdrettsfisk.*' [There is now great optimism related to cod-farming. The government shares this optimism and will now make a strong effort to develop cod as a farmed fish.]

commercial or part of the market. Simultaneously and deliberately, the notion twists and turns this concept of commercialization and alerts us to the practices and work involved in *modifying* entities, both market and biological entities, hence *co*-modification. The notion implies that this is not a pre-given, straightforward, linear, or uniquely social process, but rather something we need to turn into the object of analysis and explore more openly;

- the notion of 'biocapitalization'. This points to the practices involved in valuation processes in the form of efforts to turn something into new forms of capital. This contrasts with such notions as 'biocapital' and 'the bio-economy' which may give the impression that such versions of the economy are a given, as well as leave unnoticed that potential capital may flow in unexpected directions and that biocapitalization practices may fail;
- acts of 'timing' and 're-timing' as integral and crucial to co-modification and biocapitalizing efforts and practices.

Value making and the modification of markets

In taking an interest in valuation processes in the economy, scholars with a background in economic sociology have sought to rewrite the approach to the value problem in neoclassical economics, implying that value is decided in the market and by the price the consumer is willing to pay—in this case then, the price I was willing to pay for the cod fillet I brought home to my dinner table. It is argued that rather than taking consumers and their motivations for granted, the valuation of goods and services needs to be investigated by focusing on the *meanings* that goods obtain for actors—as well as the social and institutional structure of markets (Aspers and Beckert 2011; Beckert 2011). How are customers attracted to the goods they purchase (Beckert 2011; but see also Callon et al. 2002; Muniesa and Trébuchet-Breitwiller 2010; Sjögren and Helgesson 2007)?

One of the main objectives in the economic sociology approach has been to demonstrate that values are socially constructed. Therefore, as Aspers and Beckert have argued (2011: 30) certain forms of markets stand out as particularly interesting. Markets such as life insurance, whale watching, and fair trade hold 'a special attraction to sociologists because value seems detached from the materiality of the commodity and in very obvious ways socially constructed'.

In important ways the above approach points in the direction this chapter seeks to move: value, or valuation, cannot simply be revealed. Rather, valuing or valuation is something that is actively done while values are something actively constructed. Valuation is a particular form of practice (Muniesa

2012). A number of studies at the intersection of economic sociology and science studies have pointed to the ways in which market action in general and valuation practices in particular are enacted by means of certain tools or devices. The notion of 'market devices' was developed in part to point to the socio-technical *dispositifs* through which actions are made economical (Callon et al. 2007; Cochoy 2007). The notion of 'judgement devices' has been suggested to provide a framework for grasping processes of singularization, hence processes that enable particular goods to stand out among others (Karpik 2010). One example is all kinds of so-called appellations, i.e. making distinct goods attractive by linking them with labels on designations of origin, brands, and certification. Related work has pointed out that valuing, or valuation, takes place in particular situations and is linked to concrete settings (Muniesa cited in Stark 2011: 334–5). Innovation documents, I argue, can be approached as settings for valuation practices.

But is there not also a problem with strands of the above approach? It could be argued that parts of the economic sociology approach stop just when the issue becomes most interesting: exploring ways in which valuation is performed in particular practices, in particular settings, and by means of a heterogeneous sets of tools is crucial. In this way the very value-construction process is made the object of analysis. However, if one then insists that the focus should be on the extent to which values are *socially* constructed, and insists that certain markets are particularly attractive to study because they seem *detached* from the materiality of the commodity (Beckert 2011), this leaves significant empirical and analytical challenges untouched. 'The social' is deliberately made into a room or a sphere of its own. This is precisely what science studies have for a long time tried to avoid (e.g. Callon 1986; Latour 1987); more recently such efforts have been pursued in relation to markets and valuations (see for instance Kjellberg and Helgesson 2007 on the co-construction of objects and economic actors). The material semiotic or post-constructivist version of science studies can be read as a response to the sociological critique of positivism which actively strived to carve out room for 'the social' up against a towering natural science which insisted that all objects could be studied with the same method (Asdal 2005). As long as the object of study is simply markets and how consumers respond and are attracted to market goods, the differences in the approaches might not be that visible. However, if the issue is to explore the value practices in the life sciences, the materialities of the objects are crucially more difficult or problematic to ignore. Valuation in processes of co-modification implies efforts to modify: the *modifications* of markets. But valuation in processes of co-modification also implies efforts to modify the very objects—for example the very life-science objects involved. Hence, the co-modification process is particularly interesting to study because of the ways in which it is not detached from, but intimately intertwined with the materiality of the object.

Value making and the modification of biological entities

A Marxist-oriented approach starts in the completely opposite direction from an approach that seeks the solution in 'the social'. In contrast to the neoclassical approach which ended by solving the value problem in the market and in consumer preferences, or the version of the economic sociology that attends to a broader, however purely social sphere, the Marxist version sees the production of economic value as inextricably linked to the production side. In the early 1980s, science-studies scholars who worked from a Marxist-oriented framework set out to re-explore and situate science, technology, and medicine in the labour process (Levidow and Young 1981, 1985). This approach is interesting in relation to the present study as the project aimed to analyse the economy with a point of departure in its relation to science and technology, including more specifically economy's relation to biology. Is nature a labour process, asked one of the scholars, Bob Young (1985).

In another contribution which sets out to trace the historical development about the move from natural history to biology as a life science, the author (Yoxen 1981) builds on Michel Foucault: 'Up to the end of the eighteenth century in fact, life does not exist: only living beings. These beings form one class, or rather several classes, in the series of the things in the world' (Foucault 1970). Foucault pointed out how, in the late eighteenth century '…life assumes its autonomy in relation to the concepts of classification…and becomes an object of knowledge among others' (Foucault 1970: 162). In Yoxen's (1981) words, 'the nature of life becomes an issue in its own right; an issue to be confronted scientifically'. More specifically, in this framework inspired by Foucault but nevertheless Marxist, biotechnology is understood as a technology controlled by capital; thus, seen as a specific mode of the appropriation of living nature, it is literally capitalizing life. The conclusion here is that life has become a productive force.

Later, related contributions, all linked to the attempt to grasp the phenomenon of the new capital that has risen from the union of biological science with profit-oriented enterprise, have given this development many names: the organic phase of capitalism, life as surplus, the bioeconomy and biocapital (see Helmreich 2008). Academic contributions on the bioeconomy have demonstrated the extent to which reproduction is constitutive to such new versions of the economy (e.g. Cooper 2008; Franklin 2007).

The aquaculture enterprise that is the object of the present study can be studied as an example of such new forms of capital conditioned upon the living: In the aquaculture enterprise the nature of the life of cod is made into an issue in its own right, confronted scientifically in order to transform cod into a viable farmed commodity. Through its very potential, the *Gadus morhua* has become part of the new biocapital in the sense that it is being subjected to explicit calculations about its life and mechanisms.

But again, as with the economic sociology approach I outlined above, we could ask if the Marxist–Foucauldian inspired approach stops right where the issue becomes most interesting because we must ask the question: how does such capital, such biocapital, come into being? Biocapital and the bioeconomy are not a given. On the contrary, just as much as consumers through a set of market or judgment devices actively construct values, life-science objects are also actively tried constructed or modified in practices of valuation. This is part of the reason why I suggest that we talk about efforts at biocapitalizing and practices of modification.

The significant and highly valuable power and contributions of the Marxist-Foucauldian approach is its way of taking seriously the ways in which materiality, biology or the living are made part of the economy and how science is a powerful practice in these endeavours to modify objects. This is in contrast to the economic sociology approach I outlined earlier that insisted on the significance of *social* constructions, exclusively. The problem with a Marxist-Foucauldian framework is rather the opposite of this latter approach: The conclusion that life has become a productive force risks being seen as an all-encompassing rationality driven by its own inner logic—without leaving space for a more open-ended empirical analysis of what the 'the economy' might entail and how it might be performed.

Several of the papers in this book, along with the introduction, grapple with the problem of how to define 'the economy' at the intersection of economic sociology and 'the bioeconomy'. It would therefore be naive to assume that one chapter in this book could resolve this problem. The approach that is chosen for this chapter is to confront these problems by attempting to carve out a site to enable the study of how the economy is performed and how attempts are made to modify living bodies to become part of the market in new ways. In doing this I suggest that we need to draw from both the above-mentioned broad traditions.

In trying to grasp this phenomenon it might, however, already be too simplistic to take for granted that there is only one such thing as 'the economy'; i.e. that there is only one economy (cf. Asdal 2014 and the introduction, Dussauge at al., this volume). One of the reasons why it might be advantageous to transform the notion of biocapital into a notion that alerts us to the fact that this is a *practice*, a biocapitalization practice, is that efforts at co-modification are something that neither biological organisms nor other actors or entities simply comply with or adhere to. The codfish for instance, may be said to live by its own economy. At least, as the ensuing analysis will show, it tends to move in unexpected ways and swim in its own directions so that it in part defies biocapitalization efforts.

As I will demonstrate in the next section, (Innovation documents), value cannot be readily defined as simply economic value or profit. Whereas economic value in the form of profit is definitely the important objective in realizing innovation strategies in cod farming, a broad set of valuation and evaluation practices is part of these strategies in complex and in part entangled ways. The innovation strategies attempt to produce value for consumers as well as other actors, such as industry and policymakers. In doing this, valuation processes in the innovation strategies involve appraisal, devaluation, the construction of kinship value, and ways of linking new objects to already highly valued versions of the past. Hence, narration processes, linking current cod-farming practices with an established (rich) history, and linking the current version of the farmed cod with an already renowned wild version of the cod, are important valuation strategies. As I will show, the latter strategy enacted a new version of cod altogether.

Such complex and down to earth fundamental valuation practices, including the relevant devices involved, are part of what we need to explore if we are to grasp the life sciences and their involvement with economies. What I will focus on and argue in particular as part of the ensuing story is that the innovation documents themselves ought to be approached as distinct valuation agents and sites for performing values.

Innovation documents as sites for value making and co-modification

As mentioned in the first section, (At the table), the codfish has lately presented itself to the public as part of a series of innovation documents and strategies with the stated objective of turning the precious Atlantic cod into a farmed species. In the early innovation documents and strategy, the bearded *Gadus morhua* is staged instantly on the cover page (see Figure 9.1). Here it is captured swimming towards the reader. If you look closely, you will soon notice that here on the front cover the codfish stands out as a quite particular form of object, a commodity, and an export product. Its forehead is stamped, or branded, with the text 'Norge—Seafood from Norway' (RCN/NIRDF 2001). In this way the cod is inscribed with a particular value, an assumed market value from and for a particular nation state.

Immediately, the front cover draws our attention to the fact that value as part of economic activities is not something we should simply take for granted. Value is not simply a fact to be revealed in the form of individual or given consumer preferences (Fourcade 2011b, 2011a), nor is value something that is intrinsic to objects or subjects. Values are actively done or enacted. The brand

Figure 9.1. Norge—Seafood from Norway
Source: From cover page of RCN/NIRDF 2001

'Seafood from Norway' can be grasped precisely as a judgment device (Karpik 2010), a guidepost for individual and collective action: the codfish is marked with a story of its origin that also appeals to a certain kind of quality. But how to cash in the potential market value and how to transform the cod from a species swimming in its own directions into a viable market product? Those are the questions.

In the innovation documents, the *Gadus morhua*, or rather the *Gadus morhua* potential is fleshed out to enable the production of future market values. The assumption that there is already a market that values the emerging new product, a sphere inhabited by consumers who are willing to repeat the market action I had already enacted, is what the reader is presented to: 'There exists a large international market for cod, and a wide range of cod-related products are in demand in large parts of the world', it is argued (RCN/NIRDF 2001: 2).

This quite general assumption is coupled with more concrete estimations of the future values that can come to be realized: In twenty years cod may produce values that will equal the current production of salmon, i.e. an export value of about kr 10 billion. Hence, the *Gadus morhua* has it in itself the potential to be transformed into 'a salmon' (see also Jensen et al. 1985; RCN/NIRDF 2001).

The reader is presented with a certain reality which is also part of a certain vision (RCN/NIRDF 2001: 2):

> Cod farming is an industry where Norway, with its geography, infrastructure and competence in fish farming will have huge competitive advantages. Norwegian research and development environments have a solid competence in cod as a farmed species. Cod is probably the fish species that can most quickly reach large production volumes in farming because modified salmon technology can be used in the phase of growth in the sea. Today there is huge interest amongst several commercial actors to commence large-scale cod farming, and there is a lot of private capital available for investments in production units. The probability of succeeding in cod farming is significantly larger today than it was 10–15 years ago.

The text envisions values in the form of future profit: There is a market out there, inhabited by consumers who are already demanding cod or cod-related products. 'Norway', for its part, is enacted as a valuable site from where to produce and supply such goods due to natural conditions, skills, and technologies. Then the codfish potential is valued because of its kinship values, in other words, its proximity to another co-modified species, namely salmon. Hence, this is about valuing the '*Gadus morhua* potential', the potential to reap what the salmon has already yielded. Moreover, there is an available cash flow ready to invest in and capitalize on this potential. Last, the very timing is evaluated as good: it is assumed to be easier to succeed in farming cod now than it was before.

But then to repeat the question I asked above: How can the *Gadus morhua* be successfully co-modified with 'the market'? Or to put it another way, how can the demands of the market be imprinted on the biological? It has been argued that contemporary biology has become expert at stopping, starting, suspending, and accelerating cellular processes, wedging these dynamics into processes that look like a molecular version of industrial agribusiness (Landecker 1999). Others again have argued that things are not that simple and that 'biotech geese cannot lay golden eggs without daily tending'. The innovation documents are examples of the latter, i.e. the need for a number of tending practices. More precisely, this has to do with more than simply tending.

The codfish is fleshed out as in need of a number of transformations, as needing to be modified and assembled. For instance, from letting the cod take its own directions it needs to be homed in, assembled in nets that hold. Also, cod fry need to be grown in sufficient quantities and on a large scale at a sufficiently low price. Another challenge is to modify the cod's preferred time to grow as sexual maturation has to be delayed so the cod will reach the optimal size. Breeding needs to be improved in order to enhance its production capacities. Diseases and parasites need to be better controlled and the cod itself must be developed into a more resistant organism. The cod's behaviour needs to be modified, as do its eating habits (there is sometimes a problem of cannibalism). The economically most important properties of the relevant cod stocks need to be charted, and so forth.

Coordinating, assembling and readjusting: innovation strategies as innovation devices

The subtitle of the first in a series of innovation plans for cod, which started in 2001, already points out the need for a so-called 'coordinated effort'. Hence, working in coordination is the way of working that is most highly appreciated and valued. Simultaneously, the document, or this genre (Miller 1984) of documents, can be read as acts of timing such coordination efforts. The innovation documents (RCN/IN 2006; RCN/IN/NSRF 2009; RCN/NIRDF 2001, 2003) are efforts in timing necessary adjustments; they are acts of assembling actors and resources that will enable bottlenecks, as they are sometimes called, to be overcome.

So how is this done? On the one hand, the documents time such efforts through (what looks like) acts of repetition. The 2003 plan can be read as close to a copy of the plan from two years previously, and the 2006 plan is close to a copy of the 2003 plan, and so forth. The publishing years are changing, but the title stays close to the same, as does the cod-origin story (how it all started more than a hundred years ago), as well as its possible future. Through these acts of (what looks like) repetition, speed is maintained, actors keep being assembled, and the future-farmed cod as a successful industry and export product is kept in realization. Therefore, in this sense, the documents act as what I have suggested we should label innovation devices—distinct devices at work in order to realize values. This is done by addressing and adjusting to the constantly emergent needs for co-modifications. Such acts of timing and re-timing are one of the aspects of the timing issue—and one of the crucial and interesting ways in which these documents act as innovation devices.

If we read closely, we can see that the strategy documents are also shifting, adjusting, and reordering to the successes as well as to unforeseen problems in the market, and in the efforts taking place elsewhere to correct, adjust, and modify the cod potential. If we continue reading more closely, we see that the *Gadus morhua* offers little but trouble.

Already the 2006 strategy document expresses the ways in which the cod resists and defies a number of the modification efforts it is exposed to: for instance, a too early sexual maturation continues to give low growth and weak profits (RCN/IN 2006: 38). Moreover, there are many of what are referred to as external threats (RCN/IN 2006: 39). Again, if the cod escapes the net, the economic potential is simply left to swim away. There are other aspects in relation to which the cod was not so easily homed in: there was too little knowledge about diseases and contagions; what about the cod's own welfare? Too little was known. At the same time, conflicts over the cod farming sites were increasing. Apparently, Norway, and its coastal areas, was not simply a promised land for the production of new species under water. Others wanted to use this available water–land differently, so that conflicts over the use of land were intensifying. Moreover, what were the potential ecological and genetic effects on wild cod from the farmed cod?

Co-modification as timing

If we align the aquaculture enterprise with the new biocapital and analyses on how life has become a productive force, the impression could be given that this is a fairly straightforward, close to mechanical process. I already pointed out ways of reasoning about the bioeconomy that may give the impression that nature is now under control, that one has become expert at stopping, starting, and suspending (cf. Landecker 1999) or in my own wording, timing nature. But looking carefully into the emerging problems in the cod-farming enterprise, such timing efforts are precisely what is regarded as one of the crucial problems. So the question still is how to, in practice, stop or time biological processes, and by which means? The innovation documents point to the many problems that researchers try to address and deal with. Early sexual maturation is one of the problems that are confronted. This is defined as one of the main problems, 'bottlenecks' to solve if profitable farming of Atlantic cod in sea cages is to be developed (Taranger 2001; Taranger et al. 2005).

In delineating this problem the farmed version of the cod is being compared and contrasted with the wild Atlantic cod. The spawning period for most stocks of Atlantic cod is between January and April, but the growth and age of puberty varies between stocks. This is also influenced by the availability of prey and the temperature in the cod's environment (Taranger et al. 2010: 487). The spawning age differs between stocks; for instance, there is a difference between the northeast Arctic cod stock that usually spawns at an age of between four and eight years, and the Norwegian coastal stock that spawns at three years and older. But not only do the farming practices eliminate these variations between stocks, the spawning time is profoundly changed as well. Under 'normal growing conditions' the spawning age in farming is two years, sometimes even earlier (Taranger et al. 2010: 487).

According to researchers in the field, early sexual maturation leads to a number of problems, such as loss of body mass, reduced harvest quality and feed utilization, often increased susceptibility to diseases and spawning associated mortality—and welfare problems related to this—as well as potentially negative genetic and ecological impact on wild stock due to spawning in cages or escape (Taranger et al. 2005; Taranger et al. 2010).

As one of the researchers explains the problem (Norberg 2002), 'A major bottleneck is early sexual maturation, which commonly occurs at two years of age in cod reared in captivity. This is before the fish has reached market size and is accompanied by weight loss due to spawning'. This takes us right to the heart of the co-modification problem: modifying the biological to provide a viable market product. In a very concrete way, then, the market is written into the body of the cod, meaning that not only do the innovation documents seek to time the necessary events in order to achieve a successful co-modification process, the scientists in the field also work to time the biological with the market, in this case so that the cod reaches market size.

The problem is then that the growth of the fish stops before it has reached a size which makes it a viable market product. Moreover, the very spawning process causes the weight of the fish to decrease. To ensure a viable cod farming industry, which according to the problem formulation depends on how fast and at what cost a marketable size can be reached, the 'onset of puberty must be delayed by at least one year' (Norberg 2002). The challenge then is to co-modify cod with the market. As I have pointed out, one of the crucial acts in this process is to time, or re-time, the cod.

One of the strategies involved in re-timing the cod has been to 'pinpoint the timing of reproductive events', for instance, the critical times for the offset of puberty at one and two years of age, as well as the critical times when growth influences sexual maturation (Norberg 2002). One hypothesis is that such assumed critical windows for sexual maturation open at particular points of time in the season; for instance, December–January for a two-year-old. This means that 'season', a particular 'nature time', has been brought to the researcher's attention.

But this is not the only point of departure for scientists who not only try to open such critical windows but to intervene at these points in time. As one of the other researchers involved has put it: 'Fish sets its biological clock according to the light, but the mechanisms are not very well known' (Geir Lasse Taranger, interviewed in Ditlefsen 2007). Hence acknowledging that the codfish lives according to its own biological time and that it is governed not by the season as such, but light; that is the point of departure in efforts at re-timing the cod with other means. Using continuous light in seawater tanks has been one of the projects enabled by the innovation strategy. In this way, experiments in the wild produce enhanced environments (Knorr Cetina 1995).

This problem is also addressed as a timing issue quite explicitly and continuously elsewhere: According to its own description, the project 'Pubertiming' has 'provided new fundamental knowledge on the cellular, molecular and endocrine events that lead to the onset of puberty in salmonids and sea bass, as well as applied knowledge on puberty control in farming by improved understanding on how various photoperiod regimes, light intensities and spectral composition of light affect the light perception and ultimately the onset and completion of puberty' (Taranger et al. 2005). The above timing and re-timing efforts as they are dealt with by the researchers is then another crucial aspect of the co-modification process.

However, despite these efforts, a follow-up innovation document published in 2009 can be read as a way of letting go of earlier timing efforts. Now the *Gadus morhua* situation is enacted as more or less out of control, out of synch, with the preferred market and production regime. The cod, it is now acknowledged, had not been so easy to control as the salmon when it came to delaying its sexual maturation. Hence, the biomass remained too small. The costs incurred from a cod which refused to step out of its own rhythm were huge (RCN/IN/NSRF 2009).

The cod was also continuously plagued by diseases. Despite extensive vaccinations against it, the disease 'vibriosis' for instance continued to be a problem. On top of this, there were a number of other diseases, as well as resistance problems due to the high amounts of antibiotics. Moreover, the cod continued to swim following its own path and escaping from the nets. Not only was this now acknowledged as a straightforward economic loss, the escaping cod were becoming a risk to their environment as well, as the escapees were threatening to pollute other species (the wild cod) with their own enhanced genetic material.

Moreover, the cod escapees did not always *appear* the way the familiar *Gadus morhua* did. Due to intensive production methods, the cod, or selected cods, changed their anatomy. Their backs and sometimes their jaws were deformed or disfigured.

Deformation, for example a bent neck or deformation of the spine, had been a problem over a long period of time, but was now under better control. Nevertheless, the cod was facing an image problem. Even if the quality of farmed cod was judged to be good in the markets, the problem remained that the farmed cod was seen as inferior to the wild cod.

Co-modifying markets with market research

To begin with, as we saw above, the market situation was thought to be good: The innovation strategy was based on the assumption of an already existing market inhabited by consumers who were providing a demand for cod or cod-related products. On a general level this was reflected in the market research. For instance, cod was described as having been an important seafood product for centuries, treasured by Norwegians for thousands of years for its quality. It was also said that Norway was the leader in the production of farmed cod (Luten et al. 2002). The market research envisioned the same results as the innovation strategies: a substantial increase in production was expected, based on a planned increase in the production of fry and with the use of technology adapted from salmon farming. Whereas the production of fry had been 500,000 in the year 2000, this was expected to rise to between two and three million in 2002. But would the consumer in practice be attracted to the farmed cod? And if not, how could this be enabled?

At this point reference was made to the fact that there was very little research on how consumers judged farmed cod versus wild cod. However, in expert panels set up to determine the sensory qualities of these two versions of cod, the wild cod was assessed in a significantly different way from farmed cod. To make the farmed cod more similar to the sensory qualities of wild cod, a longer starvation period before slaughter seemed to be an important factor (Luten et al. 2002). This means that not only did the experts suggest modifying the farmed cod to achieve an optimal size in the market (as in the previous example), they also suggested modifying the cod to increase its sensory value: its praise value.

One report suggested that consumers judged the quality of the farmed cod fillet to be just as good as the wild cod fillet, or even slightly better when it came to particular attributes. However, what this research pointed out as striking was that consumers, if they were informed about the cod's origin, seemed to value the products more. Hence, the research suggested making what others have called judgement devices (Karpik 2010) integral to the innovation strategies: 'Additional information could provide the name of the catching vessel, farming location, processing and so on' (Luten et al. 2002: 59).

However, follow-up research gave a somewhat more troubling picture: 'Apparently, farming of fish is associated with less favourable characteristics and perceived as such, which has to be taken into account in the market focus' (Kole et al. 2003: 4). Consumer research argued that price was not the most important factor for the consumers; both when buying and eating fish, price was deemed by the consumers to be only moderately important. However, the information the consumers were given about the product stood out as important—and the label 'farmed' 'induced more negative evaluations' (Luten et al. 2002: 59).

This kind of market information about the consumers and their valuation practices was written into the innovation documents. The fact that the quality of the farmed cod was judged to be good but despite this was seen as inferior to the wild cod showed, it was argued, that 'it is not always the technical quality that is decisive, but rather properties that are external or added to the product, but which cannot be accounted for as easily as the pure technical qualities of the fish itself' (RCN/IN/NSRF 2009: 11). Interestingly then, the division between the technical or assumed objective character of the codfish and so-called added or what seems to be understood as subjective or socially constructed aspects of the cod was established as part of the innovation strategies. As I have pointed out above, this is the same division that forms the point of departure in economic sociology. This shows that such divisions between the social and the object itself is something that is produced in concrete practices, hence, something that we as analysts ought to empirically and analytically study rather than take for granted.

Due to the poor image the codfish apparently had for consumers, market researchers proposed two strategies: (1) to literally disassociate the farmed cod from the 'farmed' label. Instead farmed fish could be positioned in a so-called 'different fish food category'. In so doing, the contrast with wild captured fish could be avoided. This repositioning could possibly be positively linked with a higher price level: 'If well marketed, farmed fish might... be able to benefit from higher price categories, featuring its own particular added values' (Kole et al. 2003: 31). Alternatively, it was suggested (2) the poor image of farmed fish could be changed. This could be done by building production that was adapted to and designed for 'positive consumer perceptions'. Producers who did so could then communicate this and present it as 'added value in the market' (Kole et al. 2003: 31). In this way, there was an attempt to make consumer perceptions, using a number of innovation devices including those for acting upon people's judgements, integral to the very co-modification process.

Taken together, the recommended strategies can be understood as addressing directly the issues this chapter is concerned with: the co-modification of markets and biological entities. Market research contributed directly to the innovation strategy. At least, this is how we can read the ensuing innovation documents.

Co-modifying the farmed and the wild: Not simply farmed but *'fresh cod'*

To co-modify the *Gadus morhua* in a way that secured a profit proved far more difficult than previously assumed. Rather than a new start, the decade from the year 2001 onwards seemed to repeat many of the earlier experiences from the 1980s when cod had, in much the same way, proved to resist efforts at being successfully farmed and fitted for market.

A 'zero-escape vision' was introduced to prevent the cod from escaping their nets. Stronger nets had to be built. In this 'zero-escape vision' we can draw parallels to the more overarching strategy: the *Gadus morhua* was to be modified to minimize costs and optimize the available biomass. The risk was the loss of potential biological capital. As such, the life of the *Gadus morhua* was the productive force, the live potential to be worked upon. As little as possible was to escape from the available potential.

But the various efforts were not sufficiently successful and the markets did not turn out as expected. The findings in the market research were repeated as part of the innovation strategies: to build a more successful product, the farmed *Gadus morhua* was set out to capitalize on *Gadus morhua* in the wild.

By way of shifts in the innovation document strategies, the notion 'fresh cod' emerges as a device for assembling the farmed and the wild harvested cod into *one* undivided commodity. Hence, the innovation documents acted as innovation devices precisely in the direction that had been suggested by the market research, to dissociate the farmed cod from being just that—farmed. Interestingly then, rather than to be made to stand out as a distinct and unique product (cf. Karpik 2010), the suggestion was to downplay the unique in the form of the farmed. Hence, the farmed cod could be made comparable to other products that were deemed more favourable. The innovation device at work in our case then was not directed towards producing the unique, but to produce a neutralized cod; the cod as a seamless part of an already highly valued product. Hence, this was a quite distinct and particular form of valuation.

Another dimension of the valuation practice and innovation strategy was to link the farmed codfish with an already much valued past.

Linking the harvesting times of the past and the emerging farmed-cod future together is in itself nothing new. On the contrary, linking Norwegian history with the future fish-farming enterprise was an integral part of the cod-plan genre from the very beginning. Norway has long traditions when it comes to harvesting, processing. and selling cod, it was stated repeatedly (RCN/IN 2006: 8; RCN/NIRDF 2001: 5, 2003: 7). Over the centuries this has had an effect on where people live and on the infrastructure and employment along the coast: 'Norwegian exporters have a worldwide market for their cod products and the fish is renowned for its good quality' (RCN/IN 2006: 8; RCN/NIRDF 2001: 5, 2003: 7).

But as available fishing quotas are being reduced, farmed cod, it was argued, might enter the picture as a substitute: farmed cod may in the longer term make up a larger part of the production volume, 'thus contributing to maintaining the delivery of raw material and the activities of the fish-processing industry along the coast, and contributing to more stable deliveries to the markets' (RCN/IN 2006: 8; RCN/NIRDF 2001: 5, 2003: 7).

Hence, 'history', or 'the past', already appreciated and much valued contexts, were written into, folded into, an emerging cod aquaculture enterprise. This is also true the other way around: the *Gadus morhua* potential as a farmed commodity was seamlessly made part of an established and valued national culture; once again therefore a distinct form of valuation, or revaluation.

In the later and updated plans, the farmed cod is not so much offered as a substitute but rather as complementary to its wild member of the *Gadus-morhua* family. As formulated in the 2009 strategy: 'Farmed cod and wild cod will complement each other and make it possible to offer European consumers fresh cod all year round. This is important in order for the fresh cod—both wild and farmed—to be given space in the supermarket shelves' (RCN/IN/NSRF 2009).

The same complementary strategy is made part of the innovation documents. Already from 2003 (RCN/NIRDF 2003) onwards, these are no longer directed towards the farmed cod exclusively, but towards the farmed and the harvested cod in combination (RCN/IN/NSRF 2009). 'Fresh', not 'farmed' cod is made to stand out as the new commodity. Hence, again, this can be read to be precisely the dissociation strategy that I touched upon earlier in this section. In order to comply with assumed consumer demand, the farmed cod is co-modified with the harvested wild cod.

So how is this done? On the one hand, 'fresh cod' rather than 'farmed' is established as the object of the innovation strategies. Moreover, as I pointed out earlier in this section, the harvested wild cod is no longer made part of 'the past'. On the contrary, the harvested wild member of the *Gadus morhua* family is presented as a valuable resource, a resource upon which the farmed cod can capitalize. To put it another way, the farmed cod is co-modified with the harvested wild cod. As is stated in the 2009 cod plan, 'The perception and judgment of the cod as a market product depend on the ability to support the farmed, as well as the wild, cod. They are both high-quality and sustainably governed products' (RCN/IN/NSRF 2009: 11).

Performing co-modifications

The neoclassical, now quite common definition of economics is that it is concerned with contributing to an optimal allocation of resources for competing aims. In principle then, economics does not have any given or natural limits; it can, in principle, be applied to any sector of society and to any action—as long as this is an action that is responding systematically to changes in the environment. However, as Michel Foucault pointed out in his *Collége de France* lectures (Foucault 2004), if this is the case, then '*homo-economicus*' is different from how we thought we knew him. Then '*homo-economicus*' is no longer the rational, sovereign character, a figure one should leave alone and '*laissez faire*' so that he or she can pursue his or her own interests. On the contrary, the economic man is a thoroughly governable character, and economics is more about analysing techniques that act upon the actions of others.

Foucault's approach here is instructive when it comes to analysing value practices in the life sciences. I have suggested the notion of co-modification as a way of opening an empirical space for exploring the techniques and the ways in which products are being acted upon—hence how bodies are being modified. But value practices also 'act back' and take part in modifying the ways in which markets are constructed and the very products that are offered. Hence, I have worked from the assumption that valuation practices are performative: they are practices that take part in enacting new versions of reality. Nevertheless, and as the present innovation story on valuation practices should be fit to demonstrate, there are limits to performativity (Asdal 2011; Fourcade 2011b) and co-modifications. Modification efforts do not always succeed and they definitely do not always succeed in enabling values in the form of profitable commodities.

What this suggests is a more open or open-ended approach than a Foucauldian approach conventionally has inspired. A number of studies at the intersection of economic sociology and science studies have contributed to opening up the space for less disciplining and more explorative approaches. Studying innovation documents as distinct valuation agents and sites for performing values, as this chapter has attempted to do, aim to add to this volume of work. In doing so, this chapter has argued for adding to such contributions by exploring further how valuation practices can be grasped not only as social constructions, but also as object formations. Hence, I have suggested the concept of co-modification, the modification of biological entities as well as market entities and how such modifications are intimately intertwined.

Despite careful tending and modification, so far farmed cod has made itself too difficult and costly to develop into a profitable business (e.g. Aftenposten 2012). Indeed, and quite ironically, the newly farmed commodity has had to lean on—to collect value from—the wild.

It never occurred to me that the cod I had bought at the fishmonger could have been a farmed product. Nor did it occur to me that the fact that I never asked could possibly be explained by a deliberate market strategy. However, as I later returned to the same store and asked, the fishmonger replied and said 'good luck'. He wished the cod farmers all the luck with their efforts, but so far he would not sell their product because he did not like the texture of the farmed cod. Hence, highly priced and little praised: values in tension that, so far, do not enact what was hoped for and invested in.

10 Norms, values, and constraints

The case of prenatal diagnosis

Ilana Löwy

Values into techniques

The life sciences, the introduction to this volume states, are saturated with values (see introduction, Dussauge et al., this volume). The position of values in scientific activity was captured by the term 'moral economy'. This term originated in E. P. Thompson's (1971) and James Scott's (1977) studies of peasant rebellions. In her influential 1995 text 'The Moral Economy of Science', science historian Loraine Daston proposed a different use of the term 'moral economy'. Daston investigated the genesis of a 'scientific ethos', the history of the constitutive elements of modern scientific practice such as striving for precision, replication, or objectivity. Her study is focused on the unique forms of moral, emotional, and aesthetic elements adopted by the 'tribe' of scientists and on the emotions and values of this 'tribe'. Daston's scientists move in a world of noble, bourgeois, or protestant values, such as honour, punctiliousness, or introspection, and they adapt these values to their specific needs, above all the production of scientific knowledge; but (in that text) they do not seem to be concerned by institutions and laws, hegemony and subordination, financial constraints and political developments, economical conflicts and political struggles.[1]

From the 1990s onwards, numerous scholars adopted the term 'moral economy'. They studied the moral economy of corruption in Africa, selfhood and caring, illegitimate births, nostalgia, French immigration policies, degeneration and drinking in Navajo society, AIDS in South Africa, ancestor worship among Chinese immigrants, the invention of the aquarium, the diffusion of daguerreotypes, the regulation of human embryonic stem cells,

[1] The economical and political variables that affect the place scientific activity occupies in a given society were studied by scholars who belonged to the 'old' tradition of sociology of science (e.g. Ben-David 1971) but this topic became less prominent in the 'new' social studies of science. The new approaches greatly enriched our understanding of science in the making, but, in some ways at least, make the study of the broader context of scientists' activities more difficult (Löwy 2010a).

and even internet file sharing (Fassin 2009a: 1240a). The multiple uses of 'moral economy' might have trivialized and weakened the term.[2] On the other hand, academic trends are not produced in a void. The popularity of the term 'moral economy' reflects, one may argue, a growing interest in the role of values and emotions in shaping of social and economic developments. The coincidental use of the term 'moral economy' in studies of popular rebellions and investigations of norms that influence scientists' activities, French anthropologist Didier Fassin argues, is a heuristically felicitous occurrence.[3] It makes the limits of both uses visible—the restriction of this term to the traditional/dominated strata of society by Thompson and Scott, and its dissociation from economic and political consideration by Daston. A combination of both approaches, Fassin proposes, can promote a study of multilevel intersections between political and economic variables, technologies, subjectivities, and emotions (Fassin 2009a: 1255).

Fassin is interested in the 'moral economy' of social relations, in particular those that produce marginality and exclusion: illegal migration, urban poverty, illegitimate births (Fassin 2004, 2005, 2007, 2009a, 2009b). Scholars who studied the sciences examined the incorporation of values into objects and techniques (Feenberg 1999; Winner 1977). A medical innovation—as the pioneer of the social study of science, Ludwik Fleck had already argued in 1935—may be especially well adapted to studying how values shape techniques (Fleck 1979 [1935]). The trajectory of prenatal diagnosis (PND), a medical technology with numerous economic, societal, cultural, emotional, and ethical implications, displays the role of conflicting values in the emergence of new technologies, and in their transformation into routine—and therefore frequently unexamined—medical procedures.

PND—a cluster of techniques and approaches that, when taken together, provide information on the foetus—is today a self-evident component of the medical supervision of pregnant women. The initial stimuli for the efforts to make foetal malformations visible were environmental risks: infections (rubella, CMV, herpes simplex, toxoplasmosis); drugs taken early in pregnancy (thalidomide); and pollution (radiation, industrial waste, pesticides). Early PND techniques were also directed at women seen as being at a higher risk of giving birth to malformed children: those with a family history of genetic disease, women who had given birth to an affected child or miscarried because of a major foetal malformation, older women, and those with health problems (diabetes, heart disease, hypertension) that can affect foetal development.

[2] The fate of the 'moral economy' (and perhaps also of 'value practices') may be compared to the polysemy and partial trivialization of the term 'social construct' (Hacking 1999).

[3] Daston (1995: 3) explicitly claimed that her use of the term 'moral economy' was very different from the one proposed by E. P. Thompson.

However, in the 1990s, numerous Western European countries extended prenatal testing to women without known risk factors. Prenatal screening (PNS) replaced PND.[4] This radical—and rarely discussed—shift was driven to a large extent, this text argues, by the aspiration of preventing Down's syndrome, a condition redefined as a public health problem. The development of high resolution ultrasounds and tests for foetal markers in maternal serum accelerated the shift to PNS for Down's syndrome. The generalization of such screening favoured in turn the detection of additional foetal anomalies.

Originally an exceptional procedure, PND was transformed into a routine medical technology (Boyd et al. 2005).[5] The rapid diffusion of PND was often presented as a non-problematic technological response to users' demand: pregnant women want to be reassured that they will give birth to healthy children. A history of PND indicates, however, that the generalization of this innovation was initially driven by the goals of professionals: researchers who developed new methods to detect foetal anomalies, producers of instruments and reagents, clinicians, and public health experts. Women had limited input into the shaping of a method that modified their experience of pregnancy, focused their attention on dangers to the foetus, and, not infrequently, forced them to make difficult choices. Their values, to be sure, affect their decisions about the future of their pregnancies, but not the complex chain of events that transformed the majority of pregnant women in industrialized countries into, to borrow Silja Samerki's apt expression, 'managers of fetal risks' (Samerski 2009). This chapter investigates how pregnancy has changed in the last fifty years, whose values shaped that change, and what are the consequences of this new configuration.

PND and at-risk pregnancies

Today, PND—that is, a cluster of techniques and approaches that provide information on the foetus—is perceived as one of the key elements in the

[4] On the other hand, as one of the promoters of the shift to prenatal screening in the UK, David Brock, explained, obstetricians focus on the well-being of the individual patient, and many of them dislike the idea that they have a responsibility to reduce the incidence of serious congenital abnormalities in populations (Brock 1982: 131).

[5] This text follows developments in Western Europe, with a special focus on France and the UK. In the US, in the absence of a national health system, practice varies greatly and, while obstetrical ultrasounds are widely diffused, serum tests for the risk of chromosomal anomalies are not employed in a systematic manner. In 'intermediary' countries, prenatal surveillance of the foetus is usually available only to middle class women and is complicated in many of these countries—such as nearly all of Latin America (with the exception of Cuba and Uruguay)—by the illegality of abortion.

surveillance of pregnancy. The term prenatal *diagnosis* indicates that this approach informs a woman if her foetus is 'normal' (that is, does not have detectable problems) or if it suffers from any given malformation. Initially, however, this technique was employed to indicate the existence of a risk of malformation rather than its presence.

In the 1950s, physicians started to perform 'amniotic taps' (trans-abdominal sampling of the amniotic fluid) on individual women whose child was at risk of haemolytic disease of the newborn, a condition produced by Rh incompatibility between mother and foetus (Bevis 1952; Cowan 2008). Children with this condition often died shortly after birth, and those who survived often suffered from severe health problems, including mental retardation. If the amniotic fluid showed evidence of the destruction of foetal blood, doctors induced labour as soon as possible and then rapidly replaced the newborn child's blood containing the destructive maternal antibodies with fresh blood. At that time the 'amniotic tap'—renamed amniocentesis—was a risky technique. There was no way to see precisely where the doctor places the amniocentesis needle and as a result some foetuses were hurt and died in utero. Accordingly, this approach was employed only when the experts believed that the foetus was in immediate danger (Walker et al. 1964).

In 1960, two Danish researchers, Povl Riis and Fritz Fuchs, were able to determine the sex of a foetus by analysing cells from the amniotic fluid of a pregnant woman. The woman's previous child, a boy, died at birth from haemophilia. When she found out that was pregnant again, she decided to terminate the pregnancy, a decision made possible by a Danish law that legalized 'eugenic' abortions for carriers of the haemophilia gene. Her obstetrician told her, however, that it may be possible to determine the sex of the foetus before birth. She agreed to an amniocentesis, learned that the foetus was female, and gave birth to a healthy girl (Riis and Fuchs 1960).[6]

In 1960, Denmark was one of the few countries that allowed abortions for the risk of foetal malformation. The thalidomide disaster of 1961 and the German measles epidemics of 1962–4 contributed to the liberalization of abortion laws in numerous Western countries (Dally 1998; Greenhoouse and Siegel 2010; Reagan 2010).[7] Many doctors felt that women who contracted German measles early in pregnancy and were terrified at the possibility of giving birth to severely deformed children should be entitled to legal abortions. They also became increasingly concerned by the mortality and morbidity from illegal abortions, especially among poor women. A shift in

[6] Riis and Fuchs did not analyse chromosomes of foetal cells but employed an older method of sex determination—the display of the 'Barr body', a structure present only in female cells.

[7] Thalidomide was rapidly taken off the market, limiting the number of women who took the drug and then learned about its danger to the foetus; but the high visibility of the thalidomide scandal heightened women's awareness of other risks to the foetus, such as German measles.

professional attitudes played an important role in the push to liberalize abortion laws (Gold 2003; Reagan 2010). The thalidomide disaster and the German measles epidemics also led to the development of a new branch of knowledge: the epidemiology of congenital malformations. Data collected by regional, national, and international registries of congenital malformations such as the International Clearing House for Birth Defects Monitoring Systems, or the Latin-American Collaborative Study of Congenital Malformations transformed the prevention of inborn defects into a public health issue (Castilla and Orioli 2004; Klingberg 2010).

Women who had taken thalidomide or contracted German measles early in pregnancy asked to terminate their pregnancies without knowing whether the foetuses they carried were affected; in all probability, they sometimes aborted healthy foetuses.[8] Similarly, haemophilia carriers who decided to terminate their pregnancies upon learning that the foetus was male had a 50 per cent chance of eliminating a healthy foetus (half of the male children born to women who carry the haemophilia gene receive a normal X chromosome from their mother, and are therefore disease free). However, in the 1960s and 1970s, new scientific developments opened the way for direct diagnosis of anomalies of the foetus and the possible abortion of affected foetuses.

Screening for Down's syndrome: The price of disability

In 1959, cytogeneticists who studied human chromosomes redefined several inborn conditions an aneuploidies—the presence of an abnormal number of chromosomes. The most frequent among these conditions was Down's syndrome (trisomy 21) (Harper 2006). In the 1960s, scientists developed ways to examine the chromosomes of foetal cells in amniotic fluid and perform biochemical tests on these cells to reveal the presence of several hereditary disorders (Nadler 1968). In the late 1960s and early 1970s, amniocentesis—the only way to obtain foetal cells for analysis—was, a risky procedure (with an estimated 3–5 per cent probability of procedure-related miscarriage), but many women from families with severe hereditary disorders believed that such a risk was acceptable. They explained that without a PND, they would never have had the courage to became pregnant (Christie and Tansey 2003; Davidson and Rattazzi 1972). At the same time, some doctors started to offer amniocentesis to pregnant women in their 40s, a group at high risk of giving birth to Down's syndrome children.

[8] Only a fraction of women who had taken thalidomide during pregnancy had affected foetuses. This argument was used in the 1960s by opponents of the liberalization of abortion (Kajii et al. 1973).

In the early 1970s, the introduction of obstetrical ultrasounds made amniocentesis less dangerous. When doctors were able to see where they were to introduce the sampling needle, the rate of miscarriages following this procedure decreased to between 1 and 2 per cent (Bang and Northeved 1972). The technical improvement favoured increased use of the technique. Until the late 1980s, PND was only offered to women who belonged to a well-defined risk group. On the other hand, in the early 1970s, some public health experts had already evoked the possibility of generalizing PND. Three US-based social epidemiologists, Zena Stein, Mervin Susser, and Andrea Guterman, explained that Down's syndrome was rapidly becoming a public health problem. In the 1940s, the great majority of trisomic children died before puberty, while in the 1960s, thanks to medical advances such as the use of antibiotics to treat respiratory diseases and surgery for heart defects, more than half of these children survived to adulthood.[9] New educational methods improved the quality of life of people with Down's syndrome, but were unable to change the basic fact that people with this condition remain in a state of dependence all their life, and impose a severe burden on their families and society.

The Down's syndrome burden could be alleviated, Stein et al. argued, by offering PND of this condition to all pregnant women (Stein and Susser 1971; Stein et al. 1973). Such a screening effort might be costly, but its societal value should not be evaluated in purely monetary terms alone. Newborn screening for phenylkentonuria (PKU), Stein et al. explained, was implemented in numerous countries despite the fact that this screening was much less efficient in preventing mental retardation than the generalization of screening for Down's syndrome was likely to be. An additional argument in favour of the generalization of screening for trisomy 21 was the discriminatory nature of existing practices. The benefits of testing for risk of Down's syndrome were already available to women that could afford it: societies could decide to offer them to everyone (Stein et al. 1973: 309).

Extension of PND: Serum makers and diagnostic ultrasounds

In the early 1970s, Stein et al.'s recommendation that screening for Down's syndrome be generalized, was not seen as a 'doable' proposal. The test was costly, stressful, and risky. A 1–2 per cent risk of miscarriage was acceptable

[9] Akin's argument that the improved survival of children with inborn conditions greatly increased the burden on the health system was made at the same time to promote prenuptial screening for thalassaemia in Cyprus (Beck and Niewöhner, 2009).

for carriers of hereditary diseases (who have either a 50 per cent chance to give birth to an affected child if the disease is dominant, or a 25 per cent chance if it is recessive) and for older pregnant women (a 45-year-old woman has a 3.5 per cent chance of giving birth to a Down's syndrome child), but not for younger women who had a much higher chance of aborting a healthy foetus than of giving birth to a trisomic child (Cuckle et al. 1987; Hook 1981). Accordingly, until the 1990s, physicians usually recommended amniocentesis only to women over the age of thirty-five. The choice of this cut off—thirty-five years—was legitimated by the claim that at that age a woman's chance of giving birth to a Down's syndrome child (over 1 per cent) was greater that her chance of losing a healthy foetus following an amniocentesis. This claim, geneticist Robert Resta argues, had no factual support. Physicians had dependable data on the effect of maternal age on the frequency of foetal chromosomal anomalies but not on the risk of miscarriage following this intervention. The latter risk was strongly related to the individual operator's skills and the quality of the ultrasound equipment; it varied greatly from one site to another (Resta 2002).[10] Older pregnant women were expected to choose between two different kinds of risk without reliable data on the magnitude of one of those risks, while younger pregnant women had no way of knowing whether they carried a Down's syndrome foetus.

The solution to this double dilemma was the introduction of a two-step evaluation of Down's syndrome risk: first a risk-free serum test and ultrasound, and then, if indicated, a risky invasive test. The development of a serum test for Down's risk was partly an accident. In the early 1970s, biochemists had noted a correlation between high levels of alpha-foetoprotein (AFP, one of the proteins secreted by the foetus) in pregnant women's serum and major neural tube defects (anencephaly, spina bifida) of the foetus. This observation opened the door to a simple and risk-free way of verifying whether a foetus had a higher probability of being affected by one of these malformations (Brock and Sutcliffe 1972; Kolata 1980).[11] The next step was the observation, in 1984, that women who carry a foetus with Down's syndrome have an unusually low level of AFP in their serum (Cuckle et al. 1984; Harris and Andrews 1988). The addition of two other biochemical markers, chorionic gonadothropin and unconjugated estriol, increased the sensitivity of 'Down's risk' detection. Scientists suggested screening all pregnant women for the blood levels of these three biochemical markers (the 'triple test') and offering an invasive test only to those at higher calculated risk of carrying a trisomic foetus, regardless of their

[10] Resta had also shown that cost/efficacy calculations played a central role in the selection of thirty-five years old as the cut-off for recommending amniocentesis.

[11] The use of AFP to evaluate the risk of neural tube defects became obsolete in the 1980s when the growing resolution of obstetrical ultrasounds made direct diagnosis of these defects possible (Watson et al. 1991).

age (Harris and Andrews 1988; RCOG 1993; Sheldon and Simpson 1991; Wald et al. 1988).

The triple test was initially performed during the second semester of pregnancy (at 16–20 weeks) but in the 1990s some experts proposed it be performed during the first semester of pregnancy instead; this view was increasingly adopted by obstetricians (Aitken et al. 1996; Wald et al. 1996). The main reason for the change in the timing of the triple test was the finding of an additional 'marker' of a risk of trisomy 21: the nuchal translucency measurement (the assessment of the quantity of fluid behind the foetus' neck). This measurement needs to be made at 10–12 weeks of pregnancy. An abnormal nuchal translucency at that stage of foetal development indicates a higher probability of chromosomal anomalies (Nicolaides et al. 1992). The combination of serum tests and ultrasound results, experts proposed, increased the efficacy of screening for Down's risk (Borrell et al. 2004; Palomaki et al. 2010; Wald and Hackshaw 1997).

Obstetrical ultrasounds were first introduced in the 1950s. They became more efficient in the 1970s, thanks to two technical innovations: linear array scanning and transvaginal ultrasound (Blume 1992; Donald 1969; Nicholson and Fleming 2000). These innovations made it possible to use obstetrical ultrasounds to establish the age of pregnancy, a parameter which became especially important when doctors started to measure pregnancy-related biochemical markers in pregnant women's blood. The normal values of serum markers vary according to the age of pregnancy: a result that is normal at 12 weeks is abnormal at 16 weeks. Before the advent of ultrasounds, doctors determined the duration of pregnancy from the woman's report of the date of her last period. Ultrasound replaced fallible human memory with mechanical objectivity, increasing the predictive accuracy of serum tests.[12] The quasi-obligatory coupling of the triple test for Down's risk with ultrasound favoured in turn the observation of structural anomalies in trisomic foetuses and led to the development of the present-time population-based approach to prenatal detection of this condition.

Looking for Down's risk: From diagnosis to screening

A quarter of a century after Stein et al.'s paper, several Western European countries and two US states, California and Iowa, introduced population-based screening for Down's syndrome, based on the combination of serum

[12] On the definition of mechanical objectivity, see Daston & Galison (2007).

tests and diagnostic ultrasounds. A specially designed computer program, which also takes into consideration the woman's age, then calculates her risk of having a Down's syndrome foetus.[13] If this risk is greater than the (presumed) risk of spontaneous abortion following amniocentesis, health professionals usually propose that she undergo the latter test (Cunningham and Tompkinson 1999; Wald et al. 1988). In doctors' public discourse, the introduction of screening for Down's syndrome was legitimated by women's wish to be reassured about foetal health. In the professional literature, such screening was also justified by public health considerations, above all the escalating costs of long-term care for people with Down's syndrome (RCOG 1993: 7; Sheldon and Simpson 1991: 302). PND of Down's syndrome was coupled with the implicit (and accurate) supposition that the majority of women who learn that they carried an affected foetus would decide to terminate the pregnancy. Health professionals assumed that the sum of women's decisions would decrease the predicted economic burden of caring for people with Down's syndrome.

The extension of PND from selected groups of 'at risk' women to all pregnant women was facilitated by the rapid generalization of use of obstetrical ultrasounds in the 1980s and 1990s. Sociologists, philosophers, and anthropologists who examined the consequences of the spread of this technology often focused on the role of foetal images ('baby's first photograph') in changing the meaning of pregnancy for women and families (Chazan 2008; Duden 1993; Palmer 2009; Petchesky 1987). Fewer studies examined the role of ultrasounds in the transformation of all the pregnant women into a single 'risk group'.[14] For example, France implemented nationwide 'good practice rules' in 2009; under these rules, doctors are obliged to offer first semester screening for aneuploidies (above all Down's) to all the pregnant women. This screening should include a diagnostic ultrasound with detailed (and specified) measurement of the foetus and a serum test. Women have to sign an informed consent waiver for the serum test. On the other hand, the first semester ultrasound is not subject to the formal rules of informed consent (Journal Officiel (France) 2009; Vassy 2006).

Sociologists and psychologists who discuss PND dilemmas often focus on women's decision to undergo amniocentesis, especially when the estimated risk of carrying a Down's foetus is close to the risk of miscarriage following this diagnostic procedure. Such debates frequently underplay the fact that the main tool today to detect foetal malformation is a routine ultrasound. They

[13] Today, some experts argue that due to the recent increase in the resolution of obstetrical ultrasounds and the refinement of visual criteria for Down's syndrome risk in the foetus, serum tests for this condition are not cost-effective any more.

[14] Alternatively, DNP can be seen as an approach that is transforming the foetus into a potential danger to the mother, the family, and the larger group. Such a view of the foetus seems to be frequent in Israel (Ivry 2009b; Remennick 2006).

also seldom discuss the unanticipated dilemmas produced by PND, such as the detection of sex chromosome anueplodies (Turner syndrome, Klinefelter syndrome, triple X), conditions linked with 'minor' but sometimes non-negligible health and cognitive problems, or the observation of anatomic anomalies (e.g. agenesis of corpus callosum), which can lead to severe disability in some cases but in other cases do not have noticeable effects on the child's health. The uncertainty as to the future child's fate blurs the boundary between dia-gnosis (through-knowing) and pro-gnosis (before-knowing), making women's decisions especially difficult (Gross and Shuval 2008).

Obstetricians and geneticists who developed PND techniques in the 1960s and 1970s explain that they were responding to women's demand (Christie and Zallen 2003). This explanation is correct, but only at a specific moment in the history of PND. At first, this approach was offered to women who already had an affected child or were aware of the presence of hereditary disease in their family, and were concerned about the possibility of giving birth to a second affected child. The 'users' demand' argument is less accurate for risks linked to maternal age. In the 1970s, health professionals had to educate women to recognize age as a major risk factor for Down's syndrome (Bermel 1983). In the US, this education was directly related to fears of litigation. Many obstetricians believed that they must persuade their patients (or at least all those over thirty-five) to undergo amniocentesis. Otherwise they may be liable for the lifetime support of a child with a birth defect (Powledge 1979).[15] Fear of litigation was probably less important in countries with a national health service, but the rapid spread of PND in these countries might also have been favoured by a mixture of professionals' desire to prevent the birth of handicapped children and their aspirations to extend their domain of expertise and professional jurisdiction (Powledge and Sollitto 1974). Empirical studies indicate that the main force behind the rapid spread of PND in the 1980s, 1990s and 2000s was a 'push' from health providers (Vassy 2005; Williams et al. 2002). The rapid diffusion of methods of investigating the foetus led to the construction of new needs and therefore new markets for technology and skills (Lippman 1991). It also led to the transformation of PND—an exceptional regime of knowledge applied to selected groups of users—into prenatal screening, a routine medical technology applied to populations (Vassy 2005).

[15] In the US, amniocentesis expanded exponentially in the late 1970s and early 1980s. In 1979, professionals estimated that of the 40,000 amniocenteses performed in the US from 1969 onwards, 15,000 were carried out in 1978 (Powledge, 1979).

NIPT: New technology, similar values

The prenatal detection of foetuses with Down's syndrome is predicted to change dramatically soon, with the expected replacement of amniocentesis by the examination of foetal DNA in maternal blood. Attempts to diagnose foetal malformations through the examination of foetal cells in maternal circulation started in the late 1970s, following the development of the cell sorter technique (Herzenberg et al. 1979). Efforts to make this method sufficiently reliable failed, however. Early attempts to isolate foetal DNA from maternal blood were similarly unsuccessful because the 'noise' (maternal DNA) was too strong to allow efficient isolation of the 'signal' (foetal DNA) (Walknowska et al. 1969). In the twenty-first century, this obstacle was overcome thanks to advances in molecular biology. Today, it is possible to examine foetal DNA in maternal circulation, starting as early as the seventh week of pregnancy (Feero et al. 2012; Go et al. 2011). The new approach is named 'Non-Invasive Prenatal Testing' (NIPT). Experts recently called to prepare for the next revolution in PND: the large-scale diffusion of NIPT (Greely 2011; Hahn et al. 2011). Debates on the potential consequences of the generalization of this new technique focus on the possibility of examining the foetal genome and testing the 'genetic fitness' of the foetus. These debates evoke the spectre of eugenic drift and the ruthless elimination of foetuses that fail to fulfil parental expectations, however trivial, making each pregnancy truly tentative (Greely and King 2010; King 2011; Rothman 1986).

The predictions that the development of foetal DNA-based tests that will provide a genetic 'profile' of the child-to-be will lead to a wave of abortions of foetuses with the 'wrong' eye colour or insufficient potential to excel in sports are currently (2014) remote speculations only. Earlier alarmist predictions, that the liberalization of abortion would lead to the frequent termination of pregnancies for trivial reasons, such as a woman's reluctance to modify her holiday plans, never materialized. Women who elect to terminate a pregnancy, all professionals attest, view this act as an important and difficult decision. There is no reason to assume that the introduction of NIPT will change this attitude. Moreover, there are no immediate prospects for the development of accurate and affordable non-invasive tests to detect specific foetal traits. The main obstacle to the development of such tests is the practical difficulty of making fine distinctions between maternal and foetal DNA. It is much easier to detect markers originating in paternal DNA and absent in the mother. It is also technically feasible to detect massive changes in foetal DNA such as the presence of an abnormal number of chromosome copies. Accordingly, initially, companies involved in the commercial exploitation of NIPT offered three main applications of this technique:

a) Detecting known mutations of paternal origin. If the father is a carrier of a dominant somatic mutation, such as the gene for Huntington disease, detecting DNA with this mutation in pregnant women's blood means that the foetus also carries this mutation. The transmission of a dominant paternal mutation is relatively rare, but when it does happen, NIPT can provide a simpler and less stressful way to test for the presence of selected hereditary diseases.

b) Detecting foetal sex. Since the mother does not have a Y chromosome, the presence in her blood of Y chromosome markers indicates that the foetus is male, and their absence, that it is female. This is important for the diagnosis of X-linked pathologies such as haemophilia or Duchene muscular dystrophy. Rapid determination of foetal sex can also be helpful in conditions such as CAH (congenital adrenal hyperplasia). Women who have already had a girl with this condition are invited to take a drug during their next pregnancy to prevent the 'virilization' of external sex organs and the development of endocrine disorders.[16] If the foetus is male, the women can stop taking this medication and avoid its unpleasant side effects (Devaney et al. 2011; Lewis et al. 2012). A non-medical—and highly problematic—application of this technology may be the rapid abortion of female foetuses in societies with strong preference for male children (Puri et al. 2011).[17]

c) Screening for aneuploidies, above all Down's syndrome. It is technically feasible to detect a triple dose of chromosome 21 markers (the mother's DNA will only have a double dose of such markers). At first, experts proposed using NIPT as an 'intermediary' method: women defined as 'high risk' for Down's on the basis of a serum test and diagnostic ultrasound, would undergo NPIT for Down's. Only those who tested positive with this second blood test would be invited to undergo amniocentesis, since NPIT results are not 100 per cent certain. A generalization of the new approach should allow for a reduction in invasive testing during pregnancy, from 5–10 per cent to fewer than 1 per cent of pregnant women. Later, if the price of the new test drops, it will probably replace the existing serum tests (Chitty et al. 2012). The use of NIPT to screen for Down's syndrome will probably make little difference in countries such as Denmark where more than 90 per cent of pregnant women already undergo serum test-based screening for Down's risk (Schwennesen and Koch 2010; Schwennesen et al. 2012). By contrast, it can promote a large-scale

[16] This treatment is, however, controversial (Dreger et al. 2012).

[17] A blood test that reveals the sex of the foetus aged 7–9 weeks—therefore, at a time when a pregnancy can still be kept secret—may provide some relief to women who are sometimes forced against their will to undergo traumatic late-stage abortions of female foetuses and who are stigmatized for their inability to produce sons (Puri et al., 2011).

diffusion of screening for abnormal numbers of chromosomes in countries such as the US where this approach is not generalized. In these countries, the introduction of the new method may speed up the shift from PND to prenatal screening (Greely 2011).[18]

The immediate future of NIPT seems to be limited to modulating the use of existing diagnostic approaches. The more distant future of this technology may be very different. Perfecting NIPT, coupled with a sharp reduction of its price—if this happens—will probably lead to the generalization of a simple, risk free 'reading' of the entire foetal DNA, as part of an early stage of routine surveillance of pregnancy. The outcome may be a significant scaling up of the level of supervision of human foetuses and the women who carry them. The recent sequencing of the entire foetal genome using free DNA in maternal circulation and the use of partial sequencing of foetal DNA for diagnosis of specific pathologies or an increased probability of such pathologies make this development more probable (Fan et al. 2012; Talkowski et al. 2012). Consequently, expectant mothers may receive a flood of confusing and potentially destabilizing information about the well-being of their future child. Even if this does not lead to the termination of pregnancy, such information may strongly affect their experience of child rearing.[19]

The puzzle of screening for Down

Making the foetus accessible to the medical gaze radically changed the meaning of pregnancy for tens of millions of women. One may argue that humans have always attempted to limit the danger of inborn defects, mainly through selective mating. Pedigrees may be seen as early attempts to 'visualize' heredity and reshape offspring, a trend amplified by the advent of eugenics (López-Beltrán 2007). There is, however, an important difference between estimating the probability of a potential problem and developing techniques that make it possible to directly display an already existing one. Scholars have investigated the rise of the entity 'foetus' and the role of ultrasounds in transforming how pregnant women view their pregnancies (Duden 1993; Morgan 2009; Petchesky 1987). There are fewer studies on the consequences of a steep

[18] An additional and uncontroversial potential use of NIPT is to test Rhesus D negative pregnant women with Rhesus D positive partners if the foetus is Rhesus D positive and, if that is the case, to administer a treatment that prevents the haemolytic disease of the newborn (Oxenford et al., 2013).

[19] This situation already exists in neonatal screening. The crucial difference is that in neonatal screening parents receive information about their newborn baby and face the decision of how to treat this baby, while with the further development of NIPT, they may receive such information about a foetus, and face the decision about possible termination (Baily and Murray 2008; Grob 2011).

increase in the level of surveillance of the foetus in 'low risk' (that is, normal) pregnancies. This text argues that the generalization of PND and its transformation into screening technology was strongly affected by two events: the recognition of the right of women at high risk of giving birth to severely handicapped children to terminate their pregnancies; and the aspiration of reducing the prevalence of a specific congenital condition, Down's syndrome.

The recognition of the right of women exposed to teratogens or those from families with hereditary disorders to have abortions was the result of joint pressure from women and gynaecologists. Women who had taken thalidomide, contracted German measles, or feared that they would give birth to a child with a severe inborn disease often desperately wanted to terminate their pregnancies. In the early 1960s, many doctors agreed that these women were indeed entitled to abortions. By contrast, the second event was above all the result of health professionals' interventions. Numerous women wish to avoid a birth of a child with Down's syndrome but the implementation of regional and national screening programmes was a top-down, not a bottom-up, process. Moreover, this process was shaped by two incommensurable 'moral economies'—or to follow Boltanski and Thevenot's expression, two 'worlds of worth' ('*mondes de valeur*')—that shaped this technology, that of gynaecologists and obstetricians, committed to ethics of individual-centered medicine, and that of public health experts and administrators, who reason in terms of the cost/efficacy of health measures (Boltanski and Thévenot 1991).

It is possible to argue that, from the latter point of view, countries that generalized screening for Down's risk did not make a cost-efficient choice. The 'prevention' of the birth of several hundreds of Down's syndrome children per year in France or in the UK is unlikely to greatly alleviate the expense of caring for children with severe inborn defects. On the other hand, many experts are probably sincerely persuaded that, when the burden for families and society is very high, even a limited improvement is much better than inaction. During the interwar era, scientists who enthusiastically promoted eugenics were often aware of the fact that the sterilization of the 'mentally defective' would only have a limited effect on the number of people with mental handicaps, because most hereditary defects are recessive and are transmitted by 'normal' carriers. They believed, however, that preventing even a small number of 'undesirable' births was, in and of itself, a worthy goal since it contributed to a reduction in human suffering (Paul and Spencer 1995).

Advocates of eugenic measures argued that selective birth control would diminish the misery of the poor as well as the state's expenses. Similar mixed motives are advanced by promoters of PND, who explain that this approach reduces individual suffering and the burden on the public purse (Paul 1995: 134–5). The latter goal is rarely mentioned in the late twentieth and early twenty-first century. Nevertheless, it was one of the main driving forces in the generalization of screening for Down's.[20] The history of PND displays the fundamental ambivalence of an approach grounded in the expectation that the sum of individual decisions will lead to a reduction in the burden of disability in populations. Moreover, scientists, doctors, and health administrators who hail the benefits of PND focus on its positive side—reassuring pregnant women—and seldom dwell on the individual and collective costs of this diagnostic technique.

Whose values? Whose choices?

Discussions of the social effects of DNP have as a rule focused on women's autonomy and their right to 'informed consent'; they have rarely taken into account the role of upstream decisions that affect 'real life' reproductive decisions. The administrative, professional, legal, and economic variables that led to the generalization of specific prenatal diagnostic tests and the exclusion of others and that made specific interventions available to pregnant women and provided agreed upon interpretations of these interventions were, by contrast, rarely debated outside the narrow circle of experts.

The termination of a wanted pregnancy is frequently a traumatic experience.[21] Professionals insist, nevertheless, that this decision is always the 'woman's choice'. The insistence on the key role of 'women's choice' in PND is not merely a rhetoric device. Pregnant women indeed have real choices and exercise them (Wertz and Fletcher 1993).[22] Obstetricians, paediatricians, and foetal medicine experts frequently explain that, since future mothers/parents will have to live with the disabled children, only they can decide if they can accept the children. Pregnant women who receive a diagnosis of foetal malformation also stress that they alone, and not health professionals, should decide if they will continue the pregnancy (Statham 2002). On the other hand, the 'choice' discourse often masks the paucity of real life options and actors' limited control over situations (Lippman 1999; Samerski 2009; Sandelowski and Barroso 2005). The scope of women's decision is shaped by material,

[20] These considerations are valid only in countries where abortion is legal. In countries where abortion is illegal, the management of 'risk of disability' remains in the private—and privatized—sphere, and eliminating the risk of foetal malformation is the privilege of the affluent.

[21] The termination of an unwanted pregnancy is frequently perceived as 'reparation' of an accident produced by the absence or failure of contraception. In these cases, women do not 'recognize' the foetus as their future child. In contrast, when a woman decides to terminate a wanted pregnancy because of a non-lethal foetal malformation, she sees it as the elimination of a child, even if the abortion takes place in the early stages of pregnancy (Boltanski 2004; Statham, 2002).

[22] Miana Meskus (2012) investigated the gradual disappearance of the visibility of PND-related decisions as a collective and communal issue and their transformation into an exclusively personal and private question.

institutional, and legal constraints: access to specific tests, to information about these tests, their cost, their interpretation. It is also strongly dependent on the country and region where a woman lives: Holland discourages intensive prenatal testing, while France strongly encourages it. Accordingly, pregnant women—with the exception of those who are opposed to abortion, especially on religious grounds—are expected to make the decisions seen as rational in a given society (Ivry 2009a).

The routine use of PNS and PND transformed pregnant women into 'responsible managers' of risks for their future children (Samerski 2009). The expression 'management of fetal risks' refers to the obligation for a woman who undergoes a screening for Down's syndrome and other aneuploidies to understand correctly probabilistic data provided by a combination of serum test(s) and ultrasound data, evaluate the dangers she is facing—raising a disabled child versus losing a healthy foetus—and then decide whether she is willing to undergo amniocentesis (Schwennesen et al. 2010). Some women reject this logic. From 2006 onwards, a higher proportion of US women referred for screening for Down's risk because of their age elected to undergo amniocentesis directly without prior assessment of their risk by other methods. Their choice is attributed to their lack of confidence in probabilistic evaluations: they do not see a 'good' risk number as sufficient to be reassured (Nakata et al. 2010). On the other hand, many women believe that a 'good' amniocentesis result (that is, the absence of chromosomal anomalies) attests that the foetus they carry is 'all right'.[23] Women more familiar with PND know, however, that the reassurance provided by direct examination of foetal karyotype is very limited. When anthropologist Faye Ginsburg was told that the results of her amniocentesis were normal, she told her colleague and friend Rayna Rapp that she then knew that the foetus was free of four anomalies among the hundreds that can occur (Rapp and Ginsbourg 1999).[24]

Women who do learn that their foetus has trisomy 21 must rapidly decide if they wish to continue the pregnancy or terminate it, and the great majority choose the latter option (P. A. Boyd et al. 2008). The high rate of abortion for Down's syndrome may be perceived as a paradoxical result in an era with a growing disability movement and increasing public support—on the rhetorical level at least—for people with physical and intellectual disabilities (Ville 2011). The paradox may, however, be apparent only. Parents of disabled children who became activists in the disability movement, often insist on the importance of high levels of parental (especially maternal) investment in the

[23] Writer Peggy Orenstein may illustrate this trend. When she became pregnant, after years of stressful fertility treatments, she did not accept this pregnancy as a fact before receiving the results of chorionic villus sampling (CVS) (Orenstein 2008: 196–212).

[24] Faye Ginsburg's daughter was later found to suffer from a rare hereditary disorder, familiar dysautonomia.

education and training of special needs children. This investment, activists stress, is indispensable to allow disabled children to reach their full potential.

The visibility of the disability movement and uplifting stories in the media about exceptional families who raise unique children might have increased the apprehension among pregnant women that they and their (unexceptional) families will be unable to provide intensive care and unfailing support to their disabled children, and will be forced to cope with the guilt of failing the children.[25] This apprehension may be coupled with a realistic assessment of the amount of support that will be provided by society, the level of energy mothers/parents often need to invest to secure access for their disabled children to the institutional help to which they are entitled, the difficulty of maintaining this help over time, and fears about the future of disabled children, teens, and adults in economically uncertain times. A pregnant woman's decision to terminate her pregnancy when she learns about a congenital malformation of the foetus may be motivated not only by the 'dream of a perfect child' (Rothschild 2005) but sometimes also by 'a perfect mother's nightmare': fear of the high cost of providing the right kind of care to this child, then to a disabled and dependent adult.[26]

The introduction of a new medical technology, anthropologist Marilyn Strathern had observed, often radically modifies the choices open to all people, including to those who reject it (Strathern 1992). The generalization of PND in numerous countries forces nearly all pregnant women who live in these countries to face complicated choices. Even women who, in principle, should be able escape the dilemmas triggered by screening for foetal anomalies, such as religious women who are expected to reject medical technologies that predict the future of their child and trust in God's providence, may suffer from stress produced by PND. Religious women who refuse serum tests for Down's syndrome risk still undergo ultrasound examinations that can detect numerous foetal anomalies, including a 'typical image' of Down's syndrome foetuses. Moreover, the existence of PND may put additional pressure on religious women who refuse this technique but at the same time are terrorized at the thought that their faith is not strong enough to protect them from the fear of giving birth to a disabled child (Ivry et al. 2011).

Technologies are initially shaped by the values and preferences of people who develop them. Later, however, they can be modified by their users. Users of the telephone, the French minitel, the microwave oven, and, in medicine, AIDS activists who criticized clinical trials of anti-retroviral drugs successfully

[25] The debate on 'bad parenting' does not as a rule, one may note, include discussions of 'bad parenting' of disabled children, a topic that may be viewed as too difficult to tackle.

[26] Akin's argument that women's apprehension that they will not be able to be 'perfect mothers', and its consequence, lower tolerance for unplanned pregnancy, were proposed to explain the persistence of a high proportion of abortions in France despite the widespread diffusion of contraception (Bajos and Ferrand 2011).

challenged the 'technological codes' of new technologies (Epstein 1996; Feenberg 2002; MacKenzie and Wajcman 1999). The intervention of AIDS activists was, however, an exception rather than the rule. In the majority of cases, patients, health-care users and activists have modest input, if any, into the shaping of medical technologies. Their main contribution is frequently to accept or reject an innovation, either through organized pressure from users, or, more often, by voting with their feet (Löwy 2000).

PND, this text argues, conforms to this rule. This technology was developed by health professionals. In spite of the omnipresent rhetoric of choice, pregnant women had little input into its content (which tests are used and how they are calibrated) or availability (which tests are reimbursed by national health systems and major health insurance providers). The discourse of choice often masks the range of constraints on people's individual choices in a state or market system (Kerr and Cunningham-Burley 2000). Affluent and well-informed women can purchase tests unavailable where they live and/or pay for tests that are not reimbursed by health insurance, but this option is open to only a small minority of women. Similarly, women can refuse prenatal testing, but in countries where PND has become part of the routine supervision of pregnancy, only a small proportion of women reject health professionals' offer to check if the child they carry is 'all right' (Schwennesen et al. 2010; Vassy 2006).

Public debates on new developments in human reproduction in the late twentieth and early twenty-first century tend to focus on rare techniques and on developments that might become possible in the future, and seldom discuss the messiness and complexity of decisions prompted by widely diffused routine techniques and approaches (Kerr 2009). The high visibility of innovations used by only a handful of women, such as pre-implantation diagnosis (e.g. Franklin and Roberts 2006), contrasts with the low visibility of innovations that have radically modified routine management of 'normal' pregnancies, such as the increase in the resolution of diagnostic ultrasound images in the late 1970s, the introduction of serum tests in the 1980s, or the crafting of software that provides an individualized 'Down's risk' number in the 1990s. If this trend continues, the next innovation in the PND pipeline, NIPT, will probably also be produced, calibrated, diffused, and regulated without any significant 'upstream' input from the values of pregnant women.

ACKNOWLEDGEMENTS

I am grateful to all the participants of the workshop 'Moral Economy of Science' for their important input, to an anonymous reviewer, and editors of this volume, Isabelle Dussauge, Claes-Fredrik Helgesson, and Francis Lee, for their help in shaping the final version of this text, an anonymous reviewer, and to my colleagues from the programme '*Les enjeux du diagnostic prénatal dans la prevention des handicaps: l'usage des techniques entre progrès scientifiques et action publique*', who continue to shape my thinking on these issues. All remaining errors are my own. This study was supported by the Agence Nationale de la Recherche (ANR) programme '*Sciences, technologies et savoirs en société*' (ANR-09-SSOC-026-01).

Part IV

Valuations and Knowledge

11 Purity and interest

On relational work and epistemic value in the biomedical sciences

Francis Lee

Boundaries, interests, and modes of purification, a pragmatic take

Actors in biomedicine constantly produce different versions of valuable science and knowledge. Certain tropes of value, such as 'science for innovation', 'translational science for clinical utility', or 'basic science for biological understanding' are endlessly being made, remade, coordinated, and hierarchized. We are all too familiar with these different stories from political discourse, social analyses, and the news media (cf. Shapin 2008). The biosciences are simultaneously assembled: 'in articulation with neoliberal, entrepreneurial modes of participation' (Hayden 2003: 29) and other tropes of value such as Merton's CUDOS norms (1973a).[1]

The problems with many contemporary approaches that purport to analyse these tropes are that they tend to contribute to reproducing and performing them. For example by calling for specific versions of entrepreneurship and innovation, as in the large literature on the Triple Helix model (Etzkowitz and Leydesdorff 2000), or alternatively decrying the corruption of the biosciences, as in the literature on Biocapital (Sunder Rajan 2006).[2] The drawback of this situation is that these analytical approaches often presuppose and reify a specific ontology of science and industry, clouding the work that goes into

[1] In Merton (1973a) the norms are universalism, communism, disinterestedness, organized scepticism.

[2] These questions should be understood in light of recent academic debates where the biosciences are argued to be shaped by economic interests. Patents and industry–science collaborations are argued to be changing the research landscape in profound manners; some even argue that there is a shift in the political economy of the biosciences (cf. Mirowski 2004, 2011). Notions such as biocapital (Rose 2007; Sunder Rajan 2006, 2012; Yoxen 1981), bio-value (Mitchell and Waldby 2010; Waldby 2002), and bio-piracy (Shiva 1997) have become increasingly utilized for describing this intermingling of the biological, the scientific, the medical, and the economic. It has been argued that there is a shift in the scientific apparatus of knowledge production and a redrawing of boundaries in (and between) science and business (cf. Widmalm 2007: 120). However, as other researchers have shown, these boundaries are not clear-cut, but are complexly intertwined (Shapin 2008).

making these entities in practice. That is: in approaches calling for 'science for innovation', science and industry are seen as separate pre-existing entities that are found in a System of Innovation (Lundvall 1992) or in relation to a Triple Helix of state, university, and business (Etzkowitz and Leydesdorff 2000). As a consequence, these approaches perform science, industry, and the value of them in particular manners, stressing how science will produce knowledge that contributes to business innovation and thus national economic growth.

Alas, the same argument is true for many critical approaches analysing the science/industry relation. For example, the literature on biocapital presumes that capital is a dominant force in the contemporary biosciences, and that capital has an (often corrupting) influence on scientific knowledge production. Biocapitalist critique thus decries leaks between two separate entities, reifying an ontological separation of science and industry where one (business) corrupts the other (science).

In both these approaches the values of science and knowledge are performed in completely different manners, albeit using the same ontological cut: ontologically separate, pre-existing entities that interact, even if their relations are seen as beneficial in one case and detrimental in the other. Like economic theory (Callon 1998), the tropes that these approaches purport to analyse are also articulated by them.

A possible inroads to rectify this analytical impasse could be to approach these tropes on science and industry by attending to the rhetorical production of the boundaries between them (Gieryn 1983, 1995). The question is then shifted from the reifying and normative question of 'how pre-existing, and ontologically separate entities *should* interact' to the analytical question 'how these entities come to be seen as separate in the first place'. A powerful example of this is Gieryn's introduction of boundary work, which calls for analysing how the boundaries of science are made in practice. The underlying assumption of this approach is that games of social interests, power, and authority explain how and why boundaries are erected. The boundaries around science are seen as performed, permeable, and contingent, but the explanation for them is sought in a presupposed power game, trying to explain 'uneven distributions of authority, power, control, and material resources' (Gieryn 1995: 441).[3] The pre-existing ontologies of the normative approaches such as triple helix or biocapital are here dismantled in favour of a powerfully contingent and performative account of the boundaries of science.

However, as is well known, the reliance on interests as an explanation has been widely criticized. For example, Susan Leigh Star (1991) has pointed out that this type of explanation assumes a Machiavellian understanding of scientists, where science is but an arena for great scientists to further their

[3] Another similar example is Callon's work to explain science as a strategic game of enrolment and mobilization, where actors attempt to align other actors' interests in order to serve their academic goals (Callon 1986).

own career interests. Another critique, developed by Steve Woolgar (1981), argues that interests become black-boxed in approaches using interests as explanatory strategy, leaving unsolved important parts of the work to understand science in action. Woolgar argues that a fruitful technique would be to utilize an ethnomethodologically inspired approach, and attend to the *making of interests* in practice. He argues that interests should become a topic for inquiry, rather than a resource for explanation. This strategy has been put into practice by Cori Hayden (2003) in her analysis of bioprospecting in Mexico, where she follows the making of interests: in research agreements, in intellectual property rights, in indigenous rights movements, etc. Rather than seeing interests as that which explains, Hayden investigates the making of multiplicities of interests in bioprospecting. Interests become something that needs to be investigated and understood.

A performative approach to both science/industry relations would shift the analytical searchlight from boundary work as a result of interests, to highlighting the making of both boundaries *and* interests. This would mean that we would avoid reifying interests as explaining action, and instead taking seriously the various and shifting 'interests' attributed to science in practice. This would sidestep the fundamental analytical problem of conflating analysis and critique of biomedical tropes: where does the empirical trope of corruption of science begin and where does a biocapitalist critique end? Thus, rather than contributing to the 'normative surfeit' (Zuiderent-Jerak, in press) often associated with analysing science/industry relations, this approach would be able to produce descriptions of the multiplicities of shifting boundaries and interests in science.

Rather than presupposing an ontological boundary between science and industry, and rather than presupposing social interests as the explanation, this chapter proposes to, in a pragmatic vein, pay attention to the actors' parallel construction of *interests* and *science/industry relations.* Similarly to Viviana Zelizer's (2012) recent work in economic sociology, this approach to analysing the science/industry relation would allow for a description of how actors engage in *relational work* to distinguish meaningful, valuable, and appropriate relations in the biosciences without recourse to pre-stabilized tropes on what science or industry 'are' and what Machiavellian scientists 'want'. An important consequence of this move would be the acknowledgement of the ontological multiplicity and complexity of science and industry: the multiplicity of tropes, the multiplicity of interests in knowledge, and the multiplicity of organizational forms. Thus, just like Annemarie Mol (2002) has shown disease to be ontologically multiple, science and industry, with concomitant boundaries, relations, and interests, are here argued to be performed in practice, and therefore to be ontologically multiple and shifting. Thereby actors' making of science/industry relations can coincide, clash, be hierarchized, and calibrated in different situations.

What is at stake in the making of interests and science/industry relations is nothing less than the making of the conditions of possibility of the biomedical sciences. The multiplicity of tropes, objects, relations, and boundaries that are produced around the science/industry nexus shape what it is possible, valuable, and desirable to do in the biosciences. The argument is that the making of these perform what is seen as productive at the lab bench, what is seen as valuable knowledge, how ownership and organization should be arranged, and which methods are seen as fruitful. Furthermore, the yardsticks—the ways of assessing value—for determining all of these things are at stake. What is valuable, and how do we determine this? Fundamentally, the question at stake is the negotiation, delineation, and coordination of several versions of 'Good Science'.[4]

This chapter introduces and highlights a specific type of relational work in the science/industry nexus: that of *purification*. This concept emphasizes how actors in their relational work constitute science and industry as separate ontological entities in specific ways. The argument is that just as nature and society are purified (Latour 1999a), science and industry are produced as ontological entities through purification: assembling certain configurations of interests, boundaries, and values. In doing this the chapter identifies *two modes of purification* that are used to separate science and industry: first, *a temporal mode* of purification; and second, an *organizational* mode of purification. This suggests a multiplicity of strategies to fashion different versions of science and industry as well as a multiplicity of possible relations. To elucidate these themes, this chapter enquires into two evaluations of a research project and how they construct different interests in the project, which includes different boundaries between science and industry, different views on what is valuable knowledge, as well as different yardsticks for the laboratory bench.

Beginnings: A commercial database to find potential protein targets

The case used to explore these themes is the peer review and evaluation of a large bioscientific research project, the Human Protein Atlas (HPA).[5] The

[4] This argument draws on the theoretical work of Laurent Thévenot and Luc Boltanski, who have worked extensively on justifications and valuations (Boltanski and Thévenot 2006; Thévenot 2007). However, the orders of worth that they produce are much too static to be able to capture the multiplicity of relations that are produced in the science/industry nexus.

[5] This inquiry started as part of a larger research project on the links between research and research policy in Sweden from 1960 onwards. It began as an open-ended inquiry into research of the political process and led to the establishment of the HPA project. The chapter is based on interviews and

HPA is a protein mapping project that today aims to chart all the genetically coded proteins in the human body to execute a 'large-scale characterization of potential protein targets' that can be 'used to understand disease and develop new and more efficient drugs'.[6] That is, the project aims at finding potential protein-based targets for identifying, diagnosing, or treating disease, with special focus on cancerous diseases.

The project is to result in a map consisting of annotated images of protein locations in different tissue types (see Figure 11.1). These images are argued to be useful starting points for the development of medical diagnosis and intervention. Echoing the familiar innovation trope in the biosciences, the idea for the HPA project was articulated at the height of the genomic boom in the end of the 1990s and was tied to a commonly occurring dream of creating a proteomic 'goldmine' through pharmaceutical development (cf. Ezzell 2002; Service 2001a, 2001b, 2001c).

The beginning of this story takes place in a Swedish company called Affibody, which was founded in 1998 as a gamble on the medical and commercial success of certain synthetic molecules, Affibodies (interviews, Fredrik Pontén and Stefan Ståhl).[7] The company had two main research trajectories: to develop artificial antibody molecules that could be used to identify proteins *in vivo* (interview, Fredrik Pontén); and to produce in the style of Craig Venter's Celera Genomics, a database and map of the proteins in the human body and offer it for commercial subscription (interview, Mathias Uhlén). The dream was to take the next step after the mapping of the genome: to go from mapping the genes to mapping the proteins (interviews, Mathias Uhlén and an anonymous researcher).

The feasibility of mapping proteins in human tissue was first explored in a preliminary study undertaken by Affibody. The study started in March 2001, and used antibodies to map 168 of the 225 proteins of chromosome 21. The preliminary study was deemed successful and was reported in a publication

analysis of documents around the HPA. The conflicts around the valuation of research, and methods that emerged as themes in the very first interview, I followed up in subsequent interviews. I conducted twenty interviews with (1) people who were affiliated with the Human Protein Atlas; (2) people who had insight into the funding process leading up to the project; (3) researchers who had utilized the atlas in different ways; as well as (4) researchers with insight into proteomics or antibody methods. The external informants had affiliations ranging from research foundations to international research organizations. The insiders (and to some extent the outsiders) were chosen based on being mentioned/recommended in previous interviews. The 'outsiders' were identified using interviews, contacts, and publications. All informants were offered anonymity: most declined, while some of the critical informants were adamant about being treated anonymously.

[6] Affibody Annual Report 2001 for the first quote, and KAW Annual Report 2006 (Knut och Alice Wallenbergs Stiftelse 2007) for the second.

[7] Affibody was named for an artificial and patented protein, an Affibody molecule, which is produced from a specific domain of 'Protein A'. Affibodies can be described as artificially constructed antibodies or so called antibody mimetics which can bind to any protein.

(Agaton et al. 2003). For Affibody, and also later the fully realized project, the interest in what was called the proteome was articulated as a commercial potential that would be realized by selling subscriptions to access a database of proteins:

> The platform is currently used to generate Affibodies™ and antibodies for large-scale characterization of *potential protein targets.* (Affibody 2001: 15)[8]

> If we could develop 22000 Affibodies, one for each human protein, it would have been a goldmine. (Interview, Stefan Ståhl)

In the quote above 'potential protein targets' alludes to finding targets for medical intervention; that is, proteins that can be targeted in medical diagnosis and treatment. The articulated dream was to find the seeds for the next blockbuster drug. One might say that the interest was in creating a treasure map that showed the location of proteomic gold.

However, in 2001, in the wake of the dotcom crash—according to project mythology—Affibody's board of directors decided to discontinue the study to focus on other ventures. The decision made it necessary for the project team to find other financial sources. This led the main scientist–entrepreneur, Mathias Uhlén, to pursue a project proposal with John Bell at Oxford University, whom Uhlén knew from a previous project. The proposal included moving the project from Affibody to Oxford with financing from the Wellcome Trust, and would entail a move from the commercial setting in which the project was born, to an academic setting at Oxford University.

Temporal purification: Or the linear model in action

In the spring of 2002 discussions between the Wellcome Trust and Uhlén got under way, and negotiations about what was to become the HPA project began in earnest. A preliminary budget of about £100 million was outlined. A time plan for a 10-year Oxford-based project was drawn up (interview, Mathias Uhlén). During the spring of 2002, negotiations were held between the Wellcome Trust, the Swedish research foundation *Knut och Alice Wallenbergs Stiftelse* (KAW)—who wanted to keep Uhlén in Sweden—and representatives from the international pharmaceutical industry (notably Astra Zeneca).

As the negotiations progressed, the Wellcome Trust set up a peer-review process that involved a large number of scientists (12–20; sources vary) in a review of the proposal. According to Michael Morgan, who was in charge of

[8] My translation. All quotes from Swedish actors are henceforth translated by the author.

the negotiations at the Wellcome Trust, most of the reviewers were critical of the project. The reviewers, and the Wellcome Trust, argued that the value of the protein map, and the reason for doing the project, was to provide a resource to the academic community, rather than seeing the project as a route to a proteomic goldmine. The treasure map, intended to find proteomic gold, changed characteristics to become an atlas that was to be used as a scientific resource. As Michael Morgan recalls the review process:

> It became clear that there were serious concern[s] being expressed by at least some of the scientific community about the feasibility of the proposals and the value to the scientific community. (Interview, Michael Morgan)

As an echo of Merton's CUDOS norms (1973a), the value of the project was seen as dependent on its usefulness to the academic community. Rather than commercializing the resulting data and tools, which was the original plan in Affibody, interest in the project had shifted to produce data and tools that could be used widely and freely. The importance of the wide-ranging usefulness of the project—not just for the economic benefits of the research group or its investors—was stressed. The importance of free distribution of tools and reagents and the free release of data was also emphasized:

> Yeah, because this being a resource we wanted to make sure that there was no proprietary tools being involved, or reagents being involved, that we would not be able to distribute freely to the scientific community. (Interview, Michael Morgan)

The project was seen as producing a *free resource* for the scientific community. This stance on ownership was also reflected in a worry about leakages between the proposed project and the commercial activities of the Uhlén research group. As a UK charity, the Wellcome Trust was concerned with the link between the proposed research project and private enterprise:

> There were issues of, around, the involvement that Mathias [Uhlén] had with... [Affibody]... And I think there were some concern[s] being expressed by our legal colleagues as to what exactly was the relationship. Because as [a] charity... in the UK, charities are not allowed to use their money to help support private enterprise. And so the legal team would have been very concerned to ensure that there was no possible leakage between, how should I put it, Mathias' academic activities and his innovative industrial activities. (Interview, Michael Morgan)

For the Wellcome Trust it was of utmost importance that they were supporting science rather than private enterprise. Legally, a clear demarcation was drawn between acceptable and unacceptable support. As the Wellcome Trust saw it, the purely academic nature of the project—i.e. ownership of data and tools, as well as the relations to Affibody—were vital for being able to finance the project. The project needed to be legally and organizationally cleansed from its ties to the commercial sphere for it to become acceptable.

However, the dichotomy between the academic project and the commercial sphere drawn up by the Wellcome Trust legal team was more complex in the actual negotiations, as the pharmaceutical industry participated in the negotiations with the express purpose to be able to rapidly capitalize on the knowledge that was generated by the project (meeting-minutes, OE). The interest generated in producing pure knowledge for the scientific community was combined with an interest in making it possible for the pharmaceutical industry to utilize knowledge after the facts had been discovered.

The project was consequently carried out as a purified scientific activity where science and business were to be separated in the quest for creating an atlas of the proteome. However, as the pharmaceutical industry was invited to partake in the project, it was not a complete separation, but rather a *temporal purification*. The imagined project was enacted using the common 'linear model of science' (cf. Godin 2006) where the academy produces knowledge, and industry becomes involved in making knowledge commercially useful. Basic science was seen as an initial stage and industrial development, it was argued, would evolve later. The linear model of science was rhetorically, legally, and organizationally enacted as the model of science.

However, eventually, and despite efforts to adapt the project to the linear model—to purify it—and despite efforts to defend the project against critics, the Wellcome Trust decided to decline to fund the HPA project. The number of interpretations for why this actually happened increased with the number of interpreters.

The goldmine of proteins returns—a hybrid science

The Wellcome Trust, however, was not the only contender for financing Uhlén's Oxford project. As indicated above, the large private Swedish research foundation, KAW, had at the outset become involved in discussions on financing the project. The KAW foundation was established by, and still is tightly connected with, the Swedish industrialist family Wallenberg. The loosely formulated goal of the foundation is to fund research that would be 'of benefit for Sweden'. The Wallenbergs can be described as *the* industrialist family in Sweden and has (over more than a century) had a key role in Swedish business, politics, and science. The board of KAW consists of researchers as well as a number of members of the Wallenberg family and business empire.

Erna Möller, the immunologist who was the executive board member of KAW at the time, explains how they saw a partially different set of interests as the reason for mapping the human proteome. As in the Affibody preliminary study, the *raison d'être* of the project was to screen for medically interesting proteins that could be used for identifying diagnostic biomarkers for cancer:

> The important thing is if they [the Protein Atlas] identify a protein that is typical for something... And perhaps later it becomes obvious that it is something incredibly interesting that only exists in one cell, or during a certain part of the cells' developmental stage, or perhaps in a specific form of tumour. And perhaps not in all tumour forms, but some that only exist in the most malign? These could become really important things. (Interview, Erna Möller)

Here, as in the Affibody preliminary study, interesting protein structures were seen as being biomarkers for cancer treatment and diagnosis. As in the Wellcome Trust case, these arguments were also accompanied by a specific articulation of ownership and intellectual property rights. However, for KAW, the linear view of science that the Wellcome Trust articulated seemed to miss the point. For KAW it was seen as a positive development that companies were founded on research output. A valuable project pinpointed interesting and valuable protein property, and founded biomedical and biotechnological companies. The atlas shifted back to a treasure map. Möller explains the stance:

> Wellcome followed the principle that everything should be free, and we accepted that. But we thought it was a shame, as the researchers had the possibility to develop and found their own companies.
>
> During the course of the investigation, when you find a new exciting antibody, then you say that this is something that shouldn't be sent out immediately, but it should be looked over. Should it be stowed away? Patented? (Interview, Erna Möller)

KAW argued that the development of patents and companies on scientific results was the model for pursuing the HPA project. Thus, Möller and KAW argued for a different version of the project from the linear model proposed by the Wellcome Trust. For KAW, just as for Affibody, the value of science was tied to the treatment or diagnosis of cancer and the establishment of commercial actors. The goal was market-oriented interventions in the biomedical realm. The gauge for a good project was joint medical and economic development. In performing this version of science, the actors made bioprospecting for proteomic gold the yardstick for science.

Yardsticks in the lab: Interests, epistemic value, and methodological assessments

Let us now attend to how the performance of interests played out in the valuation of knowledge. How 'value for the scientific community' or finding 'medically interesting proteins' entered into the assessment of knowledge and yardsticks for the proposed methodologies. As is customary in any scientific evaluation, peer-review processes were put in place by the two research foundations involved. The evaluation of the proposed methodology revolved around the usefulness of different types of antibody molecules, the criteria for usefulness, and how appropriate they were for achieving different articulated interests. What I highlight here are not the inherent characteristics of different types of methodologies, but rather the making of interests, measures, and values in the review process and how they pertain to the valuation of lab work. The focus is on how the relational work of pursuing interests in the science/ industry nexus intertwines with the values of the lab bench.

Let us first consider the Wellcome Trust and their articulated interest in producing a protein map 'for the scientific community' and how this played out in evaluating the lab work of the proposed project. As is commonly the case, the reviewers' critique was not solely focused on organizational forms or on the ownership of data or tools. The map was deemed interesting as a resource for unknown ventures. Thus, the review and evaluation of the project were predicated on an understanding that *the entire map* was potentially of scientific interest. Deciding on which proteins (and parts of the map) were valuable was left unarticulated in the review process—the interest was left for future researchers to decide. It was the atlas view of the project that carried the day. Further, the idea that the project should become a widely used resource was tied to specific yardsticks for evaluating the proposed methodologies. These arguments point us towards the assessment of different types of antibodies—which were the tools of the project. Michael Morgan of the Wellcome Trust again:

> First of all there was a question as to whether or not the approach was the most appropriate one. I remember there was discussion about monoclonal antibodies [produced in cell-cultures] versus antibody being raised in rabbit... And the question about monoclonals of course is that it becomes a *permanent source of material*, whereas antibody raised in rabbit is sort of *a one-off exercise*. (Interview, Michael Morgan)

The tension highlighted here was between the wish for a 'permanent source of material' versus doing a 'one-off exercise'. The argument from the Wellcome Trust reviewers was that interest in the project hinged on the production of identical batches of antibodies that could be used within the scientific community as a permanent resource for further research. This articulation of the project tied into a long understanding of so-called monoclonal antibodies, which can be produced in laboratory cell-cultures and thus produce an eternal supply of identical antibodies, as well as to the common view that monoclonal antibodies make it possible to ascertain, through repeated experiments, that they do what they should. They are then said to be specific or to bind specifically. As one of the informants, Hans Wigzell—who is also the former Vice-Chancellor of the Karolinska Institute and one of the founders of Affibody—expressed it in the 1984 Nobel Prize presentation speech for monoclonal antibodies:

> Köhler's, and Milstein's development of the hybridoma technique for production of monoclonal antibodies have in less than a decade revolutionized the use of antibodies in health care and research. Rare antibodies with a tailor-made-like fit for a given structure can now be made in large quantities. The hybridoma cells can be stored in tissue banks and the very same monoclonal antibody can be used all over the world with a guarantee for eternal supply. (Wigzell 2012)

The value attributed to monoclonal antibodies in the biosciences was tied to the possibility of producing an 'eternal supply' of identical antibodies which allowed an experimental replicability, and a possibility for standardization and packaging. Monoclonals were supposed to 'revolutionize' the 'tinkering' with antibodies into a tool 'that could yield standardized, reproducible results' (Cambrosio and Keating 1992: 369).[9] Thus, the value of the antibodies in the Wellcome Trust's review process was tied to an eternal supply of monoclonal antibodies. Furthermore, as many of the proteins in the body are unknown, that fact that the project was to produce monoclonals for all genomically coded proteins in the body was articulated as of immense value to laboratory research: both as a location map of proteins in tissue, and as an eternal source of tools (so-called reagents) for further lab work. The map, from a value perspective, was seen as a homogeneous entity, an atlas of the proteins. The map, the unknown proteins on it, and the antibodies were given homogeneous epistemic value.

As I have shown above, KAW's reviewers, on the other hand, tied the project to a completely different set of interests: that of rapidly screening for interesting proteins. The project was to do a first pass through the proteome in order to identify proteomic real estate for patenting, and only go into depth for certain interesting proteins. The last step would then be to execute the costly process of developing monoclonals. Recalling Möller's words:

> I was completely floored by it being possible... This [project] is insane! If this [methodology] works it is a hundred times faster and better than... making monoclonals en masse. And if you have an antibody and know what its target is: it's as easy as pie to take the protein and make a monoclonal. But then you only do it for the maybe one per cent of all proteins that are interesting (Interview, Erna Möller).

[9] The use of polyclonal antibodies was sometimes described as bordering on an uncertain and unscientific 'black art practiced by immunologists'. According to Cambrosio and Keating this division between monoclonal and polyclonal antibodies echoed a division of immunologists into 'those who believed in immuno-chemistry and those who believed in "immunomagic"' (Cambrosio and Keating 1995: 74).

Thus, in KAW's assessment of the project, the proteins were given a heterogeneous epistemic value. Interesting parts of the proteome were ascribed higher epistemic (as well as medical and economic) value. The enactment of this particular interest in the project made other yardsticks for assessing antibodies salient. John Bell who was supposed to host the HPA project in Oxford, and who championed the project at the Wellcome Trust recalls:

> [Mathias Uhlén] had enough data to suggest that you could get a *pretty good* monospecific reagent out of more than half of the genes you look at, *which was enough*, and that's from the *first pass*.
>
> But, the trouble is that if you try to think in high-throughput terms—that's the way you think—you'll immediately not get there because [monoclonal antibodies are] too *cumbersome*, it's too *slow*, it's a hell of a lot of *screening* you've got to do well. This [polyclonal] methodology is much, much more powerful. (Interview, John Bell)

The idea of doing a high throughput, 'pretty good', first pass of the proteome in order to discover medically interesting proteins was tied to a completely different epistemic valuation involving completely different yardsticks for the assessment of the antibodies. In the review process, the project's production methods became articulated as speedy, easy, and efficient. KAW and others argued that it would be inefficient and costly to produce monoclonals for all of the proteome, and that it would be better to produce them for the economically and medically interesting parts of the proteome. *The project was on a treasure hunt, not a topographical survey.* Briefly, KAW and the project team contended that the methods made it possible for the project to quickly and efficiently produce polyclonal antibodies that in the next step could be used to identify proteins in specific tissues—to find the coveted location of interesting proteins in the human proteome. A good map was a treasure map, and a good antibody was no longer articulated as being replicable and eternal—rather it needed to be simple, fast, and cheap.

Thus, the differing interests attributed to the project were tightly connected to the articulation of both epistemic value and methodological yardsticks. In sum, the arguments were: 'a map to be used as a resource for further scientific discovery needs to be replicable and specific' vs 'a map for bioprospecting needs to be efficient in identifying valuable real estate'. The yardsticks deployed to assess monoclonal antibodies, eternal replicability for the scientific community, was contrasted with antibodies that were sufficiently accurate, cheap, and fast for bioprospecting for proteomic gold.

Performativity and purification: Redrawing the boundaries of science and industry

In the fully funded and running project, KAW's and the researchers' articulation of speed, ease, and efficiency was highly performative. For example, the production of antibodies was outsourced to a factory in China to get the speed up and the production costs down:

> I had suggested to Mathias [Uhlén] that he needed to look at very high throughput methodologies for producing antibodies and he talked about this centre . . . in China as a way of really producing polyclonals *very rapidly*. So, since then he started by making antibodies in Scandinavia and I said 'No, you're never going to get there.' And he's now developed I think very good collaborations with the Chinese. That's really *got the price down* on making the antibodies. (Interview, John Bell)

Furthermore, the annotation of the images resulting from the atlas (see Figure 11.1) was outsourced to Indian pathologists. The argument for this was that foreign organizations were cheaper and more motivated to perform the type of monotonous work necessary for producing high-throughput analysis.

Moreover, the interest in commercializing the intellectual fruits of the HPA mapping project was performative in the complex ownership relations developed for executing the project (see Figure 11.2). The patents and intellectual property that the university-based HPA project generated were transferred to a holding company, *Atlasab Intressenter*. The IP holding company in turn owned a stake (32 per cent) of Atlas Antibodies, which was founded to commercialize the IP generated by the project. Atlas Antibodies was also partly owned (32 per cent) by a research foundation (controlled by the HPA-researchers) which was established to finance research at the participating universities as well as two university holding companies (6 per cent). Two venture capital companies held the last stake (30 per cent) of Atlas Antibodies; the Wallenberg family controlled one of the venture capital companies, Investor Growth Capital.

> The plan was that the results should be freely available. Now this didn't happen as they [the group] patented some interesting finds. And that was good. Because this is a possibility for Atlas Antibodies to sell. (Interview, Erna Möller)

Nevertheless it is important to underline that KAW did not view science as being the same thing as business. The separation between science and business was rhetorically and practically upheld. For KAW the complex organization of investment and ownership led to an undesired ambiguity, where the relations between science and industry needed to be clarified. Here, the connection between the KAW and the Wallenbergs' industrialist legacy was important in that the separation between the foundation and the venture capital arm of the Wallenberg family empire, was stressed. Just as for the Wellcome Trust, the

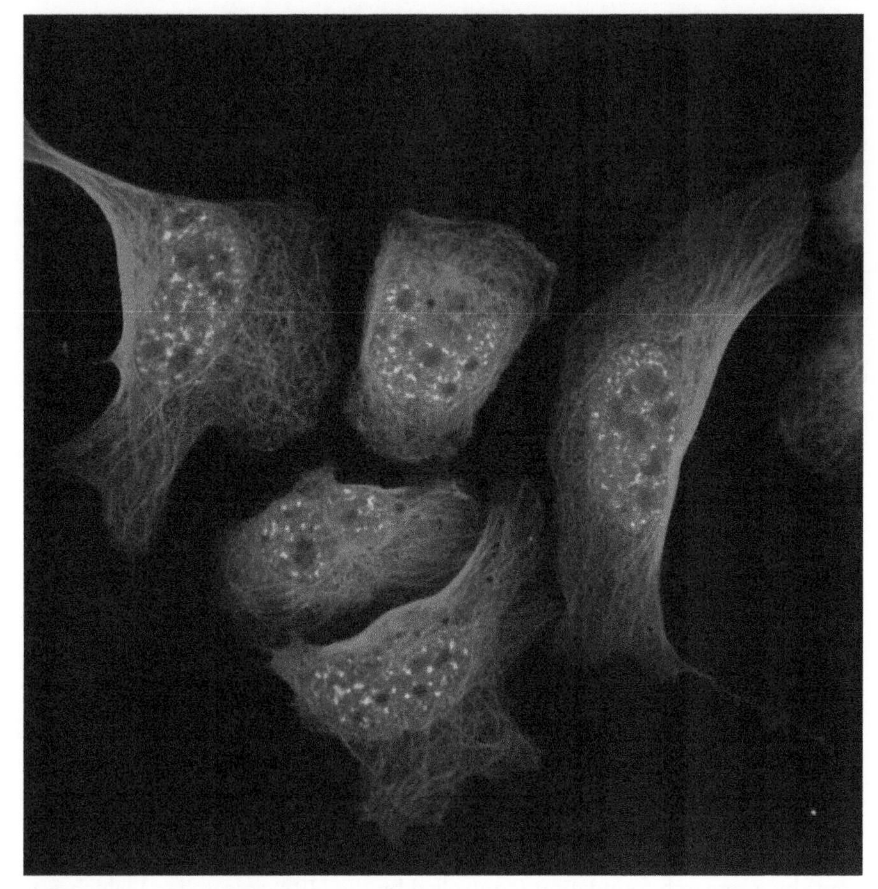

Figure 11.1. Immunoflorescence in human cells

Image from the Human Protein Atlas. RBM25; HPA003025: U-2 OS. See <http://www.proteinatlas.org>, accessed 27 May 2014.

project was still articulated as scientific—not primarily a matter of industrial development:

> We just think it's nice if we get some nice business out of it, but never, no interference. And no money back. On the contrary. That's why there was some discussion when Investor Growth Capital [the commercial sister of KAW], wanted to invest. And then we said that they could not be lone investors. And they [Investor Growth Capital] had a hard time understanding that. Because they don't think of themselves as the foundation... They are completely separate. They never ask us for advice. But I thought it was very good that there were other investors.
>
> I don't know if Health Cap and Investor Growth Capital will get back their money on this. I don't know... I know very little about Atlas [Antibodies] because for us it's a clear-cut case: We give money. You develop. We have absolutely no... we don't want to interfere. (Interview, Erna Möller)

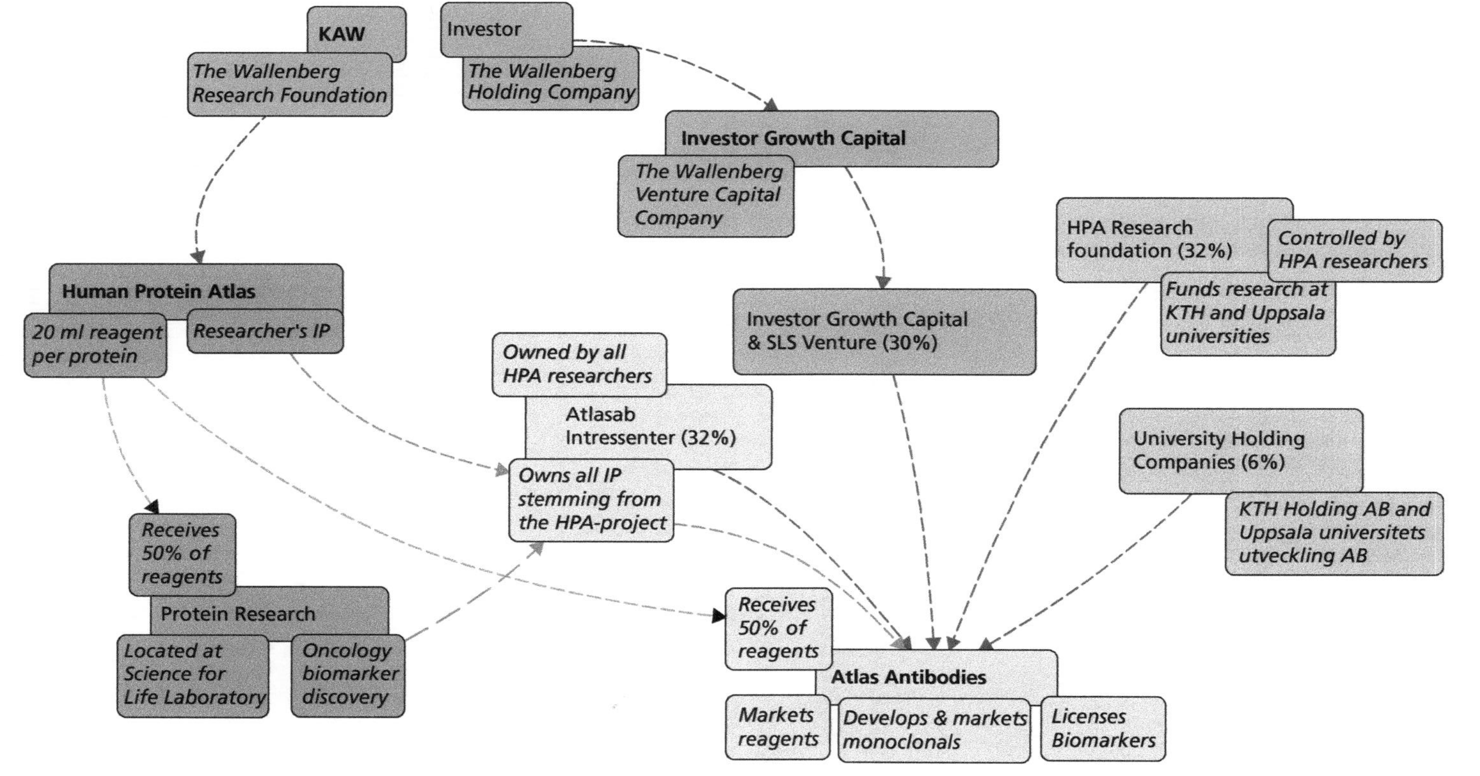

Figure 11.2. A schematic representation of the complex organizational structure that was developed

Thus, Möller and KAW articulated both a hybridized science/industry relation, and a simultaneous purification of science and industry. Investments in science should be kept separate from commercial development. The purity of the two spheres was upheld. However, rather than enacting a temporal purification, as the Wellcome Trust, Möller and KAW stressed an *organizational purification*, where the Wallenberg's science foundation, KAW, and their venture capitalist arm, Investor Growth Capital, did not mix their business. Rather than pursuing the linear model of scientific development, KAW enacted a division of labour founded on different roles in developing and commercializing science. The boundary between science and industry was enacted with different boundaries for what was deemed acceptable relations, what was seen as interesting knowledge, and what was seen as productive methodologies.

Relational work, interests, and epistemic value: Some concluding remarks

This chapter makes a lateral move in analysing the tropes of the science/industry nexus. How do we understand that the biosciences are constantly articulated in a multitude of conflicting manners? At one point they are articulated as being for innovation and national economic innovation; at another time and place they are articulated as a fundamental science by the same actors; and elsewhere, further down the road in a new situation, they are pronounced a production factory for providing new means to clinical ends.

This chapter has argued that by attending to the performance of these tropes, the performance of these interests, in the biosciences, we might provide an opportunity for avoiding the all too familiar narratives of self-interested scientists; a space for fashioning empirically sensitive stories that attend to the multiplicities of interests and values that scientists produce and relate to in practice. The tropes of medical development, economic innovation, and scientific progress are all present in the biosciences, and scientists perform and relate to all of them at different junctures.

By exploring the prolonged valuation of a large-scale protein mapping project, the HPA, the chapter has shown how the performance of different interests in science produce profoundly different valuations of science: with effects on what is seen as epistemically valuable; which yardsticks should be used for evaluating methodologies; and how laboratory work should be organized.

The mapping of the proteins—and by extension this may apply to any bioscientific project—does not have unambiguous value or interest. Is it a treasure map that marks proteomic real estate that has potential for medical intervention and economic development? Is it an atlas, a topographical survey, where all proteins are potentially valuable? What is an interesting and valuable map? And how should it be produced?

This chapter has proposed to attend to the making of interests and values in terms of *relational work* and different *modes of purification*. Thus, it attempts to understand how actors produce science through their interest-work and value-work. Following this line of inquiry, the chapter proposes two modes of purification in order to understand the strategies that actors use to uphold the difference between science and industry: one *temporal mode of purification*, and one *organizational mode of purification*. Inspired by Annemarie Mol (2002) the chapter suggests that it might be fruitful in the future to attend to further modes of relational work such as coordination, clashes, hierarchies, and calibrations.

By looking closely at the valuations of science we can begin to understand the epistemic shifts, breaks, and reorientations of the academic hierarchies of university research. How should science be done? For whom? With what purpose? The HPA is not unique in this manner, but rather one of many that question the disinterestedness of science and belies the boundaries between scientific work and commercialization.

ACKNOWLEDGEMENTS

My first thanks go to Mats Benner and Sven Widmalm who made it possible to undertake the investigation of the HPA, and who contributed both to the empirical investigation and to this chapter through generous discussion and commentary. I am also grateful to all my informants who generously shared their time, and made it possible for me to gain an insight into the secret world of peer review. Thanks also go to readers and collaborators. Thanks to Peter Keating and Aant Elzinga who reviewed this chapter, and helped me see much more clearly the relation between the valuation of antibodies and epistemic value. Thanks to my co-editors C. F. Helgesson and Isa Dussauge for rewarding discussions and careful readings. Thank you to Ilana Löwy, Morten Sager, Shai Mulinari, and all participants of the 'Moral Economy of Life Science' workshop in Vadstena who helped shape this paper. Finally, thanks to Jenny Lee, Anna Tunlid, and the ValueS Seminar at the Department of Thematic Studies—Technology and Social Change, Linköping University, for perceptive comments and valuable advice. Research for this paper was carried out within the project *Det forskningspolitiska laboratoriet*, financed by the Swedish Research Council (*Vetenskapsrådet*).

12 Data transfer, values, and the holding together of clinical registry networks

Claes-Fredrik Helgesson and Linus Johansson Krafve

Introduction

Large-scale research necessitates coordination. This chapter takes an interest in what counts as valuable in the coordinative practices needed for large-scale research. This focus is motivated by how values are understood when examining the organizing of research. Some scholarly contributions emphasize the role of shared values, rules, and norms for holding together collective research endeavours, such as when referring to an identifiable and stable *moral economy* in a domain of research (see Daston 1995; Kohler 1994, and the introduction (Dussauge et al., Chapter 1) to this volume). Other scholars have explored the practices, procedures, and tools necessary to achieve large-scale research (see, for instance, Keating and Cambrosio 2012; Shrum et al. 2007) while showing less interest for the values in play in these practical arrangements.

This chapter brings the topic of values to the surface without presuming those values to be fully shared and part of what is holding the overall endeavour together. We pursue this aim by taking a pragmatic view on values, where we see them as enacted in practice (see, for instance, Dewey 1913, 1939; Joas 2000, as well as the introduction (Dussauge et al., Chapter 1) and conclusion (Dussauge et al., Chapter 14) to this volume). It is reasonable to anticipate that any large-scale research endeavour relies on some justification as to how it is valuable and worthwhile. It is, however, less clear what to presume about the values enacted in the many coordinative practices that hold the larger endeavour together. In other words, we want to probe into how the enactment of values relates to the overall holding together of the endeavour.

The empirical site chosen is a type of large-scale clinical research that centres on the analysis of observational data for a particular group of patients. Such clinical registries regularly involve a number of clinics that see the same type of patients and collaborate in the collection of data about these patients.

This makes it possible for them to investigate epidemiological questions such as the incidence of the disease within the overall catchment area as well as the different treatment strategies used, etc. Our focus is on how the amassing of data is accomplished in the dual context of clinical practice and medical research. The registries are a form of large-scale research that produces ordered data and analyses about specific patient groups. This is made viable through the participation of many researchers and clinics that are simultaneously situated in care as well as other research endeavours. Researchers participating in the same registry may at the same time be competing with one another, making the sharing of data sensitive (Ceci 1988). Given clinical registries' highly embedded quality, they provide an interesting site for examining the arrangements that make it possible for them to produce comprehensive results and in that specific sense appear as a unified 'macro-actor' (Callon and Latour 1981). This further echoes the long-standing interest in science studies in how the practices involved in the production of knowledge are varied and convoluted in ways not fully reflected in the resulting knowledge they produce (see e.g. Collins 1985; Latour and Woolgar 1979; Pickering 1995).

The coordinated transfer and pooling of data is a prerequisite for large clinical registries. Here, we will focus on the different arrangements used to transfer data from clinics to the main registry hub where analyses are made, rather than following the steps transforming patients into data and aggregated analyses (see, for instance, Helgesson 2010). Of specific interest here are the different arrangements used and what counts as valuable in these different arrangements. The topic of values is particularly pertinent since the work to transfer data often involves an element of exchange where the recipient provides something in exchange for the data and hence explicitly articulates certain values in relation to this work. Examining three specific clinical registries, two European, one Swedish, we ask: what different arrangements are used for the gathering and transferring of data within the clinical registries and what counts as valuable in these arrangements?

Clinical registries as a specific form of large-scale research

The central feature of a clinical registry is that many centres, usually clinics, repeatedly collect the same data for patients suffering from the same (often chronic) condition. The patient data from the different centres are then gathered at a main registry hub where the analysis of the data and other overall tasks are coordinated. However, this work is not confined to those working at the main hub. Some researchers or research-oriented clinicians in other participating centres may be involved in the analyses and usually also partake in the governing bodies of the registry. Governing bodies might for instance be tasked with deciding what analyses and research proposals should be allowed to use the registry data.

A clinical registry can be multi-tiered in the sense that larger hospital centres operate as local hubs gathering data about patients treated in smaller centres. In addition, a large international registry may involve national or regional registry hubs that in turn relay their data to a main registry hub. A clinical registry is thus constituted by a large set of relations between different centres at hospital clinics and clinics in university hospitals. Some of those participating in the work primarily gather and transfer data about patients, while others may also be more or less involved in the analysis and publication of results based on the data as well as in the governance of the overall registry.

Depending on how a registry is set up it can be valuable for investigating differences in treatment strategies used, for examining how frequent is a disease (its prevalence), and for studying how patients respond to treatments over time. A central feature of a registry is that it is observational, gathering data on ordinary patients that have a condition within the area of the given registry. Questions related to issues like prevalence contribute to the general imperative of clinical registries to have a comprehensive coverage, since the usefulness of a registry for answering such questions relies on it including a majority of the patients that fall within its scope. This means that a registry depends on the participation of all relevant clinics regardless of how they are situated in clinical care as well as in other research endeavours. The imperative of comprehensive coverage further means that a registry presupposes participation among research-oriented clinics that simultaneously might compete in the domain of research.

It is useful to compare the clinical registry with the large multi-site randomized controlled trial (RCT), which is another form of large-scale research endeavour highly prevalent in the nexus of health care and clinical research. Similar to registries, the large RCT hinges on data gathered about numerous patients who have given their consent for data collection. Yet, in an RCT, data are only collected for patients meeting certain specified criteria and patients are usually randomized to receive different treatments. An RCT further has a regular terminal point whereas a clinical registry is more permanent and follows patients as they repeatedly interact with clinical practice. Data collection in a registry is hence more akin to the recurrent entering of information into patient records.

Another important difference relates to coverage. Large RCTs are multi-sited to enable the recruitment of a large number of patients for the purposes of statistical analyses. There is however, given the randomized design, no imperative to include a certain proportion of the broader implied population of patients. This stands in stark contrast to clinical registries where comprehensive coverage constitutes a huge benefit. A clinical registry thus requires an inclusive and more permanent collaboration between specialists in the field than do even very large and longitudinal RCTs. Even a large RCT can be conducted involving a more limited sub-set of specialists than is possible in the case of registries. Finally, there are differences in the research questions asked. A registry as a research form is, for instance, necessary for answering questions related to the prevalence of a disease. Yet there is some overlap regarding certain questions related to response to treatment. Such questions are largely considered to be best answered by RCTs, but in cases of rare diseases and the use of expensive and rare pharmaceuticals, it has increasingly been considered valid to pose such questions to registry data.

The comparison between RCTs and clinical registries not only brings out some defining characteristics of registries as a large-scale research endeavour situated in health care. It is further relevant to bear in mind the differing characteristics between clinical registries and RCTs, since RCTs may overlap with the operation of a clinical registry.

In summary, a clinical registry is a form of large-scale research endeavour that is intimately enfolded in the practices of everyday care. It is a form of research that can coexist and overlap with other forms of clinical research, including large multi-site RCTs. A registry further entails the involvement of a large variety of clinics that can be differently situated in clinical care as well as in other forms of research endeavours.

Arrangements holding large research endeavours together

EXAMINING THE HOLDING TOGETHER OF LARGE-SCALE RESEARCH

As we posited in the introduction, a clinical registry can be seen as a large-scale research endeavour with significant differences between its overall outcomes and the practices constituting it. This broad starting point rejects giving *widely shared* values, rules, and norms a favoured role in our understanding of what it is that holds together such a large-scale endeavour. Conceptualizing a large-scale research endeavour as a macro-actor provides an emphasis on how a large-scale, and apparently well-held-together, entity might be constituted by a large and diversified set of entities (Callon 1986; Callon and Latour 1981). It allows us to appreciate how an ordered outcome can be an effect of the catering to and accommodation of a diversity of devices and practices rather than of an outright homogenization of the diversity (see, e.g. Helgesson and Kjellberg 2005; Star 1991).

We will, as mentioned in the introduction, focus on the practices related to the transfer of data in the registry networks, a work that often involves an element of exchange. Our focus not only provides a concentrated site from which to observe some crucial practices in the holding together of large-scale research, it also provides a vantage point for critically engaging with the notion of moral economies in science in the realm of large-scale research (Daston 1995; see also the introduction (Dussuage et al., Chapter 1) to this volume). A central feature of the notion of moral economies in science is the emphasis on how a scientific domain might be characterized by a particular set of values. This leads down an analytical route emphasizing shared values that we do not aim to take here. (Look for widely shared conceptions of what counts as central values and you might find part of what coordinates and holds large-scale research endeavours together.) There is, however, another aspect of the notion of moral economies in science that introduces a critical engagement with the macro-actor approach chosen here: the attention to exchange practices that sometimes comes with the notion of moral economies in science.

FRUIT FLY SCIENTISTS, EXCHANGE, AND MORAL ECONOMIES IN SCIENCE

Robert Kohler's (1994) work on the community of fruit fly (drosophila) scientists in the early twentieth century emphasizes the role of an exchange system of specimens between scientists for sustaining a very specific set of values and rules within the their research field. He notes how there were certain rules guiding behaviour within the community when it came to sharing different stocks of flies as well as sharing information about yet unpublished results. Among the rules he identified were *reciprocity*, in that receiving stock also entailed obligation to provide stock; *disclosure*, in the sense that recipients of stock were expected to disclose their intended use of that stock; and that *tools developed were the property of the community* (whereas claims could be made to specific research problems).

Kohler notes, clearly alluding to the Thompsonian conception of a moral economy (Thompson 1971) that the exchange system 'was not unlike the preindustrial moral economy of provision, in which landowners took responsibility for feeding the poor in hard times and enjoyed the privilege of deference and social status in times of abundance' (Kohler 1994: 141). In the case of the 'fly people' this meant a mode of sharing and the avoidance of competitive behaviour:

> Sharing in the means of production was held as a golden rule, and individualistic competitive behaviour was discouraged at the same time that inequalities were accepted as natural. (Kohler 1994: 141)

This exchange system and its moral rules, as Kohler calls them, were established when the community was small, but the principles for exchanging stock did not change when the community grew in the 1930s:

> It was these rules of exchange, as well as the rituals of group work, that gave the fly people their distinctive identity. These were defining traits on their view: the fly people were a people who took as standard the tools and practices of their founding group. They were a people who discouraged monopolies, shared tools and trade secrets and did not poach and ambush. (Kohler 1994: 168)

The route of analysis taken by Kohler emphasizes a widely shared set of arrangements for interaction and exchange as central to fostering the community of fruit fly scientists. These arrangements were in his analysis not only reflecting the values of the community; they further sustained them and thus participated in holding the community together. In other words, this analytical route highlights how the overall community was sustained and shaped by interactions and exchanges that repeatedly enacted certain values.

Kohler's analysis of the community of fruit fly scientists is a helpful point of reference in highlighting the importance of arrangements for interaction and exchange in larger research endeavours. Yet we want to refrain from turning Kohler's observation of exchange sustaining *widely shared* values into an assumption for our analysis. This makes the notion of a prevailing and unified moral economy a less helpful analytical *starting point* for the empirical examination of the enactment of values in coordination of large-scale research and the overall holding together of the endeavour. Equipped with the notion of macro-actors, we will not presume that widely shared values, rules, and norms are necessarily what hold large-scale research endeavours together.

At least two traits of the registry networks are similar to the community of fruit fly scientists. First, just as the fruit fly research community, the registry networks clearly rely on practices of interaction among several researchers and groups. In the registries, it is patient data rather than stocks of flies that are the primary object of exchange. Second, from a research point of view, the registry networks share a certain form of inequality with the fruit fly community. In both cases, some researchers are more centrally situated than others. For instance, in the case of the registries, researchers more closely related to the main registry hubs are in a significantly better position to make extensive use of the data gathered. As large-scale research endeavours, a major difference between them concerns the coordination of the scholarly output. While the many exchanging fruit fly scientists published individually or in small groups, the registry networks predominantly produce results in publications using data from the participating centres and where the authorship is shared among several of the participating parties. Hence, while the unity of the fruit fly scientists was most clearly manifested in their community and the shared arrangements for interaction and exchange, the unity of registries is perhaps most clearly manifested in the coordinated output and the formal governance of the registries.

EXAMINING ARRANGEMENTS FOR TRANSFER OF DATA

Enabling the acquisition of certain kinds of data has been put forward as a central explanation for large-scale collaborations in scientific communities (Shrum et al. 2007). Yet, the sharing of data in science is a sensitive issue. Several scholars have observed how large-scale scientific endeavours might differ in terms of how data are produced and shared. Vertesi and Dourish's (2011) study shows how there were quite different patterns of data sharing characterizing two collaboratories (Finholt 2002) within planetary science. They furthermore stress that these differences in the practices of data sharing should be understood in terms of how they were situated in a larger 'data economy' encompassing the production, use, and circulation of data:

> Articulating different economies of data, involving production alongside use and exchange, may help us to better understand how data acquires meaning in scientific communities and beyond. (Vertesi and Dourish 2011: 541)

The notion of economies of data ties in to other observations of interactions in scientific communities as constituting economies, such as the previously discussed notion of moral economies in science or indeed cycles of credit in science (Latour and Woolgar 1979). The topic of who is awarded an authorship on scientific publications and on what grounds is a central part of these discussions. Patterns of co-authorship have even, as Shrum et al. note (2007), in some research trajectories on large-scale research collaborations been used as the key descriptive element of the collaborative endeavour. We will, however, follow a more restricted route and take note of the observation that the assignment of authorships is a part of what goes on in large-scale research endeavours (see, for instance, Biagioli and Galison 2003). To this, it should be added that it is also a debated and contested topic within different disciplines, not least within medicine (see, for instance, Bhopal et al. 1997).

The notion of different economies of data suggests it would be fruitful to direct our attention towards the differences in data transfer and exchange practices. This is important given how the participating parties are differently situated in the registry network as well as in clinical practice and other research endeavours. Taking an analogy from the study of market exchange in retailing, there is the possibility of identifying different 'modes of data transfer' (compare 'modes of exchange' in Kjellberg and Helgesson 2007), including the different ways in which transfers are organized, who the parties are, as well as what are transferred.[1] A mode of data transfer thus denotes an identifiable set of arrangements for the gathering and transfer of data. Such a mode includes a configuration of those involved, what data are transferred, how and with what qualifications, and what is being considered valuable in this work. The notion, modes of data transfer, consequently directs attention to the different ways in which data may be gathered and transferred.

Highlighting how the practices to gather and transfer data are configured in different ways opens the way for appreciating how transactions might be interwoven into a broader and more varied pattern of interaction, much as Zelizer (2005) found when examining economic transactions in spousal and other intimate relations. Developing a 'connected lives' approach that allows for investigating how economic and other values can be simultaneously at play even in intimate relationships, she proposed the notion of 'relational work' to denote how the involved parties deal with the compound nature of relationships:

> For each meaningfully distinct category of social relations, people erect a boundary, mark the boundary by means of names and practices, establish a set of distinctive understandings and practices that operate within that boundary, designate certain sorts of economic transactions as appropriate for the relation, bar other transactions as inappropriate, and adopt certain media for reckoning and facilitating economic transactions within the relation. All these efforts belong to relational work. (Zelizer 2005: 35)

The notion of relational work is useful here since it adds another analytical route for examining the setup of different modes of data transfer, by highlighting such matters as the work to establish and sustain distinctive understandings and practices regarding what is appropriate in specific relations.

In the following, we will use the notion of modes of data transfer as a device to aid in investigating different arrangements for the gathering and transferring of data. For a given arrangement, this includes examining: who are the involved parties; how the arrangement configures those involved; what is the distinctive understanding concerning appropriateness in their relations; what is appropriately to be considered as valuable; and so on. Moreover, on a more distanced level, it allows asking whether the arrangements examined appear at all as orderly and patterned along the lines of a single major mode, or indeed along the lines of a few discernible and distinctively different modes.

[1] Compare, for instance, the 'mode of exchange' in self-service retailing relying on pre-packaged goods and an active consumer cruising the aisles, with the 'mode of exchange' in manual service retailing that does not have to rely on pre-packaged goods but on a much more active salesperson (on qualification in exchange, see Callon et al. 2002; Cochoy 2009; Kjellberg and Helgesson 2007).

The gathering and exchange of data in three registry networks

INTRODUCING THE THREE INVESTIGATED REGISTRIES

Two of the registry networks examined are pan-European while one is a Swedish network. The three registries investigated are: (1) SMSreg (Svenska MS-Registret), a Swedish registry for MS (multiple sclerosis); (2) Eutos, a European registry for CML (chronic myeloid leukaemia); and (3) Eurofever, a registry within paediatric rheumatology collecting information about patients affected by the major auto-inflammatory diseases (AID) in childhood. The three registries involve different specialties within medicine, which also relates to what kinds of clinics are engaged in the treatment of patients. A further important difference between the registries is the relative rarity of the focused diseases, where MS is by far the one with the highest prevalence and where AID are rare or so-called orphan diseases. This has implications for how many patients may potentially be registered, and how typical are the relevant patients for the various clinics involved. Moreover, for the very rare diseases it can be difficult to conduct traditional large-scale RCTs due to a shortage of patients. Table 12.1 depicts some key features for each of the three registry networks investigated.

The selection of the registries was done to ensure variety as well as to focus on some specific traits such as the registry-based collection of drug treatment data. The Swedish Medical Products Agency (MPA) made the selection, since our study was supported by MPA as part of a larger project at the agency (Assessing Drug Effectiveness—Common Opportunities and Challenges for Europe).[2] The registries are not necessarily selected to be representative of medical registries. Yet we maintain that the varying rarity of the diseases covered and the registries' differing reliance on support from pharmaceutical companies make them an ideal set for exploring different arrangements for data exchange and the holding together of large-scale research.

Our empirical material consists primarily of interviews with representatives of the main hubs of each registry as well as with representatives of some centres, usually clinics, participating in each registry. When selecting centres to interview, we have tried to include both research oriented and less-research-oriented clinics. Research-oriented physicians do however dominate our sample based on snowballing and convenience, which consequently makes the perspective of other clinicians somewhat under-represented. Seventeen semi-structured interviews with informants based in nine different countries were

[2] The larger project had a particular interest in the registry-based collection of drug treatment data, particularly where multinational data collection is needed in the case of orphan diseases. Such registries further operate to assist in fulfilling regulatory requirements to produce follow-up data provided for drug companies at the time of approval of new drugs.

Table 12.1 Some basic data about the registries

Network	SMSreg	Eutos	Eurofever
Disease	MS	CML	AID (primarily among children)
Number of patients	11,893	1,098	1,880
Patients included per year	1,000	n/a	1,880 recruited in first 18 months
Number of centres	71	n/a	67
Number of countries	1	14	31
Main registry hub	Stockholm	Heidelberg/ Mannheim and Bologna	Genova
Most important source of financing	SALAR	Novartis	European Agency for Health and Consumers (EU)

Sources: SMSreg: Årsberättelse 2009–2010; Eutos: Presentation Hasford at ELN workshop in Mannheim, 1–3 July 2011 (concerns population-based part); Eurofever (Toplak et al. 2012).

conducted between March and October 2011.[3] We have refrained from disclosing in what specific countries our informants operate to ensure their anonymity and have for the same reason identified these using generic Anglo-Saxon pseudonyms in quotes.

Each of the three registries is related to some form of professional society or association and their operation is further governed by one or several committees as well as bylaws. The organization most closely tied to *SMSreg* is the MS association, which is separate from other professional bodies in Sweden. The MS association welcomes not only neurologists and other physicians, but also other professionals interested in MS. The MS association appoints the governing board of the registry. The major financial support for SMSreg comes from the SALAR (Swedish Association of Local Authorities), an association for the municipalities and county councils in Sweden.[4] *Eutos* is founded with the support of grants from the European Union, the European LeukemiaNet (ELN), and the pharmaceutical company Novartis. Eutos is built up from several regional and national registries within Europe that each collects and administers data from their respective regions. Furthermore, Eutos is partly the achievement of a professional association, in the sense that it was inspired by the Swedish leukaemia registry, which was run by a professional association

[3] The interviews ranged from 23m to 2h32m. All interviews were recorded, transcribed, and entered into a CAQDAS software where they have been coded using primarily a descriptive coding strategy (Saldaña 2009). Representatives from four centres from four different countries were interviewed for Eurofever and representatives from four centres from three different countries were interviewed regarding Eutos. For SMSreg, representatives from three different clinics were interviewed.

[4] County councils are the public sector organizations responsible for the provision of health care in Sweden.

of haematologists. However, ELN is an association of physicians, scientists, and patients with interest in leukaemia, rather than a professional association for haematologists. Finally, *Eurofever* is an extension of a network of trial collaborators that host many clinical trials. Eurofever could in that sense be seen as being facilitated by the prior collaboration of a group of clinical researchers conducting trials. Central to registry governance is a steering committee consisting of a handful of experts in paediatric diseases, but a meeting involving a larger group is held once a year. The first publication with results from the registry was published in 2012 (Toplak et al. 2012).

All three registries are situated in settings where other forms of clinical research take place. The coexistence with clinical trials might raise tensions, since a trial can at times require that data for patients recruited to the trial be excluded from the clinical registry. This causes gaps in the registry data from patients recruited to clinical trials. Competition for patients between clinical trials is not new. Yet the temporary exclusion of patient data appears to be an effect with wider implications given the broader and more longitudinal scope of registries.

There is a large variation in how data in a registry are seen as useful for clinical practice. Sometimes historic data from a given patient are referred to for decisions regarding his or her current treatment. In other registries the feedback to clinical practice is more remote, so that, for instance, clinical practice is informed by publications based on studies on the aggregate of registry data. There is thus a variation as to how, and to what extent, the registry is considered to be valuable in informing day-to-day clinical practice. Some informants report on how they use the locally gathered data for informing and benchmarking their clinical practice against other clinics, whereas others say that the registry has a value to clinical practice through the publications that stem from the research based on the registry as well as the guidelines developed on the basis of such research.[5]

The following sections explore different arrangements used for the gathering and transference of data within the clinical registries and what counts as valuable in these arrangements.

DIRECT CLINICAL USEFULNESS AND POSSIBLE PAYMENT FOR USING THE REGISTRY

There is variation both within and between the registries as to how the actual registration is performed in relation to the patient visit. In some cases patient data are entered into the registry at the time and point of care. However, it is

[5] See, for instance, Lachmann et al. 2013 regarding Eurofever and Baccarani et al. 2013 in relation to Eutos.

more common to enter the data into registry forms at a much later date, thus relying on patient records as the primary source. There is also a large variation in the frequency of data collection for a single patient, from updates at each visit to biannual or annual updates.

The notion of direct usefulness clearly relates to how data are entered into the registry in relation to the time of care. The gathering of data at clinics for SMSreg relies heavily on the data being useful in clinical practice. The usefulness of the registry in clinical practice was heavily emphasized by a representative of the main registry hub:

> The idea is that you as a clinical active physician work with the registry. You open the registry in parallel with having the patient record open. You then enter your data directly at the point of care. The registry is designed to make it efficient to work with. You get a good overview of the patient and all his data by using the registry. You are thus spared much of the flicking around in the patient record. (Our translation, interview with Thomas Mason, physician and researcher at SMSreg main hub)

Another informant working in a research-oriented clinic at a university hospital confirmed this usefulness of the registry:

> It is very useful to be able to bring up compressed data during a patient meeting in way that is impossible with the patient record... It is simply much more convenient to get an overview over the patient's situation with the MS-registry menu. (Our translation, interview with Sophie Thomson, senior physician and researcher at university hospital)

However, the comments from an informant at a non-research oriented clinic indicate that relying on usefulness can be a double-edged sword when it comes to enticing clinics to enter data into the registry. The informant repeatedly stated that they often did not have the time to enter information at the point of care and that the lagging behind of registrations was a constant worry. In turn, less rigorous entering of data made the registry less useful when it came to the treatment of individual patients:

> You can of course in the registry view what happens to an individual and what has happened historically with treatments and so on. But this relies on how many parameters one enters and the frequency with which one updates the registry. It is not optimal if you do not make the registrations... We have furthermore to also enter information into the patient record... We can hence not neglect to document in the patient record just because there is a registry. I have understood that some colleagues to a large extent rely on what they see in the MS-registry and perhaps put less energy into the patient record.
>
> The MS registry is a perpetual worry of mine. We have to keep the work at the clinic moving along... My attitude as an ordinary clinician is that for the moment I do it for the greater good, to allow researching colleagues in neurology to do their research, and with the prospect that we ourselves some time when we feel less behind can also use it for [clinical] feedback. (Our translation, interview with Stephanie Harper, neurologist at non-research oriented clinic)

FINANCIAL COMPENSATION IN EXCHANGE FOR DATA

There are instances where financial payments are made in exchange for the delivery of data about patients. Of those studied, the most clear example is that of Eutos where €300 was mentioned as the payment for a completed form.[6] These payments are very much related to the financial support given to the registry from Novartis and are made to national or regional hubs from the main registry hub, which in turn receives payments from Novartis. At times, the regional or national hub uses this financial support to pay the clinics that report cases to them for further delivery to the main registry hub:

> Well they get [financial amount omitted], which is what we're left with after we've taken all the bank charges and only if the form is complete. We can't, we're not interested to have all the things missing that we want. (Interview with Daniel Powell, physician heading a regional hub within Eutos registry network)

It is, however, far from always the case in Eutos that the clinic seeing the patient is involved in these financial transactions in exchange for completed forms containing patient data. On several occasions we were informed that the per-form payment to a national or regional hub was converted into other arrangements for their subsequent engagement with different centres delivering patient data. A representative of the main registry hub emphasized that the money was given to the national registry hubs, and that it was up to them how best to use the money to get the data. In Italy, for instance, the money is used to reimburse a central laboratory for providing physicians all over the country with free molecular and cytogenetic analysis.

Another line of reasoning maintained by representatives of hubs receiving the money from the main Eutos hub was that it was they who actually did the heavy lifting. The compensation should therefore stay with them even if the data concerned patients from other clinics:

> CFH: Do you pay a specific hospital for reporting a case?
>
> Informant (heading a regional hub): No. No, because we send our stuff to get the data, we don't pay anything to doctors. (Interview with Jack Roberts, physician heading a regional hub within Eutos registry network)

The situation for Eurofever is a bit different when compared to Eutos. Eurofever has no pharmaceutical company funding such a large proportion of the work as does Eutos. Instead Eurofever relies on a mixture of EU grants and

[6] Here, a completed form can mean both the form completed when the patient is first entered into the registry as well as each of the subsequent follow-up forms about the same patient that are completed after one year and then again after another year. Each such completed form thus results in a payment of €300.

support from pharmaceutical companies. This translates into an even more mixed picture of how the gathering of data is financially compensated. As mentioned previously, Eurofever covers several very rare diseases. The arrangements involving financial compensation for the gathering of data can therefore differ not only due to contingencies such as those indicated above regarding Eutos, but also depending on the specific kind of rare disease indicated in the patient data. Entering data about patients with TRAPS (tumor necrosis factor receptor-associated periodic syndrome) was, for instance, financially reimbursed based on an EU grant:

> So the bulk of the patients I entered were TRAPS patients and that was funded through an EU grant, the non-TRAPS patients I've entered have been in my own time, and at the moment unpaid for. (Interview with Kelly Clarke, heading a national hub in Eurofever)

For another rare condition, there was money from a pharmaceutical company involved that allowed the Eurofever main hub to provide some compensation to clinics that gathered data about patients with this particular condition:

> [I]f you enter ten patients into the registry we are going to give you, I can't remember how much, €20 per patient that you enter into the registry. This, I will call [a] symbolic reimbursement, [and] is done to facilitate the work. (Interview with Mark Davis, part of team heading Eurofever registry network)

There are then a number of different ways that financial compensation appears in relation to the gathering and transferring of patient data for building clinical registries. Several contingencies appear furthermore to influence when and how financial compensation is made and how it is distributed.

Several informants who worked with financial compensation that relied on funding from pharmaceutical companies voiced concerns about the source of the funding. They emphasized, however, at the same time that it was important to be able to distribute financial resources to the centres so that staff could be paid to allot time to the gathering and transferring of data. One representative of the Eutos network lamented that support was needed to assemble a good registry, but that such support also carried with it a risk that it could be withdrawn:

> I'm not so happy with this that we . . . have to rely so much now on Novartis. The good thing is . . . that we now get the money to really start the registry. We can pay at least €300 or €400 or €500 for a complete recording of a case, which we couldn't do with the European leukaemia net money . . . We have in some areas I think a median observation time of let's say five to six years, so it's really analysable the data. But, with a three-year perspective, I'm not sure whether we are still existing. (Interview with Harry Parker, part of team heading Eutos registry network)

CO-AUTHORSHIPS IN EXCHANGE FOR THE GATHERING OF DATA

One regular arrangement in registry networks is to acknowledge the work put in to gather data by granting co-authorships to representatives of at least some of the participating centres. This practice sometimes appeared very naturalized. For instance, when we asked a researcher involved in Eutos about the number of centres participating in a given study, the number calculated and provided was the number of co-authors on the latest publication. The granting of co-authorships for delivering data is, however, interwoven with other principles regarding who is to be acknowledged as an author and where in the list of authors different contributors ought to be located. This is, for instance, laid out in the PRINTO bylaws (PRINTO (Paediatric Rheumatology INternational Trials Organisation) is the investigators' network that lies behind Eurofever):

> Primary Results of a PRINTO Study: In most cases, the lead authorship of a journal article that reports the primary results of a PRINTO study will be the PI of that study. Additionally, several other individuals who played key roles in the study's development, conduction, or analysis may be included as co-authors... The PRINTO Senior Scientist will determine prior to beginning a study a minimum number of patients that must be enrolled in each centre. If the minimum number of qualified patients is enrolled at one centre, then a single investigator will be included from that centre as an author on the subsequent publication reporting the primary results of the study. A centre that enrols patients, but does not meet the minimum number as set forth by the Senior Scientist, will be recognised in the cover page of the paper. Centres not listed in the primary publications will be listed in secondary publications that might arise from the same project. (Excerpt from Article VI of the PRINTO bylaws dated 1 October 2003; last revision 21 November 2011)[7]

Eutos, the registry concerning CML, also has rules for granting co-authorships to recognize the work done in gathering data:

> We have some rules and basically the rules are, they're not very strict. Let's say everybody who contributes data, at least if it's not just one patient or something ... can function as an author and a big centre maybe two authors and the authors are somehow ranked with regard to the number of evaluable patients that have been submitted... The first authors are of course usually those who have really done the writing and the design of the study, but that is usually not so difficult to decide... The last position again is somehow reserved for a senior author. (Interview with Harry Parker, part of team heading Eutos registry network)

The possibility to offer co-authorships is also present for regional hubs doing particular studies. One informant at a research-oriented centre at a university

[7] <http://www.printo.it/bylaws.asp>, accessed 14 August 2012.

hospital connected to Eurofever, for instance, saw the granting of co-authorships as an important part of the arrangement to get data about patients from other centres:

> Here comes the politics, and then what we do is, we buy them with co-authorships. This is how life works with academia and for most centres. In our institute it only counts if you are first or last author, you know those rules probably? Many of the smaller units they are already happy if they are author number twelve or thirteen. This is how it works. And then, you have, for each paper you have about six important places and the rest is, you know, in the rota in between. (Interview with Daniel Adams at a centre being part of the Eurofever network)

Co-authorship thus appears as a particular currency for rewarding the delivery of data in the sense that you can mint it yourself, while at the same time there are policies and guidelines for how they ought to be awarded. Several informants deliberated on the principles and practices for granting co-authorships and bore witness to the management of co-authorships being a delicate matter. One informant, for instance, reported that it was one of the 'sensitive issues' that had to be managed by the registry steering committee. We further learned that decisions on the awarding and ordering of co-authorships might result in conflicts where someone might feel inappropriately excluded or incorrectly placed in the list of authors:

> Occasionally... you have experienced that someone was not really happy, then sometimes you think he's right that he's not happy, you say my goodness, I'm sorry, we will, next time we will pay attention and that happens of course too. (Interview with Harry Parker, part of team heading Eutos registry network)

On one occasion an informant also more broadly contemplated how the registry rules for co-authorships sometimes inadvertently disadvantaged small centres participating in Eurofever:

> I mean the question is always how do you protect the very small units who put in one or two patients, who are always going to be bottom of the list for publications for everything and you know one has to be a little sensitive about that. (Interview with Kelly Clarke, heading a national hub in Eurofever registry network)

The notion of a co-authorship being a customary compensation for providing data even extended to our own interview situation. At one point, for instance, one informant stressed that our cooperation with him needed to include the possibility of awarding him a co-authorship on our publications using information about the regional registry he managed:

> But if you would like to have more sophisticated information or even look at that, you have to come to us and we have to discuss also what, what actually will be a profit for us to do this kind of work, to be co-author or something like that, and definitely you have some financial coverage for that. Just... take this [interview]

> as very informal information, but if you would like... to see how it works, you have to come and we have to discuss on a more I would say professional way, the cooperation. (Interview with James Edwards, heading a regional registry network and a hub within the Eutos registry network)

The notion of granting a co-authorship in exchange for receiving information from an informant appears uncanny to most social scientists. Yet it is worth stressing that it might be far less so in clinical research, since it appears to be such a common practice in the context of registries. When informants politely suggested a co-authorship as a possible result of the interview, they simply drew on the disposition that participating in work producing valuable data might be acknowledged and rewarded in terms of a co-authorship.

OTHER RELATIONAL ARRANGEMENTS FOR GATHERING DATA

The arrangements described above for encouraging the gathering of data have centred on the directly clinical usefulness of using the registry locally and on ways to compensate for the work done. These arrangements do, however, not exhaust the list. We have also come across other ways to obtain data about patients. One such arrangement just briefly mentioned above involves free provision of laboratory services to physicians. For regional and national hubs in Eutos, partnering and compensating laboratories for their services appeared to be such an arrangement. Here the hubs could, through the participation of the laboratories, receive notice about potential patients eligible to be registered as well as access to some of the data about them. Thus, agreements with laboratories appear to be an arrangement for the population-based registry to ensure that patients in the catchment area who become diagnosed with CML are not missed regardless of which physician they visit.

> In [a major European city] it's... complicated, but... we have established an alliance not only with the public laboratories but also with the private ones. So the public... and private laboratories are members of our group... They warn us of a given case because [at] this moment every suspicion of CML comes with a sample to a molecular laboratory. (Interview with Jack Roberts, physician heading a regional hub in Eutos registry network)

We further observed another arrangement related to the identification of patients relevant for the registry. This arrangement was most prominently articulated in Eurofever with its very rare diseases and was connected to the involvement of regional or national hubs for their expertise regarding certain patients. Such involvement as experts then becomes a route for receiving information about new cases as well as having access to patients for the purposes of gathering data about them. One clinician/researcher said this when asked about whether smaller clinics received authorship in exchange for gathering data about their patients:

> For us no, they don't [get co-authorships] because I've entered them [the patients] and because to be honest we've usually made the diagnosis, made the management decisions, done the monitoring, so we're not, we're not creaming off the work they've done. What we tend to be saying is: 'Look, here's this patient who we largely sorted out, they live seventy miles away, will you do the routine monitoring, we will keep seeing them and we'll make the big decisions.' (Interview with Kelly Clarke, heading a national hub in Eurofever registry network)

As regards the Eurofever registry we were also at times informed that participation in the registry was considered valuable since it provided a means to get into contact with other clinicians seeing patients with a similar condition. Again, the fact that this was raised in Eurofever can be related to the registry's scope to gather data about patients with a number of very rare diseases.

Another theme that came up every now and then in relation to the gathering of data was that of informants considering the necessity of nagging colleagues to report cases to the registry. This articulates that the need to remind colleagues to report cases could appear in conjunction with the other arrangements for getting data such as payments and/or co-authorships. One informant concluded that his unit could do more of this work; but that this chore would have to fall on him and that he found it difficult to find the time to chase his colleagues:

> We're not good at that and I think that needs to be someone a bit more senior. I think that is really my job, and I just don't have the time to do that... You can't just have quite a junior data manager ringing up professors in other cities saying: 'you know, why haven't you put that patient in', otherwise we'll just make, we'll just lose friends. (Interview with Daniel Powell, physician heading a regional hub within the Eutos registry network)

This need to persist in getting others to report cases is the flip side of situations where clinicians are behind in entering data, as was exemplified previously in the case where a clinician expressed that the being behind was a chronic worry. The themes of nagging and having a bad conscience emphasize that interactions within the registry networks may well take place in multifaceted relations. Moreover, the observation in the above quote to the fact that the informant would lose friends if he delegated the nagging, testifies to the tact needed to handle these relations. Another example of the need for being tactful concerned the attention sometimes needed when granting and ordering co-authorships (see section on Co-authorships above).

There are, as the above illustrates, a number of relational arrangements apart from direct compensation involved in requiring centres to gather and transfer patient data to a registry hub. One such arrangement sometimes used in Eutos is to ally with laboratories to get indications about possible new patients eligible for registration. Another relational theme concerns how the sustenance of relations between clinics within the registry network also enabled interactions regarding the care of specific patients. (This was most explicitly expressed in Eurofever.) We should also mention here that the informants at times further added that they valued their participation in the registry network since they could thus support efforts that could improve treatment guidelines and provide evidence to be used against cost-cutting health policy officials limiting the use of effective but expensive drugs.

Several modes of data transfer

A DIVERSITY OF ARRANGEMENTS

There are clearly several different arrangements used for the gathering and transferring of data in the three clinical registries examined. These include exchange-like arrangements where the gathering of data is compensated financially or through the granting of co-authorships and the use of laboratories to detect potentially eligible patient cases. There are, in the first place, three marked differences regarding the arrangements upon which the three registries rely. First, an arrangement using financial compensation to clinics for their gathering and transfer of data is by far most widespread in Eutos. Second, the direct clinical usefulness of the registry at the clinic is most clearly expressed in SMSreg. Third, the benefit from participation for gaining access to colleagues with similar rare cases was most clearly articulated by participants in the Eurofever registry.

It would, however, be too simplistic to conclude that each of the investigated registries relies on a singular arrangement for gathering and transferring data that sets it apart from the other two. First, some arrangements are present in two or all three of the registry networks. Each registry has, for instance, some principles for how co-authorships are to be granted to compensate participating clinics. Second, there is simply too much variation in the arrangements used *within* each of the registries examined. In Eutos, to take one example, some clinics reporting data from patients received financial compensation from the national or regional hub, whereas such compensation in other instances was kept at the national hub or indeed used to compensate laboratories.

When reflecting on the various arrangements we can identify a form of relational work (Zelizer 2005) in operation within the registries to the effect of differentiating between individual relations. This can, for instance, be seen in the work to articulate to whom it is appropriate to grant a co-authorship as compensation for data gathered and to whom it is not. The way co-authorships are distributed erects a boundary between those granted and those not granted co-authorships and it establishes an order of importance among the circle of researchers and clinicians who do become included as co-authors. The differentiation of what is appropriate in other relationships is similarly discernible in articulations about where it is appropriate for a hub to forward the financial compensation for data gathered to the clinic seeing the patient and where it is not. Such relational work thus establishes and sustains distinctions between different relationships. In our words, the relational work designates relations to specific modes of data transfer. This then highlights the simultaneous presence of a few distinct such modes within a single registry. Moreover, each such mode is characterized by the types of parties involved, the configuration of their relations, how and with what qualifications data is transferred, what is being considered appropriate compensation, and so on.

In *Eutos* one such mode of data transfer involves clinics, registry hubs, and the main registry hub through exchange-like arrangements using both financial compensation and co-authorships. Another discernible mode of data transfer within Eutos involves somewhat more peripheral clinics, and relies only on financial compensation. Yet another mode involves registry hubs, laboratories, and clinics and relies on financial payments to the laboratories rather than to the clinics, which instead are offered free laboratory tests. In *Eurofever*, one discernible mode of data transfer involves the granting of co-authorships, whereas another discernible mode in this registry centres on hubs taking on some of work from other more peripheral clinics related to the treatment as well as registration of patients. Finally, in *SMSreg*, the dominating mode of data transfer relies on the rather immediate gathering of data in clinical practice for the dual purposes of clinical practice itself and the registry. Yet another mode in SMSreg, and judging from one informant a less successful one, relies on the postponed gathering of data 'for the greater good'.

WHAT COUNTS AS VALUABLE IN DIFFERENT MODES OF DATA TRANSFER?

Our informants have articulated diverse things as valuable. This includes: aggregated data on the incidence of the disease; guidelines based on registry data; an overview of individual patient data at the point of care; to be noted as an author; and time; to name but a few.

Each mode of data transfer involves the enactment of certain values. Beginning with the more exchange-like modes: a mode involving the granting of co-authorships clearly enacts that it is valuable to be a co-author on an academic paper; that data reporting a patient case are valuable; and consequently that the work done to gather the data is valuable. A mode involving only financial compensation similarly enacts a slightly different set of values. Furthermore, a mode relying on the usefulness of having access to patient history at point of care enacts that access to such an overview is valuable. In all instances, patient data are enacted as valuable, but what else counts as valuable varies in different modes.

Stating that a mode of data transfer enacts certain values is not the same as arguing that it performs a commensuration (Espeland and Stevens 1998). An instance where the gathering of patient data is compensated with a co-authorship enacts both authorships and patient data as valuable, but does not equate them. What we are simply suggesting is to appreciate that a mode enacts a multiplicity of values and, furthermore, to recognize that the multiplicity of values enacted vary somewhat between different modes. We would also like to stress the variety of values articulated in a specific mode. To take one example already mentioned: the expressed need for a senior to personally spend time nagging friendly colleagues was sometimes necessary to obtain the completed forms, even in the presence of financial compensation. Hence, in that particular case, this mode involving compensation also enacted the value of time and the relationships with colleagues through the act of nagging in a respectful way.

Several things are manifested as valuable in the work of gathering and transfer of data in clinical registries and the values enacted vary somewhat between different modes of data transfer. These range from such things as the value of co-authorships and ease of access to experts seeing patients with the same rare disease, to greater goods such as the value of having the clinical guidelines that are one result of analyses performed on the registry data, or indeed the value of having aggregate data revealing under-treatment or proof of effectiveness in discussions with skinflint health policy officials.

FRICTIONS AND ORDERING IN THE ENACTMENT OF MULTIPLE VALUES

The simultaneous presence of many different values can cause friction. One angle for examining this is to consider the notion of the relational work (Zelizer 2005) and in particular how it emphasizes the deliberations of what is to be considered appropriate and inappropriate in different relations. We have, to this effect, for example encountered expressions about how it is *inappropriate* to cream off the work of others, to allow junior administrators to nag colleagues to fill in forms, and to fail to make proper entries in patient records.

Voicing reflections about the appropriate and the inappropriate appears to be particularly articulate in relation to instances where many different values are simultaneously at play. Such deliberations reflect difficulties in aligning the enactment of many divergent values. This is how we can show the dual articulations encountered in Eutos about funding—it was simultaneously articulated as possibly inappropriate to rely on funding from a pharmaceutical company and appropriate to be able to financially compensate (some) clinics for their work to gather data. And here, the independent registry governance was often mentioned as the device that made this friction tenable. Such frictions are also what appear to be in operation in frequent deliberations about what is to be the appropriate allocation and ordering of co-authorships. Here, a mix of principles, negotiations, and pragmatic adjustments were mentioned as means for grappling with this.

The work to distinguish between different relations can in this respect be understood as assisting in the ordering of the values enacted. The coexistence of a few different modes of data transfer in each examined registry thus reflects how they order and displace the many values in play.

Values and the holding together of large-scale research

The existence of large-scale clinical registries is not justified as a means of providing researchers with opportunities to receive co-authorships, nor are they justified as a means for producing work at clinics that can be financially compensated. In their unified macro-actor (Callon and Latour 1981) and collaborative appearance, these registries are related to such things as the value of producing knowledge about the prevalence of a disease and differences in treatment strategies, as well as the value of clinical guidelines that can be a direct result of analyses performed on the registry data. Yet it is equally clear that it is not such greater goods alone, or even primarily, that hold these large-scale endeavours together.

We found a rich and diversified set of arrangements when focusing on the gathering and transferring of data in three clinical registries. Each registry accommodates a few different modes for the gathering and transfer of data. Each mode further entails the enactment of somewhat dissimilar values. Hence, the registries examined here are not held together by exchanges repetitively reinforcing a single set of widely shared values. This clearly sets them apart from what Robert Kohler (1994) observed in his study of the community of fruit fly scientists. Relations appear to be central in the holding together of the registries examined. Yet they do not all operate as manifesting the values of a single unified moral economy. Indeed, there appears to be something akin to ongoing relational work that distinguishes between different relations and which values are appropriately enacted as part of them.

We introduced this chapter with an interest in large-scale research and the ambition to bring to the surface the topic of values without presuming them to be widely shared and part of what holds an overall research endeavour together. In this we have focused on the arrangements used for the gathering and transfer of data in three clinical registries, and on what counts as valuable in these arrangements. Our study suggests that large-scale research endeavours can indeed be held together by several different and partially overlapping arrangements that furthermore enact a variety of different values that also entail frictions. Moreover, our examination is further suggestive that any value attributed to the entirety of such an endeavour can only be a fraction of all the values simultaneously in play and grappled with in the coordinative practices holding them together. In terms of the various values in play, the whole can appear far more orthodoxly homogenous than the diverse coordinative activities that constitute this whole. Such a refinement of the values enacted by the entirety of the large-scale research endeavour is indeed a precious achievement, given the many and diverse value practices that participate in making it possible.

ACKNOWLEDGEMENTS

Several persons have contributed to this study. First thanks go to Nils Feltelius at the Swedish MPA. He presented the idea of studying a set of European disease registries to analyse their working procedures and ability to collect drug effectiveness data. He made the initial contacts with the registries that made the subsequent interviews possible. It was he who commissioned us to examine the organizational and financial aspects of registry networks, and he has also taken great interest in following the subsequent work.

A second thanks goes to our informants. Setting up interviews with those involved in the registries was far from always easy given their busy schedules. We are therefore very grateful to those who found time to be interviewed and then generously shared their experiences about registries as well as clinical research and clinical practice more broadly.

Finally, we would like to thank Isabelle Dussauge, Karin Svedberg Helgesson, Francis Lee, Tiago Moreira, Vololona Rabeharisoa, Ebba Sjögren, Teun Zuiderent-Jerak, and the ValueS seminar program at the Department of Thematic Studies—Technology and Social Change, Linköping University, for insightful comments and helpful suggestions on previous versions of this chapter.

13 Valuation machines

Economies of desire/pleasure in contemporary neuroscience

Isabelle Dussauge

Prologue[1]

This morning is an uphill one. On the radio, while I am making breakfast, the speakers of my favourite programme are welcoming their guest, an established neuroscientist, who has come to speak about the reward system of the brain. And I who expected a relaxing start for the day... The reward system, *the neuroscientist says after lining up a series of brain regions,* sends us pleasure when we've done something that is good for us.

I go to the shower, turn the hot tap to a maximum, and let my tiredness drip away. I stand there for a long, long while. I think: This is good for me. Then I suddenly remember the porridge I set to boil for breakfast. I turn the heat off under the small morning pan, but it's too late. My cats don't care. They are all over the kitchen looking for their share of porridge, especially the little one. The reward system is there for a reason. Evolutionarily speaking, it helps us optimize our behaviour. The brains of humans, like other mammals, have evolved with the reward system to provide us with a sense of what is good for us.

Upon leaving, I greet my cats goodbye. The little one wants to play, as every day and every night, with its favourite shoestring with an improvized glitter tail. As usual, I don't really have time. The radio goes on: We get used to rewards, and we want more of them. Research has shown that addictions to gaming, sex, or food involve the same brain mechanisms as addiction to drugs.

I reach the university and watch my colleagues run around, setting aside the books they're longing to write and read, to produce short publications instead, like small thinking machines on an assembly line, to advance their individual careers; or rather, to not lose the chance of getting yet one more short-term contract when this one is ending. This, I recall from previous readings, is what sociologists call a reward system.

[1] This section is fictional but inspired by a radio programme actually sent out on the morning radio in Sweden in 2013. The radio excerpts featured here are not quotes.

TOWARDS A CRITICISM OF NEURAL HAPPINESS

There is something I find existentially disturbing about the idea that we act for rewards. Can't we do the things we do for their own sake? Can't we be non-calculative beings? Aren't our lives worth living without the accumulation of rewards?

The popular and scientific figure according to which our behaviour is highly directed by a pleasure-seeking 'reward system' of the brain enjoys a thriving social life in contemporary culture. The salience of the reward system of the brain, I want to argue, is part of the contemporary neuroculture (Vidal and Ortega 2011) of happiness—the shared practices and meanings of happiness made neural.

A scientific description of life and happiness entails not only epistemic claims but also claims that pertain to values. The claim that our hedonic lives revolve in specific orbits around rewards, via the reward system of the brain, foregrounds certain behaviours (choices), situations (decisions), and relations (consumption, exchange) as the fabric of life. Such claims reflect assumptions about what is a good life, and which aspects of life are worth a scientific description. Besides, such claims have a cultural and moral impact in terms of what lives they treat as worth living, and with which human skills.

Moreover, since the reward-based neuroscientific account of life tends to conflate pleasure with consumption, it refers directly to a political background—contemporary expectations of happiness as a result of choices and consumption.

This chapter sets out to explore the values enacted through neuroscientific models of reward. I develop my analysis through three sections.

I begin with the observation that the neurosciences have come to describe many different registers of human life, such as economy, sex, and pleasure, under the umbrella of rewards. In the section 'Neuroframing all values as rewards' I ask: What happens to economy, sex, and pleasure when described as rewards?

Second, in that overview of the neurosciences of desire/pleasure it becomes clear that the neurosciences believe that people are driven by different desires, different registers of value. Some of those are explicitly economic values, others are not. In the section 'Neuro-relating values to one another' I ask: What relation between economic and non-economic values is enacted by the neuroscientific models of desire/pleasure?

Last, I argue that the neuroscientific models of desire/pleasure make statements of value as they describe what motivations, situations, and pleasures human life consists of. In the final section 'Happy without neural happiness?' I ask: What values are enacted by the neuroscientific models of human life as pleasurable?

A backdrop to these models, and to my interrogations, is a late modern conception of the subject as a naturally economic being, a decision maker, and an agent of their own happiness. At stake are no less than the dynamics of desire, pleasure, and happiness made neural in late capitalism. Through a journey into the neuroscience of rewards, this chapter aims to highlight a powerful contemporary scientific account of what life consists of, and thus to question the neurosciences' production of values about what makes a desirable life, a life worth living.

Rewards

Although not historically new, 'reward systems' seem to have gained renewed prominence as a popular buzzword, as a social–scientific concept, and as an institutional practice of late-capitalist Western states.

In sociology, Robert Merton (1957) developed theories of the reward system of scientific institutions, which also opened up a range of subsequent studies of the dynamics of science. In Mertonian sociology, the reward system refers to the institutional incentives, such as recognition, which come to encourage scientists to act in ways that may, or may not, conform to the norms of scientific conduct (Merton 1973b).

It has also been a century since behaviourist psychology was established as a scientific programme, in the wake of extremely popular psychological theories of reinforcement and control of human and non-human animal behaviour. Rewards, in behaviourism, denote reinforcers of behaviour.

In the neurosciences—a programmatic alliance between neurobiological, neurological, and behavioural sciences in the 1960s (Abi-Rached 2012)—the notion of brain-reward system has its self-referred origin in a paper published by two postdoctoral psychologists contemporary with Merton, James Olds, and Peter Milner (Olds and Milner 1954).

By what is usually considered chance, Olds and Milner had identified parts of the brain involved in rewarding processes. They were conducting experiments involving rats that had electrodes implanted in their brains, which received stimulation when either experimenter or rats pressed appropriate levers. When given the possibility, the rats with electrodes implanted in certain parts of the brain would abandon eating or other activities to press the lever instead. Olds and Milner concluded that they had 'perhaps located a system within the brain whose peculiar function is to produce a rewarding effect on behavior' (Olds and Milner 1954: 426), and similarly proposed the existence of a system of punishment.

Olds and Milner's discovery thus built on behaviourist notions of rewards as reinforcers and directly addressed some current debates between behaviourist schools of thought; at the same time, they contributed to a direction of research critical of behaviourism's approach.[2] The notion of reward integrated into what would eventually be widely referred to as the reward system of the brain was thus ambiguous, already drawing on different theoretical traditions.[3]

Within three decades, the notion of the reward system of the brain, including the search for pleasure centres, mediators, and mechanisms for pleasure, desire, and the role of experience in decision making, would become increasingly central in the psychology of motivation as well as in the growing neurobiology of addiction. Nowadays, as neurobiology stands for a culturally powerful framing of behaviour, desire, pleasure, self-control, and decision making, the figure of the reward centre of the brain extends far beyond the laboratory.

The notion of reward conveys fundamental assumptions about human action. In all accounts reviewed here, rewards or the reward system stand for a partial explanation of why people do what they do. Rewards and reward systems provide a justification (behaviourist, functionalist, or, as we will see, evolutionary, or else) for where we individually choose to go.

The Merriam-Webster online dictionary defines reward first as 'something that is given in return for good or evil done or received, or that is offered or given for some service or attainment' (<www.merriam-webster.com>, accessed 9 October 2012). Similarly, the sociological, behaviourist and neurological notions also equate receiving a reward with getting something enjoyable, or desirable, in exchange for one's actions.

In other words, in systems of reward, there is an exchange taking place, in which some actions are attributed a positive value and rewarded, some are not rewarded and are not part of that system of value.[4] Reward is thus an intrinsically economic metaphor, as it introduces a notion of exchange and value into its description of human actions.

What kind of story of human action is being told through the figure of the reward system of the brain? What happens to economy, sex, and pleasure when framed as rewards?

[2] Olds and Milner's work was on the one hand situated within a behaviourist paradigm, and their concept of rewards denoted reinforcers of behaviour. On the other hand, Olds and Milner were exploring questions beyond behaviourism's scope of analysis of behaviour; they were working in the lab of Donald Hebb, a physiological psychologist whose research had, in part, become directed against the current behaviourist black-boxing of the motivational and emotional 'mental processes' behind decisions (Brown and Milner 2003).

[3] For a more detailed exposé of the complex notions of drives for behaviour in twentieth-century psychology, see Fradelos (2008).

[4] This definition echoes the behaviourist definition of reward recalled by Olds and Milner: 'In its reinforcing capacity, a stimulus increases, decreases, or leaves unchanged the frequency of preceding responses, and accordingly it is called a reward, a punishment, or a neutral stimulus' (Olds and Milner 1954).

Neuroframing all values as rewards

Neuroeconomics, the neurobiology of addiction, the neuroscience of sex, and evolutionary psychology, are disciplines which all deploy a vocabulary of rewards to theorize and experiment on human pleasure and desire. How do these respective fields frame human action as centred around rewards? And according to their models, how do human beings attribute a different value to different rewards?

ECONOMICS

First, the vocabulary of rewards permeates neuroeconomic research, a growing field, described as a 'marriage of giants' (Maasen 2010) between economics and the neurosciences.

According to an important figure in the field, Paul Glimcher, neuroeconomics is a research programme with the ambition to experimentally and theoretically integrate the three different explanatory/modelling levels of economics, psychology, and neurobiology (Glimcher 2008).

Neuroeconomics emerged from behavioural economics in the 1990s before it became widely institutionalized in the 2000s. According to sociologists Natascha Dow Schüll and Caitlin Zaloom, behavioural economics and neuroeconomics have aimed to address, experimentally, the paradoxes which cannot be accounted for through the figure of the rational *homo economicus* (Schüll and Zaloom 2011). Instead of working with the rational and egoistic *homo economicus*—who is aware of what is good for them and maximizes their self-interest—behavioural economics works with a figure of the human as rational and emotional, capable of empathy and not always aware of their own best.

Neuroeconomics studies both economic and non-economic behaviour. Many experiments study financial behaviour (see e.g. Camerer et al. 2005). But often the questions explored by means of neuroscientific methods and money are not financial. Studies of decision making in various settings use money as a quantifier of how much e.g. trust, fairness, risk, short-term gratification, or conformity to social norms matter to the participants (e.g. de Quervain et al. 2004). In such experiments, rewards function as a general denomination for the outcomes positively valued by the participants.

Yet another body of neuroeconomic experiments is interested in the neural mechanisms of the valuation, expectation, and consumption of rewards. Here again, rewards may be monetary (e.g. Breiter et al. 2001) or of another order (social, emotional).

Two aspects characterize neuroeconomics' reward-centred description of human behaviour. First, the notion of reward is a floating placeholder: it can be filled with virtually any content and therefore it is used to relate different scenarios, needs, or things to each other which otherwise do not have a straightforward interrelationship (Harrison 2008).

Second, the consumption of rewards is universalized as the general figure of pleasure, and decision is made the general figure for action. To neuroeconomics, human behaviour in the world consists of a cognitive identification of alternatives and valuation of what they are worth, a weighing of those against each other, and a choice of one alternative. In other words, neuroeconomics models the world as made of rewards and life as a series of decisions demanding consumption choices. Things acquire value in relation to these decisions: They appeal to us as 'incentive value', and we enjoy their subjective value (utility) upon consumption.

In short, and as a whole, the neuroeconomic view entails not only a description of how people make decisions; it entails a description of life as made of such decisions. It also models a direction for our actions, and a relation to the things described as rewards: What we want from rewards is consuming them. The reward-centred descriptions of behaviour in neuroeconomics produce a figure of the human as a reward hunter, a goal-directed consumer of pleasurable value.

PLEASURE

Reward-centred notions of human behaviour are also central to the huge and expanding neuroscience of behavioural addictions (Elam 2010), and the closely related neuropsychology of pleasure.[5]

In the neurobiology of addiction, the grammar of reward focuses on the distinction of liking (pleasure) from wanting (desire), and on processes of learning, from experience, the mismatches between wanting and liking (Berridge and Kringelbach 2008; Elam 2010).

The neuropsychology of pleasure, or 'hedonic psychology' (Kahneman et al. 1999), sets as its mission to highlight the role of pleasure in directing human behaviour and conceptualizes pleasure as a complex brain mechanism. Main figures of the contemporary field position themselves against earlier

[5] The neurobiology of addiction is an older scientific field than neuroeconomics; and it has been a precursor to certain themes of research in neuroeconomics. Scientific concerns and concepts have been borrowed and translated between the neurobiology of addiction and neuroeconomics. The human subject of the neurobiology of addiction is envisaged as autonomous 'individual volitional units' without a social context, 'who, in modern society, must maintain self-control in the face of ubiquitous inducements to seek pleasure through consumption' (Acker 2010: 72). Historian Caroline Jean Acker argues that this model casts the addicted subject as perverse and paradoxical, acting against their best interest (Acker 2010: 72). This interest for the paradox of addiction was therefore in line with earlier trends in the sub-disciplines of behavioural and experimental economics who tried to destabilize mainstream economics' insistence on a rational subject (cf. Schüll and Zaloom 2011).

psychology which, they argue, has overemphasized negative conditioners and neglected pleasure and the seeking of pleasure as motors of human actions and central aspects of life (Berridge 2003).

About the neuropsychology of pleasure, a group of neuroscientists conducting research on female orgasms with Barry Komisaruk writes:

> By observing what we can, we have the sense that we are getting somewhere in understanding how our brain generates pleasure . . . What we lack is a concept of how any neurons produce any bit of cognitive experience . . . We know pleasure when we feel it; we just do not know which or how our neurons generate it. (Komisaruk et al. 2010: 176)

The promissory horizon of the field is thus to unveil the neural mechanisms of pleasure itself. Without an actual theory of how pleasure is generated, what neuroscientists explore is how pleasure takes place psychologically, and which areas of the brain are recruited.

Two figures in the field, Morten Kringelbach and Kent Berridge, explain that '[p]leasure and reward are at the heart of affective neuroscience and the psychology of well-being'. At stake in the neuroscience of pleasure are 'fundamental insights into our own nature' and 'better ways to enhance our quality of life' (Kringelbach and Berridge 2010: 3).

Hedonic psychology gives pleasure a distant explanatory power. For instance, Michel Cabanac de Lafregeyre writes that 'To minimize displeasure and to maximize pleasure is both the result and the aim of behavior' (Cabanac 2010: 123). Via notions of reward, hedonic neuroscience explores how pleasure directs human behaviour. The explanations proposed draw heavily on theories of motivation, especially theories positing that the search for pleasure goes through the hunting for things that 'function as rewards' and that this rewarding function is signalled to us as 'incentive value' (Kringelbach 2010: 204).

Where psychoanalytical theories have foregrounded *desire* as fuel of human actions, and where behaviourist theories have emphasized learning through *reinforcement*, hedonic psychology lifts up *pleasure* as a goal and motor in its own right. Rather than being the endpoint of desire, pleasure becomes a human process distinct from the dynamics of desire. This is reflected in the theory promoted by Berridge and others, that pleasure ('liking') and desire ('wanting') have distinct 'underlying neural circuitry and pathways' (Anonymous 2010: 11). At stake is, among others, the reinterpretation of Olds and Milner's reward system as a system of wanting rather than one of pleasure.

In a psychology of wanting, the notion of valuation is crucial. Hedonic neuroscience emphasizes that valuation is the main brain tool with which people (and other animals) navigate in a world of potential rewards (e.g. Kringelbach 2010: 204). As a result of the evolutionary pressure, when confronted with a situation of choice, brains compute a 'cost–benefit analysis' of the values and costs of different potential rewards for humans (cf. Rolls 1999:

266–87). Concomitantly, life is often implicitly described as a long series of situations demanding a weighing of alternatives, decisions, and actions.

Thus valuation is central in the neuropsychological models of motivation/wanting. However, the act of valuation is also put forward by hedonic psychologists as the very essence of *pleasure*. Pleasure is often defined as a valuation in itself (Kringelbach and Berridge 2010)—the valuation of things as pleasurable.

Hedonic psychology relates explicitly to happiness, although without consensus about what neural happiness is made of. Consider for instance the following statement by Morten Kringelbach:

> Kringelbach: Pleasure is but a fleeting moment in the state which is happiness. It is possible that 'true' happiness or bliss might be a state of 'liking' without 'wanting', which with the current available neuroscientific evidence is becoming a testable hypothesis. (Anonymous 2010: 22)

Experimental neuroscientific studies which address the question of how we should live our lives in order to be happy are, for instance, the neuroscientific studies of short-term satisfaction vs long-term benefits, featuring a neural version of the fable of the grasshopper and the ant (Ainslie and Monterosso 2004; Camerer et al. 2005; McClure et al. 2004; see Schüll and Zaloom 2011). The moral of the neural story rehearses the traditional one: rational long-term investment is for our own best, whereas short-term satisfaction is fun and emotional, but doesn't last. But it adds a neural twist to the story, which articulates an implicit humanizing reading of the fable, as demonstrative subtitles would. Yet we—our brains—sometimes act against our own best:

> The Platonic metaphor of reason as a charioteer, driving twin horses of passion and appetite, is on the right track—except reason has its hands full with headstrong passions and appetites. (Camerer et al. 2005: 56)

New moral of the story: Our brains are conflicted. We are conflicted beings. Happiness is a complicated thing.

In short, valuation is central in neuroscientific accounts of pleasure (itself an act of valuation) and wanting, which presuppose a prediction and weighing of possible rewards in the style of a cost–benefit analysis. However, this depiction of the human insists that we are conflicted, rather than rational, calculative beings. It also makes happiness an object of neural management (Elam, personal communication, 8 January 2014), by fundamentally calling its reachability into question, and by promising to yield the means to better achieve happiness.

SEX

Neuroscience of sex illustrates how pervasive the figure of the brain reward system is and how its scope reaches far beyond neuroimaging laboratories. Let us consider an example. In France, a new national educational curriculum of sex education entitled 'Femininity/Masculinity' (SVT Féminin/Masculin, known as SVTFM) was adopted in 2010 for high-school students, and educational materials were posted on an official website (Anonymous 2013b).

More surprising than their rampant sexual conservatism, an unexpected feature of the SVTFM online materials is that 20 per cent of their third biggest theme 'Living one's sexuality' (Vivre sa sexualité), and actually over 10 per cent of these sex-educational materials, are devoted to . . . the brain reward system. Beginning with the rat experiments by Olds and Milner illustrated with recent 3D-models of the human brain, the document describes the human neuroanatomy of the 'pleasure system'. Then the text explains that like the rats in Olds and Milner's experiments, humans may also become caught in behavioural addictions such as gaming and sexual addictions. The overall message is clear:

> Human sexual activity is associated with a feeling of pleasure. Neurobiologists have discovered that *the experience of pleasure during sexual activity, as well as in other human actions, relies in part on biological phenomena by the activation of 'reward systems' in the brain . . .* The reward system . . . shares in providing positive or negative information in the realization of a behaviour that is subsequently integrated with the person's other cognitive and psychological processes (by the cortex . . .) The activation of these cerebral structures must then be seen as the translation of a pleasure favouring a recurrence of that act which is necessary for the reproduction of the human species. (Anonymous 2013a) (my translation, emphasis added)

In short, according to these French materials, sexuality is one instance of, in part biologically driven, a human search for pleasure. Evolution is what we have to thank for this pleasure system.

Of course, the authors of the French educational materials have not come up on their own with the idea that the brain reward system is central for sex. In the published neuroimaging studies of human sexuality, neuroscientists introduce quite frequently (in about one out of three articles) (Dussauge 2014) notions of reward and the reward system—and the notions of valuation that follow from there.

The sexual arousal experienced by participants in specific neuroimaging experiments (where participants are asked to grade the arousal they feel upon viewing more or less erotic pictures) is usually interpreted by neuroscientists as the enjoyable assessment and anticipation of an upcoming pleasure (Dussauge 2013). For instance, Redouté et al. write: '[T]he cognitive component [of the sexual response in the brain] comprises a process of appraisal through which a stimulus is categorized as a sexual incentive and quantitatively evaluated as such' (Redouté et al. 2000: 171). The *valuation* of potential sexual rewards is thus described as a crucial part of that neural expectation of sexual pleasure.

Using work in neuroeconomics and the neurobiology of addiction in order to interpret their results, neuroscientists describe and study sexual arousal as part of a linear chain of events: identification, valuation, expectation, and, possibly, consumption of rewards (Dussauge 2014). This linear chain of events speaks both to mainstream sexology's model of human sexuality as a linear physiological process in four stages (from desire to orgasm), and to the models and metaphors used from neuroeconomics reviewed in the previous section, 'Pleasure' (Dussauge 2013).

Less often, but significantly, the neuroscience of sexuality intersects with the neuroscience of pleasure in a reversed way, where sexuality functions as a figure of pure pleasure/desire. There, sexuality comes to stand for a 'model' of the psychology of human pleasures, due to the 'acuteness and intensity' of sexual experiences (Georgiadis and Kortekaas 2010: 178). This is for instance the case in the experiment by Knutson et al. (2008) who use sexuality as one instance of desire/pleasure disrupting men's financial risk-taking behaviours.

Now, when sexuality is made the example of reward-driven behaviour par excellence, the direction of sexual desire is redescribed. In the short term, the direction of desire is the consumption of sexual rewards—pleasure and orgasm. This is congruent with sexology's inflection of desire towards a goal: orgasm.[6] But in a long-term reward-centred description of sexuality, the arrow of desire also becomes aligned with that of evolution.[7] The neuroscientific literature on rewards is heavily influenced by a normative evolutionary–psychological paradigm according to which our brain systems of motivation, cognition, and emotion must be the result of selection. In short: they assume that the brain reward system (as everything else) is or has been adaptive at some point, and has therefore been selected. This is the object of the next section.

EVOLUTION

Many of the examples mentioned above illustrate that according to the evolutionary inflected psychology of emotions, motivation and pleasure,

[6] The orgasm-centred account of sexuality has been criticized, among others, by a movement of feminist sexologists championed by Leonore Tiefer (2002). This neural model is also criticized by psychologist Nico Frijda on the pleasure-theoretical grounds that there are more pleasures at play than orgasm in sexuality; especially '[m]aking progress along the chains of the [sexual motivational] system's competences' such as 'the recognition of [a] sexual object' (Frijda 2010: 105).

[7] For a criticism of the evolutionary explanations of the female orgasm, see Lloyd (2005).

humans live in the world with the help of adaptive brain systems of emotion and motivation, and particularly, of neural rewards and punishments (cf. e.g. Rolls 1999). A broader version of that conservative argument is that putative brain systems of emotions and behaviour, rewards and punishment, exist because they are good for us. We can feel pleasure because pleasure leads us to seek to fulfil our needs, which is good for the survival of ourselves and our genes.[8]

When mobilized, the evolutionary frame is applied to both behaviours and the biological systems putatively involved in those. For instance, in the French SVTFM materials, evolution is featured as a justification for the existence both of the reward system of the brain, *and* of pleasure itself:

> Evolution has thus installed regions in our brain, whose role is to 'reward' with a pleasant sensation the execution of certain vital functions. (Anonymous 2013a) (my translation)
>
> The reward system is necessary to survival because it provides the motivation necessary to carrying out adaptive acts or behaviours which allow the preservation of the individual and the species (searching for food, reproduction, avoiding dangers . . .) . . . The reward circuit is the origin of our behaviours: Food behaviour; sexual behaviour aiming at the continued existence of the species; social relational behaviour. (Anonymous 2013a) (my translation)

Thus, these public documents argue, pleasure is evolution's tool to guide us, free individuals, towards our own survival.

Neuroscientific publications feature the same kind of presuppositions. Often, evolutionary interpretations are brought in to frame research problems (e.g. in Redouté et al. 2000). In other publications, evolution is mobilized to justify the existence of desire and pleasure: they assume that if pleasure/desire exists, then it has an evolutionary function (Cabanac 2010).[9]

For instance, Morten Kringelbach writes: 'Pleasure must serve a central role in fulfilling the Darwinian imperative of survival and procreation (Darwin, 1872)' (Kringelbach 2010: 202). Sometimes the evolutionary function of pleasure is further specified in psychological terms, e.g.: 'Yet we know that pleasure fulfills the same function in animals and humans: optimization of behavioral decisions' (Cabanac 2010: 10). Karli Watson et al. write: 'The

[8] The evolutionary framing of behaviour, of emotional life, of motivation, and of rewards owes much to the discredited sociobiology and its successor, evolutionary psychology, which are both to a high extent speculative sciences (Rose and Rose 2000). In their framework, brain structures and behavioural functions are the result of evolutionary processes (see Martin 2000). This mode of evolutionary reasoning, sometimes referred to by cultural scholars as 'pop-Darwinism', is very popular in contemporary Western culture (cf. e.g. McCaughey 2008).

[9] For a classical criticism of the general argument, see Gould (2000) and Rose (2000b) who argue that evolution has not designed human beings to any optimized purpose, that most behaviours and features of a species are the result of chance, and that biologists and laypersons must exert utmost caution before claiming that behaviours may be a result of evolutionary pressures.

rewards we evolved to enjoy continue to motivate our behavior' (Watson et al. 2010: 85).[10]

The figure of the brain functions as an important interface between the reward-psychological and the evolutionary description of behaviour. We do not act simply for the immediacy of desire/pleasure, but for the greater purpose of the satisfaction of our brain, itself shaped by our unwritten evolutionary destiny.[11]

Thus desire, pleasure,—and as I will come back to, the pursuit of happiness—are instrumentalized when framed as rewards in evolutionary psychology and evolutionary-inspired neuroscience. Whether this intrumentalization should be read as a degradation of desire and pleasure or as their cultural promotion, is an empirical question beyond the scope of this chapter.[12]

VALUATION MACHINE

To sum up so far: in the neurosciences, as in other cultural arenas, the notion of reward is crucially entwined with that of valuation in human behaviour. Neuroscientific vocabularies of rewards are deployed to explain the desires and pleasures of humans, mostly of their brains. They perform various 'neuroframings' (Dussauge 2013) of desire and pleasure, which affect these objects they describe.

Especially, the contemporary neurosciences of emotions describe the desiring brain as what I call a valuation machine—a machine which attributes value to different scenarios or possibilities of action, and which processes these values and leads to a choice.

Valuation is central to the notion of reward in neurosciences, because in most neuroscientific accounts of desire/pleasure, valuation of different kinds (weighing of alternatives, cost–benefit analyses, etc.) is what enables the human to operate in a world of rewards.

[10] In yet other researchers' theories, pleasure becomes the very footprint of evolution: 'In general, behaviors that increase the ability to conserve one's own genes (COG) are recognizable by pleasure' (Georgiadis and Kortekaas 2010).

[11] As Fradelos observes, Skinner's behaviourism was also aligned with what was to become an important trope in sociobiology and evolutionary psychology, the sociological, Spencerian, notion of the survival of the fittest: '[Skinner] also applied Herbert Spencer's evolutionary notion of the survival of the fittest in his work; arguing that the psychologically fit, those who responded well to a system of rewards, were best suited for civilized society' (Fradelos 2008: 175).

[12] The criticism of evolutionary tropes formulated, for instance, in science studies, feminist studies, or anthropology, identifies the equation of behaviour to evolutionary optimization as an unjust, degrading reduction of human phenomena, including a loss of the social (Martin 2000) and a justification of social inequities (e.g. Bleier 1984; Dussauge 2010). But we must also acknowledge that the evolutionary instrumentalization of behaviours gives those a beautiful shine, a glimmer of desirability, and opens for the production of meaning and identity; see McCaughey (2008); Nelkin and Lindee (1995) about genetics, and a similar argument by Martin (2010) about the neurosciences.

Valuation has also come to describe pleasure itself (as seen in the previous sections); and some neuroscientists use valuation even to describe what happiness is:

> Frijda: There is no happiness without pleasure; there is much pleasure without happiness. Pleasure is a core evaluative process; happiness is an emotion or long-term evaluation. (Anonymous 2010: 22)

This anticipates what is at stake in describing humans as valuation machines: happiness, or which values make up a good life. But first, the next section asks: if the notion of rewards flattens the differences between kinds of action, i.e. between orders of value (economic, sexual, other hedonic), how does neuroscience relate these orders of value to one another?

Neuro-relating values to one another

I have argued, so far, that a range of widely different moments of life culturally experienced as pleasurable are described under one neuroscientific umbrella of rewards. This 'rewardization' of different pleasures enacts, in turn, an economization of desire. What I have analysed is a redescription of desire in economic terms which privileges consumption and choice as main figures of human action, and consumption as both the motor and the goal of life (Dussauge 2014).

However, this economization is but one part of the story of neural desire. Indeed, neuroscientific models of economic, sexual, or addictive behaviour make a double gesture. On the one hand, they describe these behaviours and pleasures as *commensurable* (Espeland and Stevens 1998) with one another; on the other hand, they distinguish between these pleasures and re-separate them as different neural phenomena. What relations between economic and non-economic values are enacted by the neuroscientific models of desire/pleasure? In this section, I will argue that the double gesture characterizes the neurosciences' conceptualizations of how the economic and other values relate to one another: comparable but different.

'NOTHING-BUT' COMMON CURRENCY

Consider the case of a study published in 2010 by three neuroscientists, the main result of which was that sex and money—erotic and monetary stimulations—are processed in different regions of the brain (Sescousse et al. 2010). In line with the official press release, the popular science site Science Daily entitled their report 'Sex, Money: Specific Brain Areas for Each Pleasure' (CNRS 2010).

'So what?', we hear the alert reader say. 'Didn't we already know that sex and money were different things? How could it be a piece of news that they are processed differently by the brain?'

Indeed, the publication—and the publicity it received through popular scientific diffusion (CNRS 2010)—illustrate how self-evident the notion of a general reward system has become. Many scientific publications have been attempting to identify or document overlaps between 'neural networks' recruited by different pleasures; see e.g. Knutson et al. (2008). It is as though it has become so for-granted that sex, money, drugs, chocolate, and social relations are rewards, and that our brain processes rewards in a common system—that an experiment is needed to disentangle sex from money in neural terms.

The official press release of the French study reads as follows:

> In our everyday lives, we often encounter various types of 'rewards': a 20-euro bill, a chocolate bar, a glass of good wine... Moreover, we must often choose between them, or trade one for another. To do this, we must be able to compare their relative value on a single consistent scale, which suggests that all types of rewards are assessed in the same brain areas. (CNRS 2010)

In more scholarly terms, the authors explain that the reward system functions on the basis of a 'common neural currency' which the brain 'use[s]' '[t]o evaluate and compare the relative value of different rewards on a unique scale'(Sescousse et al. 2010: 13095).

Michel Cabanac is usually credited to have proposed that *pleasure* is the common currency of different motivational systems.[13] Cabanac argues that '[t]he common currency makes possible... a trade-off between motivations and their ranking according to the urgency of the need to access behavior' (Cabanac 2010).

In Cabanac's theory, the currency is psychological (pleasure). Other theories propose instead tentative, neural-material versions of a common currency, often featuring dopamine or dopaminergic systems (Berridge 2003 and Smith et al. 2010 for 'wanting'; Schultz 2002 and Camerer et al. 2005 for 'liking'). Yet another model posits the existence of two common currencies, one affective and one cognitive, which interact with each other (Dickinson and Balleine 2010).

The use of the term 'common currency' articulates that these neuroscientific models of behaviour envisage all human behaviour as an economy, a system of exchange. Therefore it comes to describe a wide range of different facts of life as inherently economic by nature. The use of the notion common currency subsumes the non-economic under the umbrella of the economic.

[13] In turn, Cabanac relates the history of the concept to a behaviourist theory advanced in 1975 by David McFarland and Richard Sibly (Cabanac 2010: 117).

I want to argue that what neuroscientists in these fields do is that they articulate the relationship of the different-things-we-want-or-do to one another: their commensurability in human action versus their integrity with regards to one another, to us, and in the world.

In that sense, the neuroscientific modelling under scrutiny here is a reminder of the analyses put forward by economic sociologist Viviana Zelizer in *The Purchase of Intimacy* (2005). There, Zelizer explores how people navigate the thin line of culturally acceptable and condemnable economic behaviours in their intimate relationships (see Chapter 1, Dussauge et al., this volume).

What draws Zelizer's work close to the present chapter is the stances she rejects, which she calls 'hostile worlds' and 'nothing-but' views, and the approach she promotes as 'connected worlds'.

Whenever we, people, do something, we relate to both economic and social codes of conduct, and we imbue actions and things with both meaning and economic value. The economic and the intimate do not take place in separate social spheres. Therefore Zelizer rejects 'hostile worlds' approaches which contend that economic and non-economic relations can be understood independently of each other and must be held separate to avoid contaminating one another. But neither can the intimate be explained solely by the economic, and vice-versa. Therefore Zelizer rejects 'nothing-but' views that reduce the economic to the social, the social to the economic, or more generally the non-economic to the economic. However, Zelizer acknowledges that people live with 'nothing-but' views through their practices.

The description of all pleasures and goals for action under the economic umbrella of rewards is a reminder of the nothing-but views criticized by Zelizer. By equalizing things and actions as rewards, neuroscientific models enact a 'nothing-but' view that reduces behaviour to a certain kind of economic behaviour, and which subsumes the non-economic under the economic.

NEURAL 'CONNECTED WORLDS'

However, no serious neuroscientific model—and certainly none of those reviewed here—would argue that sex, cakes, and money are the same thing, feel the same, or mean the same to anyone. Part of the ongoing discussion in the neurosciences consists in articulating the singular identities and integrity of the different kinds of things we want, like, or do.

Psychologist Nico Frijda writes:

> Are all pleasures the same or are they not?... Certainly, the pleasure experiences differ importantly. Their phenomenology is inseparable from the nature of the pleasant object or event, and what one can or cannot do with it... But at the same

> time, ... [i]t appears a safe bet that all pleasures share a neurohumoral final common pathway, as several findings ... suggest. (Frijda 2010: 105)

Zelizer's 'connected worlds' approach acknowledges that we live and deal with both economic and non-economic relations. It also contends that we constantly attempt to make culturally appropriate relations between the economic and the non-economic (Zelizer 2005: 35). Where neuroscientists debate whether humans (or their brains) process all rewards in similar, economic ways, or to which extent different rewards retain some integrity, they articulate a neuroscientific concern reminiscent of Zelizer's questions: How are the economic and non-economic related to one another in people's lives—and brains?

Contemporary neuroscientists, as did their twentieth-century predecessors, navigate between 'nothing-but' views which reduce things and deeds to their value in too-unitary economies of common currencies; 'hostile worlds' views which keep aspects of psychological life immune to each other; and 'connected worlds' views which attempt to make space for relatedness-with-integrity between economic and non-economic aspects of behaviour.

The universalist notion of reward functions as a 'boundary currency' in these neuroscientific models and explorations—to paraphrase the already economic concept of 'boundary object' coined by Susan Leigh Star and James Griesemer (1989). This boundary currency has a function precisely because the different realms of desire/pleasure *are* different.

Should the critical scientist feel at peace now, if neuroscience is simply engaging with the same kinds of models, questions, and disputes as contemporary sociologists deal with? Well, no. Primarily, there is a political canvas on which the scientific landscapes analysed here emerge. This canvas has to do with science: which sciences count as science, and which sciences are legitimate producers of discourses about the human? When neurosciences produce neural theories of the social—at the same time disregarding the social, cultural, and critical theories of the same—they continue, willingly or not, a programme of colonization of the social sciences (Rose 2000a) championed, in the twentieth century, by sociolobiology and evolutionary psychology.

Therefore we need to focus somewhere beyond—and below—the well-foundedness of neuroscientific models and reflect instead on the place they occupy in contemporary concerns: happiness.

Happy without neural happiness?

The neuroscientific models of desire/pleasure make deeply ethical statements as they describe what motivations, situations, and pleasures in which human life consists. What values are enacted by the neuroscientific models of the pleasurable human life?

I have attempted to show that the models of pleasure and desire articulated in the neurosciences cast the human, through the human brain, as a natural economic agent pursuing rewards. This economic agent is, in turn, either articulated as following economic–evolutionary–utilitarian principles of maximization all along ('nothing-but' view), or as conflicted between different hedonic values (the neural 'connected worlds' view). But in both cases, the economic agent acts for an optimization: of their experienced pleasure, and/or their satisfaction, and/or of their own or their ancestors' fitness.

Sociologist Mark Elam argues that 'neuroscience is helping us to imagine the contemporary addict as someone unable to handle the unprecedented (mind-blowing?) freedoms and responsibilities of choice and opportunity made available by late capitalism' (Elam, personal communication, 8 January 2014), leaving all of us in the situation of struggling to handle these unprecedented opportunities and requirements. Similarly, neuroscientific accounts of pleasure/desire help us envisage the human as struggling with the possibilities and imperatives of happiness on the terms of late capitalism.

The world in which the values calculated by the brain are worth something is a very special world, made of very special things: choices, choices that are good for us, choices that are not so good for us. What is positively valued here is not, for instance, human beings' capacity to feel empathy, or their capacity to produce and create things in the world. What is valued is, instead, human beings' capacity to make good choices. Furthermore, human beings are essentially envisaged as consumers, not producers, of values. Thus, through their epistemic selection of what makes a good life, these models enact a very specific set of values.

The teleological orientation of the models of the neuroscience of desire/pleasure makes them an ethical concern, a concern of value, as it articulates a direction for life. Another normative dimension is also the direction of the exchange inherent in the metaphor of rewards: through our behaviour we should win (rewards, fitness, pleasure, satisfaction), not lose.

The models are normative without needing to be prescriptive: although they do not acknowledge themselves as such, they describe the ultimate meaning of life and what paths are worth pursuing. The normative direction of the economy of desire articulated around pleasure makes these neuroscientific models relevant in terms of values, and important cultural objects of critique—matters of concern (Latour 2004).

Indeed, happiness enjoys a renewed social and cultural salience. This is visible not only in the size of the happiness industry but also in the renewed criticism of discourses of happiness and positivity (Ahmed 2010; Ehrenreich 2009). In their review of philosopher Sara Ahmed's critical book *The Promise of Happiness* (2010), Sean Grattan writes: 'The injunction to be happy manifests itself as a normalizing stricture that when questioned, according to Ahmed, allows us to ask "other questions about life, about what we want from life, or what we want life to become"' (Grattan 2011). The critique proposed in the present chapter thus appears as one of the values enacted by the neural models of desire/pleasure and happiness: their agents, cosmology, lines of action, and long-term directions preclude these 'other questions about life' which are not subsumed under an economic–evolutionary–utilitarian scheme.

The neurosciences call happiness into question and treat it as a condition continuously under threat, a 'biobehavioral insecurity' (Elam 2010: 13) calling for sustained neural governance—for there is a choice at every corner and certain outcomes may well be worth more than others. The promissory horizon offered by the neurosciences is that neural management of happiness contributes to more happiness.

Elsewhere (Dussauge 2014) I argue that the economization of behaviour is the problem with neurosciences of desire/pleasure/almost-happiness. Here I want to qualify that claim and insist instead that the problem may well be that the neurosciences' claims reinforce the view that there is a truth about human life and happiness that only they can unveil and formulate, which they, in turn, do from an economic point of view.

At stake for Sara Ahmed is an ethics of non-happiness. Ahmed shows how the cultural imperative of happiness conceals many lives and meanings, precludes alternative routes not only to happiness but routes of meaningful or enjoyable life. The happiness imperative excludes positions and directions from which and towards which meaning can be created, enjoyed, and/or killjoyed. Critically at stake in relation to the value-making in the neurosciences is the production of values outside the narrow imperative of economized happiness.

ACKNOWLEDGEMENTS

Warmest thanks to Mark Elam for generous help and comments, especially for pointing out the parallels between the biomedicalization of health and the cerebralization of happiness. Thanks also to my co-editors Claes-Fredrik Helgesson and Francis Lee, Petra Jonvallen, Mika Nielsen, all participants to the workshop 'The Moral Economy of Life Science', and the STS seminar at Uppsala University, especially Ylva Hasselberg and Alexandra Waluszewski. The Swedish Research Council *Vetenskapsrådet* has been funding the research project 'Brain Desires', of which this chapter is an outcome.

Part V

Conclusion

14 Valuography

Studying the making of values

Isabelle Dussauge, Claes-Fredrik Helgesson, and Francis Lee

Proposing valuography

Many profound concerns in the life sciences and medicine are linked with the enactment, ordering, and displacement of a broad range of values. In the introduction (Chapter 1, Dussauge et al., this volume) we put forward three concerns regarding how stakes are made, the intertwining of values and the epistemic, as well as the relationships between economic and other values. We further articulated a pragmatic stance on the study of value making in society at large.[1] In this conclusion, we propose a number of analytical and methodological means to deal with these concerns.

We propose the word *valuography* to indicate a programme of empirically oriented research into the enacting, ordering, and displacing of values. We think of *valuography* as the study of value practices following the same broad 'ethnographizing' and estranging move that several scholars have previously suggested. For example, Peter Dear (2001) has suggested 'epistemography' to designate 'an enterprise centrally concerned with developing an empirical understanding of scientific knowledge' (Dear 2001: 131) and Steve Woolgar has suggested 'technography' to refer to the analytically sceptical study of technology (Woolgar 1998). In a similar vein, Michael Lynch recently proposed 'ontography' for 'historical and ethnographic investigations of particular world-making and world-sustaining practices that do not begin by assuming a general picture of the world' (Lynch 2013: 444).[2] We take valuography to be an

[1] By the notion of pragmatism, we refer principally to John Dewey (1913, 1939) and his followers and to their tying of values to practices, specific situations, and subjectivity. In Dewey's view, values do not precede action, but are instead inseparable from their active articulation, for example, in the form of practices of valuation; see, for instance, (Dewey 1913, 1939; Joas 2000; Muniesa 2012) as well as the introduction (Chapter 1, Dussauge et al.) to this volume.

[2] Other similar examples are Steve Woolgar's proposal of 'scalography' for studies of the cultural specificity of the concept of scale (Woolgar 2012: 36) and Michael Lynch's proposal of 'ethigraphy' for the study of how some things come to be considered ethical (mentioned in Lynch 2013).

empirically oriented and analytically sceptical research programme of values as enacted.[3]

We argue that a valuographic programme makes it possible to take an interest in values while moving away from the question of what values 'really' are. This stance remedies some central problems that arise with approaches depending on stabilized understandings of value and values.[4] In this chapter, we will also discuss why our proposed form of methodological agnosticism for the study of values does not translate into a programmatic nihilism.[5] At its heart, a valuographic research programme encourages us to examine how certain things come to be considered valuable and desirable, as well as how certain registers of value are ordered and displaced. We further argue that a valuographic research programme as envisioned here has to symmetrically examine whatever is included as well as excluded as pertinent values in a given process (cf. Galis and Lee 2014). Finally, the sceptical attitude encourages a sense that these matters could be, and sometimes indeed are, otherwise (Woolgar 1988).

Valuographic tactics: Multiplicities and instabilities matter

Our valuographic research programme embraces the idea that values do not exist as transcendental entities, impinging themselves upon our actions. We take as a starting point the fact that we can examine the taken for granted and putatively stable and demonstrate how things can be otherwise. This entails moving into positions that enable us to see values as enacted, ordered, and displaced rather than as fixed and constitutive forces. In short, our programme

[3] Using the word 'enacted' is, as Lezaun and Woolgar note regarding the study of ontologies, helpful in that it 'emphasizes the generative power of the practices involved in the constitution of reality' (Woolgar and Lezaun 2013: 324).

[4] Here, approaches using the lenses of, for instance, 'capital' or assets come to mind, as well as some uses of the notion of moral economy. In these approaches, values are primarily used to denote a given set of values that guide action.

[5] Central tenets of science and technology studies (STS) articulate a broadly agnostic stance on the study of science and technology. This agnostic stance entails, among other things, a reliance on the principle of symmetry (Bloor 1976), which implies a neutral examination of what is considered scientific knowledge. Rather than trying to identify various sources of bias only in what are later identified as subjective knowledge claims, this principle encourages the symmetrical examination of whatever participates in establishing what come to be considered objective or subjective knowledge claims. This does not, however, imply a value-free approach to science; instead, Bloor and the SSK (sociology of scientific knowledge) programme often endeavour to include an analysis of society and science that rehabilitates weak actors (Pestre 2013: 208 ff.). The repurposing of such tenets in developing a broadly pragmatic approach to the study of values thus brings with it a certain agnosticism vis-à-vis values.

is an invitation to study the making of values. Below we outline a few approaches to apprehend values-in-the-making. We list a number of approaches, sites, and situations in which what are considered values can be rendered unstable. We wish to grapple with ways of attending to value articulations provoked by such instability, to ask questions about how these practices are configured, and to address how the configurations of practices in turn shape the values in play.

Attending to the multiplicities of values, the frictions that arise from them, and the concomitant articulation work performed by involved actors offers a key entry point to examining the instability of values.[6] The simultaneous presence of various values, and the articulations that this multiplicity evokes, can be explored in a wide variety of sites and situations.

What appears to give a particularly strong foothold for the making of valuographies is the drawing of attention not primarily to the ordering consequences of any stabilized values, but rather to the numerous and multifaceted frictions that come into view due to simultaneous efforts to enact different values.[7] One example here is the frictions between efforts to enact biocapitalist values pertaining to 'production, profit, and novelty', and efforts to enact genetic values pertaining to 'reproduction, management, and responsibility' in the conservation of endangered species (see Chapter 8, Carrie Friese, this volume). Furthermore, attending to the multiplicity of values in play exposes frictions between efforts to enact different notions of what is considered the proper method and goal of research (see Chapter 11, Francis Lee, this volume). Below we give a few tactical tips on how to attend to the multiplicities of values: where to look; how to investigate; what to look for.

HEEDING THE SETTLING OF COMPENSATION

The establishment of various forms of compensation provides a framework that allows the examination of values. Compensation could concern goods pricing (e.g., cod pricing as discussed in Chapter 9, Kristin Asdal, this volume) or the rewarding of individuals or organizations (e.g., physicians, researchers, and/or hospitals and clinics, as examined in several chapters in this volume) for services rendered. Compensation could also take the form of promises of future settlement (e.g., efforts to create a market for malaria vaccine described in Chapter 7, Daniel Neyland and Elena Simakova, this volume); or non-financial

[6] There is a clear affinity here with the methodological tool of ontonorms proposed by Annemarie Mol (2013).

[7] The notion of moral economies in science (see the introduction, chapter 1) is, in several ways, a helpful sensitizing mechanism for appreciating that many values are simultaneously in play. However, Daston's (1995) notions of values and moral economy instead evoke attention to stability rather than unsettledness.

forms of reputational reward such as 'the good guy benefit' (Chapter 7) and co-authorships (see Chapter 2, Sergio Sismondo, and Chapter 12, Claes-Fredrik Helgesson and Linus Johansson Krafve, both this volume).

Another fruitful line of inquiry would be to attend to regulatory struggles to settle what 'fair market value' might be, as in delimiting bribes and fair compensation for so-called Key Opinion Leaders (Chapter 2) or in providing Dutch health care (see Chapter 6, Teun Zuiderent-Jerak, Kor Grit, and Tom van der Grinten, this volume).

We argue that the struggle over compensation provides an excellent opportunity to examine many different values. Although compensation at times involves financial considerations, it never exclusively concerns economic value. Clearly distinguishing between the financial and non-financial rarely appears to make analytical sense when it comes to trying to understand the particular practices of settling compensation. Even establishing the prices of goods for market exchange is far better appreciated if we break Parson's pact (for an illustration, see Beckert and Aspers 2011; see also introduction (Chapter 1, Dussauge et al., this volume) and Stark 2000).

Highlighting situations in which compensation is negotiated and established alludes to valuation as a promising 'flank movement' (Muniesa 2012), countering the notion of values as transcendental and fixed entities that impinge upon actions. In brief, the practical grappling with compensation appears to epitomize a situation in which what are considered values might become unsettled. Investigating such situations directs attention to the agencies, relationships, settings, procedures, and devices involved in enacting, ordering, and displacing values.

STUDYING DEVICES AND THEIR CONSTRUCTION

Devices provide a second type of approach in making valuographies. This draws on the primary assertion that devices are far more than mere neutral props in the making of values (see, e.g., the algorithms for allocating transplant organs discussed in Chapter 5, Philip Roscoe, this volume).[8] Examining the

[8] One central tenet of STS has been to make epistemology and ontology into empirical research programmes by drawing attention to the role of technical practices. The detailed scrutiny of the role of such practices and of the required mechanisms has provided another generalized and central way to substantiate the notion that 'things could be otherwise'. The examination of the role of mechanisms in transporting and transforming observations has, for instance, been central in studying the shaping of scientific knowledge (see, e.g., Latour 1999a).

The role of devices in shaping reality has also been widely addressed in STS. This is particularly clear in the study of markets, where it has been argued that the use of calculative devices might help shape the market to conform better to the theory that informed the device (i.e., the performativity thesis of MacKenzie and Millo 2003). More broadly, devices have been invoked to help us understand how markets and market actors are shaped (Callon et al. 2007). In this, the notion of *agencement* has been

deliberations on the construction of different valuation devices can position a study to make sense of how various values are articulated, translated, and transplanted into devices, and of the subsequent practices of which they would become part (this is largely in line with the notion of market devices suggested in Muniesa et al. 2007).

An infinite number of devices can be explored in making valuographies. Devices can appear in the guise of kinship charts and studbooks for breeding that construct the 'genetic value' of endangered species at zoos (see Chapter 8, Carrie Friese, this volume). They can be bureaucratic devices for establishing a malaria vaccine market (see Chapter 7, Daniel Neyland and Elena Simakova, this volume). Even policy documents can act as devices guiding the innovation of both markets and biological species (see Chapter 9, Kristin Asdal, this volume).

For valuographic purposes, it is crucial to note the *translational*, or reconfiguring, quality of devices. To cite an example, devices for allocating transplant organs reconfigure values, where notions such as good clinical practice, equity, and utility are shifted through models, simulations, and algorithmic protocols (see Chapter 5, Philip Roscoe, this volume). A device, as Roscoe puts it, 'not only dictates who receives, but also what matters' (see Chapter 5). A further illustration is how market devices for allocating hospital care have actually enacted values other than those intended by the policy (see Chapter 6, Teun Zuiderent-Jerak et al.). While the policy aim was to increase incentives for quality improvement through diversifying Dutch hospitals, deploying the device produced the very opposite result, producing a set national standard for care quality. The devices seemed to undo the very values that they were supposed to strengthen.

EXPLOITING CONTROVERSIES

Controversies are prime arenas for surveying the articulation of various conflicting values, simply because central registers of value often are at stake in such situations.[9] They provide access to conflicting articulations of what will serve as

proposed to envisage actors as made up of and shaped by assemblages of human bodies, tools, devices, algorithms, etc. (Callon 2005). The notion of *agencement* directs attention to the socio-technical arrangements in actors' capacity to act and to the attribution of meaning to action (MacKenzie 2009b). In STS-informed social studies of finance, this notion has been used to examine how actors' actions and meanings are shaped by their being equipped by material and conceptual devices embedded with notions and algorithms from financial economics: 'At its most basic, a human being equipped with a financial calculator is a different actor from one without one' (MacKenzie 2009b: 23).

[9] Using controversies as an evocative resource is akin to how the study of controversies has been used in STS to examine the making and unmaking of knowledge claims (see, e.g., Collins 1981). Regarding values, this approach is related to the notion of examining disputes to access different modes of justification (Boltanski and Thévenot 1999).

central values and—through the dynamic course of the controversy—settle what come to be the most important values.

Controversies about values can treat topics ranging from what ought to be proper conduct and the proper use and distribution of resources, to what information and knowledge is valuable. Conflicts can range from disagreements in medicine about what are considered proper interactions between physicians and the pharmaceutical industry (as explored in Chapter 2, Sergio Sismondo, this volume), and conflicts about how to properly use available coastal water in cod farming (as examined in Chapter 9 by Kristin Asdal), to disputes about gathering information about foetuses that were part of the development trajectory of prenatal diagnosis (Chapter 10, Ilana Löwy, this volume).

The study of controversies also provides insight into the mutability of values. Both the values enacted and the topic of controversy can undergo significant changes during a controversy. The unstable quality of the values enacted in a controversy is well illustrated by conflicts over publication priority, for example, as examined in Chapter 4, Sven Widmalm, this volume. Widmalm describes a large and prolonged conflict over what should be considered proper conduct for scientists. The controversy was settled in a manner that saved the face of the purported violator while demonstrating the vigilance of the scientific community. In short, while the unfolding controversy enacted values related to the proper and improper conduct of a scientist and a journal editor, the settlement enacted the value of a trustworthy and vigilant scientific community.

This example illustrates, first, how conflicts evoke rich articulations about the values at stake and, second, how the course of the controversy might shift the most important registers of value. Shifts in the values that appear central in controversies underline the general valuographic point of seeing values as enacted in social processes rather than as fixed and transcendental forces.

TRACING THE TEMPORAL INSTABILITY OF VALUES

Most things change over time. Another approach to making a valuography is accordingly to scrutinize enactments of values over time, how they change, come into conflict, and are reordered.[10] The instability of values over time

[10] This approach echoes a recurrent strategy in STS, in epistemological explorations of how knowledge is historically situated, which examines how what constitutes scientific knowledge changes over time. Nuclear missile accuracy, to cite one example, is, as Donald MacKenzie demonstrated in his historical sociology *Inventing Accuracy* (MacKenzie 1990), very much a contingent and precarious achievement shaped by a complex set of processes of diverse political forms. Steven Shapin's *A Social History of Truth: Civility and Science in Seventeenth-century England* (Shapin 1994) provides another telling digest of such a position.

powerfully counters any notion that values are transcendental and fixed entities.[11]

Longitudinal studies are the obvious gateway to examining changing enactments of values over time. These can, for instance, examine the development and transformation of prenatal diagnosis and the values enacted in the shaping of this technology by various groups (see Chapter 10 by Ilana Löwy). Another temporal gateway to examining the enactment of values is to study efforts to shape the present by projecting possible and desirable futures (see, e.g., Brown and Michael 2003). For valuographic purposes, this can be explored by examining attempts to create a market to stimulate medical development (see Chapter 7 by Daniel Neyland and Elena Simakova); by studying efforts to shape the future role of genetics in endangered species conservation (see Chapter 8 by Carrie Friese); and by exploring efforts to realize a future and market for farmed cod (see Chapter 9 by Kristin Asdal).

An appreciation of temporality attuned to studying values in practice can expose the various shifting values in play and how their enactment and displacement shape and solidify programmes of action in health care and life science. 'Gerundizing' the notion of moral economies in science (Daston 1995)—studying the actors' 'moral economizing' (discussed previously in Chapter 1, the introduction to this volume)—is in our view precisely well adapted to examining such shifts in the enactment and ordering of values.

MAKING COMPARISONS

Employing comparisons, such as examining the enactment of values in different settings or situations, is another valuographic approach. Giving a study a comparative outlook enhances the possibility of appreciating multiplicities of values.

One exemplar study that implements a nearly valuographic research programme, using a comparative approach, is Marion Fourcade's (2011b) study of how the (economic) value of damaged nature was determined in oil spill catastrophes, one in Alaska, USA, and two in Brittany, France. Fourcade traces, in detail, the valuation practices, actors, and techniques involved, and demonstrates not only that damaged nature was attributed quite different economic values at the two sites; more importantly, she demonstrates how and why different registers of value and techniques were in play in settling these values. The benefits of comparative studies for making valuographies is more generally emphasized in a recent general review and discussion of the emerging field of the comparative sociology of valuation (Lamont 2012).

[11] In history, the Annales School springs to mind as particularly emphasizing the need to attend to the beliefs and values of the time examined in order to understand the past (see, e.g., Burke 1990). However, their view of values emphasizes fluidity much less than does the approach proposed here.

Using a variety of sites and situations could provide comparative leverage in valuographic work. One can compare different research endeavours, for example, examining the variety of arrangements used to transfer data in different large-scale clinical registers (see Chapter 12 by Claes-Fredrik Helgesson and Linus Johansson Krafve). Comparative leverage can also be gained by comparing how long-term research collaboration between a leading researcher and a pharmaceutical company is presented differently by the two collaborating parties (see Chapter 3, Christer Nordlund, this volume). Finally, one can compare how the same large-scale research endeavour is assessed differently by two funding bodies (see Chapter 11 by Francis Lee). Comparative leverage allows one to examine how and why different registers of value can be in play in similar or related sites and situations.

The main advantage of the comparative approach is that it provides a powerful way to investigate how value enactments can indeed be 'otherwise' in different settings. It provides material for making a valuography that challenges the idea that certain values are ingrained in the very nature of the setting investigated. In short, comparisons invite us to think about sites and situations that are similar, yet evocatively different.

A REFLECTION ON VALUOGRAPHIC TACTICS

The central valuographic starting point is to investigate values as enacted in specific sites and situations, rather than assuming that they are fixed, constitutive forces. Emphasizing ways to examine values has offered a method to move beyond the question of what values 'really' are. However, two matters related to such a move merit comment.

First, it must be stressed that the list of tactics useful in making valuographies is certainly not exhausted by the five mentioned above. Indeed, there is a great need to develop additional ways to examine values-in-the-making. These additions would increase our appreciation of how values come to be settled in various sites and situations. Such work on a valuographic toolbox is important for increasing the ways we can examine the multiplicity of values and the variety of situations in which this can be done.

Second, it must be stressed that reliance on various valuographic tactics provides for an appreciation of values as enacted. Exploiting these tactics in examining the unsettledness of values in controversies or in the settling of compensation will doubtless provide important insights into values-in-the-making. However, we must avoid the temptation to too eagerly transpose what such valuographic inquiries yield into grand theories about values in general. A certain degree of modesty is not only becoming, but is helpful in truly appreciating or grappling with the multiplicity of values in various sites and situations.

Valuographies and matters of concern

The valuographic approaches suggested above facilitate the investigation of values in a number of sites and areas pertaining to the production of scientific knowledge or the making of highly specialized valuations through employing sophisticated calculative devices. In the introduction (Chapter 1, Dussauge et al.) to this volume, we formulated three main areas of concern in relation to the making of values in the life sciences and medicine. These were concerns related to how stakes are made; the intertwining of values and the epistemic; and the relationships between economic and other values. We can now think of these as providing direction and meaning to the development of a strong critique, given the possible weakness that might come from a purely pragmatic stance.[12] It is time to return to these concerns, and to highlight how a valuographic research programme might help direct and stimulate attention to these pressing concerns.

POWER AND POLITICS

Enacting values is one way of producing *stakes*—i.e., matters of concern or care. What is supposed to be at stake, and what is at stake, in the life sciences, techno-science, and society at large is the object of intense politics. Being attentive to the multiplicity of values that are enacted provides critical and analytical leverage regarding issues of power and politics in the life sciences, techno-science, and society. Switching to emic registers of values fosters sensitivity to the myriad ways in which values are implicated in politics and power struggles. What seem to be coherent and rational processes of valuation are always already political battlefields. The aim with our programme is to show how a number of different values may be at stake in seemingly technical and mundane decision making regarding matters such as efficiency, safety, and quality of care.

Furthermore, addressing stakemaking in practice allows us to understand how matters of concern or care are constructed on at least two levels: first, there can be *conflicts about what the concern is*; second, there can be *conflicts about the correct way of assessing* a stake along a settled register of value. Here we acknowledge that actors will have varying stakes in different issues, as well as different ways of assessing these issues. For example, in the 2011 Fukushima disaster, the stakes differed greatly between actors—families worried about

[12] Luc Boltanski (2013) has discussed the various merits of a *critical sociology* and a more pragmatic *sociology of critical practice*, arguing that the latter, while well attuned to appreciating the struggles of ordinary actors, 'did not succeed in fostering a form of critique of more salient potency that could supply actors with the resources needed to reinforce their critical will and their critical efficiency' (Boltanski 2013: 48).

radiation doses; tour operators worried about the dropping number of tourists; and local governments worried about the depopulation of their constituencies (see Chapter 1, introduction to this volume). These are worries of the first kind, concerning what is at stake. But in Fukushima there were also worries of the second kind, concerning what constituted a safe radiation dose. Should the inhabitants of Fukushima stay in their area and continue with their lives, or should they go, leaving behind their lives and starting anew?

The performance of values is also a matter of stakemaking. This makes the valuographic approach to examining the unsettledness of values a gateway to examining the making of stakes and hence the politics of life sciences, techno-science, and society at large.

ECONOMIC AND NON-ECONOMIC VALUES

Distinctions between what are deemed economic, cultural, and medical values are made in practices. It is therefore pertinent to examine how certain matters come to be considered economic values and how they might come to be juxtaposed or subordinated to—or indeed take the place of—other values. Using a valuographic approach, we can address how economic value is *made* through complex entanglements with values that are not 'economic'.[13]

Our point is that the study of values-in-the-making simultaneously allows us to appreciate the enactment of different registers of value. With a valuographic approach to the practices of life science, techno-science, and society, value hierarchies are not seen as predetermined. The importance of different values is something settled as part of the action, rather than being something that exogenously settles the action. This further implies that the practices of making (or unmaking) distinctions between different registers of value in themselves become subjects for empirical investigation. For example, how are financial, medical, or scientific values related to drug development? Are they differentiated or seen as interlinked? (See Chapter 7 by Daniel Neyland and Elena Simakova for a study of this.)

The notion of *relational work* proposed by Viviana Zelizer (Zelizer 2005, see also Bandelj 2012) to address the varied practices by which people in intimate relationships differentiate and maintain social ties can provide inspiration here. It is precisely this kind of pragmatic attention to such practices that can allow a valuographic research programme to examine the enactment of economic values alongside other values. In the life sciences, techno-science, and society at large, the composition of values—i.e., the making of distinctions

[13] Of course, the very categories 'economic' and 'other' are clumsy placeholders used in studying the making of these categories in practice. See the introduction (Chapter 1, Dussauge et al.) of this volume for more discussion of this matter.

between values, the hierarchical ordering of them, and the establishment of possible commensurations—provides a good position for examining how the economic is made to matter in society. With inspiration from Stefan Helmreich (2008), the valuographic approach allows us to move beyond notions wherein economic value, capital, and assets are seen as stable signifiers of value, to instead compose empirically sensitive and analytically sensible accounts of how economic and other values are made and differentiated in practice.

KNOWLEDGE AND VALUES

Different registers of value are enacted and ordered alongside any project of knowledge production. New knowledge and technologies might inspire efforts to articulate and enact certain values and to displace others. There is an urgent need to challenge the idea, cherished in many instances, that scientific knowledge is unconnected from values.

A valuographic research programme entails asking how values are enacted in sites of knowledge production such as laboratories, government statistics offices, and market-research enterprises. While retaining an interest in knowledge production and an emphasis on knowledge as a precarious and contingent achievement from the sociology of scientific knowledge (SSK); (see e.g., Bloor 1976), a valuographic research programme needs to depart from the traditional inclination of SSK to treat the values (and interests) of a certain set of powerful actors as determinants of the knowledge produced. Our concern directs our attention to how values and knowledge are precarious outcomes of contingent practices rather than assuming that values are the impetus shaping knowledge (see Chapter 13, Isabelle Dussauge, this volume, for an example of this).

A valuographic approach can, in our opinion, sensitize investigations to the multifaceted relationships between facts-in-the-making and values-in-the-making. In this way, we can benefit from the contribution of SSK to establishing scientific knowledge production as a crucial area for critical empirical study, extending its scope to examine how knowledge production involves enacting values and making distinctions between what are deemed values and what are deemed facts. In short, a valuographic research programme permits the empirical investigation of value enactment in matters of epistemic concern.

VALUOGRAPHIES AND WORLD-MAKING

Annemarie Mol (2013) recently discussed how the enactment of particular versions of objects simultaneously enacts particular 'ontonorms', normativities, and registers of values. Using the case of dieting, she examines how food is treated so as to affect the registers of value enacted: activities that treat food as calories enact foods as fuel (calories) and valorize restraining oneself from supplying more to the body than is consumed. Activities pertaining to the 'food pyramid' cast food as mere nutrients, enacting a register of values related to the cognitive control of ingestion. In short, what food is made to be mobilizes various metrics and values concerning what constitutes proper and good dieting behaviour.[14]

The 'turn to ontology' (Woolgar and Lezaun 2013) opens avenues for exploring how the enactment of ontologies is interrelated with the enactment of values (Mol 2013). In short, this turn lets us explore the intersections between an ontographic (Lynch 2013) and a valuographic research programme. Such links can appear in the form of overt interventions, for example, when devising devices for allocating transplant organs (see Chapter 5 by Philip Roscoe); in more mundane forms, for example, in tending tomato plants in an effort to make them productive (Heuts and Mol 2013); or indeed as emergent in co-modification processes that transform something into a valuable commodity (Chapter 9 by Kristin Asdal). Practices enacting and shaping certain realities can in many ways be linked to the enactment of values. This means, we believe, that an empirically oriented ontology research programme would benefit from an empirically oriented research programme examining values.

Commitments and challenges: How can a valuographic research programme matter?

We have illustrated above how addressing contingency—i.e., matters that 'could be otherwise'—is a compelling research strategy for examining how values are enacted. It should be increasingly clear that a valuographic research programme addresses classical concerns of politics in the life sciences as well as in techno-science and society at large.[15]

[14] A key part of turning ontology into an empirical research programme within STS has been to examine how things and realities may be multiple and far from as unified as is often assumed. Entities not only 'can be otherwise' but can, in various practices, simultaneously be otherwise—that is, they can be enacted to be distinctly different versions of themselves. Arteriosclerosis, to cite one example studied in detail by Annemarie Mol, can in this way be treated as something that is done, stabilized, and constrained differently in different clinical practices (Mol 2002). The basic idea here is to investigate how objects (both mundane and scientific) are made multiple and the consequences of doing so. This, then, is an ontological version that stands in contrast to a more epistemologically oriented approach investigating how objects are perceived differently from different perspectives.

[15] To us, it seems as though the devising of constructivist perspectives on facts and technologies can only prise open the black box of politics to a certain degree. Calls to address the politics of techno-scientific work abound, addressing matters of concern (Latour 2004), matters of care (de la Bellacasa

Now we wish to ask what a pragmatic *commitment* might look like—without reifying stable value categories stemming from the nineteenth century. This question relates to our three broader concerns with *stakemaking*, the *economic*, and the *epistemic* outlined in Chapter 1, the introductory chapter of this volume and above. Our starting point has been the life sciences, but we believe that a pragmatic take on our three concerns—our three commitments—is relevant to understanding a society saturated with debates and controversies pertaining to values more broadly.

We propose an analysis of the making of values—perhaps guided by our three concerns—to gain analytical purchase on the slippery politics of today's society. We suggest that the analysis of values-in-the-making provides a tool useful in apprehending the complexities of politics and power in a society increasingly characterized by intricate webs of relationships—often economic or scientific—alongside opaque valuations that increasingly depend on 'judgement devices' that materialize credible knowledge (Karpik 2010). We also propose that, by approaching values-in-the-making, we could begin to understand the proliferation of valuations in society (Kjellberg et al. 2013)—even in popular cultural phenomena such as reality shows and competitions.[16] Our point is that value practices are a diffused and important aspect of social life and how politics are done. We propose that analysing value practices in light of our concerns—in stakemaking, in the economic, in the epistemic—could allow us to engage more fully with the politics of a society increasingly obsessed with values.

Characterizing and understanding values in a society that works through assemblages of contracts, machines, and networks is a complex and difficult matter. Classic social analysis applying a Marxist reading of values becomes increasingly problematic in a society where local workers are subcontractors for global franchises, pension funds (for the same workers?) are ravenous capitalists—mediated via diffused ownership, and an identity of belonging (to the working-class?) fills people with discomfort (on the discomfort of belonging, see the introduction in Karpik 2010).[17]

As outlined in Chapter 1, the introductory chapter, we wish to carefully approach values-in-the-making without succumbing to our own—or others'—pre-stabilized analytical categories. By following the value practices of contemporary society, by

2011), concerned groups (Callon and Rabeharisoa 2008; Galis and Hansson 2012), standpoint epistemologies (Harding 1991), situated knowledges (Haraway 1988), and ontological politics (Mol 1999). It seems as though looking at the construction of facts and artefacts does not satisfy the thirst to understand the politics of techno-science.

[16] We want to thank Jonas Bååth for bringing to our attention how widespread the performance of valuation has become in popular culture, as seen in shows such as *Hell's Kitchen* or *American Idol*; (on valuation as a voyeuristic spectacle, see Muniesa and Helgesson 2013).

[17] This uneasiness with Marxist social analysis corresponds to uneasiness with approaches attempting to grapple with biocapital in the contemporary life sciences (Helmreich 2008).

following the making of stakes, the drawing of boundaries, and commensurations of value, we might gain a different understanding of the politics of power that complements a classical analysis of labour, capital, and assets (cf. Birch and Tyfield 2012). This move would entail tracing organizations, processes, devices, the shaping of actors, and translations of values. It would further entail describing the multifarious conflicts over values that penetrate and permeate welfare systems, global corporations, NGOs, and public discourse.

REFLECTING ON THE ONTOLOGICAL POLITICS OF VALUOGRAPHY

What does an analysis attempting to eschew pre-stabilized analytical categories hide? What are the consequences of an emic valuography as a 'social and political posture' (Pestre 2013)? Having an emic outlook on the world is not innocent. Doing emic analysis—backgrounding the agency and categories of the researcher—is a choice rife with politics. The risk of an emic perspective is that we might fashion our analysis in the form of Boyle's 'modest witnessing' in attempting to efface our agency from the analysis (Haraway 1997; Shapin 1984). As Dominique Pestre has argued, and as has been debated in social science for decades, in performing an emic analysis we are creating a certain story:

> [...] we select our players (and forget a lot) and endow specific properties (at the expense of others) because we cannot not simplify things, because we have limits, because we have an idea of... what 'understand' means—because, as humans, we cannot grasp everything (and even less in a narrative, which is necessarily linear) and we want to, consciously or not, emphasize certain points, some connections, some reconfigurations. (Our translation, Pestre 2013: 210)

What Pestre points out is the need to acknowledge our situated and partial perspectives (see also Haraway 1988). In any analysis, researchers need to strategically find 'the right balance between the views of the people they study and their own' (Löwy 2010b). We advocate treating the agnostic approach as a critical device for empowering the weak and revealing the constructed nature of weakness (cf. Galis and Lee 2014). In this sense, our valuographic approach to contingency is a call for questioning the power of taken-for-granted values in the world. In liberal democracies, there seems to be an increasing technocratization of values (e.g., the increased use of cost–benefit analyses or algorithms for political decision making) and thereby of choice. By questioning the taken-for-granted value enactments in society, we wish to bring to light the political nature of valuation, desire, and values.

The second issue we want to raise is that an agnostic approach to values-in-the-making must not embrace nihilism. Our argument for a methodologically emic stance on a valuographic research programme has deliberately been combined with emphasizing concerns related to stakemaking, the economic, and the epistemic. We care deeply about the liberal democratic technocratization and depolitization of values, i.e., the moving of choice, value, and politics to arenas dominated by notions such as efficiency, optimality, risk minimization, and market competition. We want to question our and society's taken-for-granted notions of value by staying sensitive to the making of value. As in all social science, choosing one's valuographic concerns—i.e., empirical area, emotional motivation, and methodological choice—will determine the nature of one's research. As Galis and Lee have argued elsewhere, 'constructing stories is a political activity. So choose your starting point wisely' (Galis and Lee 2014).

The third issue concerns the risk that a pragmatic and emic approach might provide insufficient resources to allow actors to perform a substantial critique (see Boltanski 2013). Our emphasis on our concerns is intended to provide direction here. In addition, we appreciate the usefulness of remaining unfaithful and only partially committed to an emic register of analysis. In our view, the researcher must move strategically between using the emic device to subvert pre-stabilized values and unfaithfully using other devices to demonstrate how 'macrologies' structure values-in-the-making—to recall Pestre's (2013) invocation of Gayatri Chakravorty Spivak. In the ideal case, a valuography goes hand in hand with the strategic demonstration of violence and of the oppression of the weak.

Modes of intervention

How could a valuographic programme add to the issues it addresses? How would it intervene in stakemaking? What repertoires of intervention, mattering, and care does a valuographic research programme allow for? Here we join a number of scholars in asking how a pragmatic stance on values in techno-science, epistemology, and ontology can intervene in the fields that we study: How can we as social scientists provide a situated intervention into the value practices of contemporary society (Zuiderent-Jerak 2015, in press)? How can we as researchers of techno-science matter (Law 2004)? How can we interfere or 'diffract' (Haraway 1992)? How do we do an ontological politics intertwined with a logic of choice (Mol 1999, 2008)? How can we care (de la Bellacasa 2011)?

For us, one way to approach these questions has been through thinking about stakemaking through values. How are stakes made? How are they agreed upon? How are they disagreed on? How are boundaries drawn around stakes? The corollary also becomes: How can we as researchers intervene—and care about—stakemaking? Below we outline four modes in which a valuographic research programme might intervene in stakemaking, in how values are made. These are not mutually exclusive, nor are they to be seen as exhaustive. Rather, they must be seen as provisional placeholders for the myriad ways in which a valuographic research programme could inform and contribute to politics in a society consumed with values and valuation. These modes are: rebalancing, caring, interfering, and inspiring.

Through these moves of intervening in values-in-the-making we also wish to make clear that a pragmatic and agnostic research agenda is not at the same time a nihilist agenda. A caring commitment to exploring the making of values in practice does not mean nihilism; rather, it means a commitment to exploring how value practices are implicated in power relations, oppression, the drawing of boundaries, commensurations, and their effects.

REBALANCING STAKEMAKING: EXPERIMENTS AND NUDGES

Balance–rebalance. Our first mode of intervention attempts to nudge, to rebalance, to redirect—ever so slightly—the delicate enactments of value that occur and recur ubiquitously. The question is how we, as researchers, can participate in stakemaking, how boundaries between different sets of values are drawn and redrawn in practice, how commensurations are made and unmade. This requires, first, the continuous identification of critical issues as well as the likewise endless work of transforming such issues into solvable problems. Such a style of civilized multi-problematization (see Callon 2009) might very well benefit from an interdisciplinary approach incorporating valuographic elements.

With Teun Zuiderent-Jerak and Georges Canguilhem (Zuiderent-Jerak 2015, in press), we might consequently ask how we can work with situated interventions to shape and produce new kinds of actors—new agencies of normativity—that point creatively and constructively towards the future, rather than conservatively to the past.[18] In line with Zuiderent-Jerak's proposal of 'situated intervention', this mode of intervening would entail getting closer to local practices and their 'normative surfeit' and attempt to nudge them in another direction through experimental interventions. Such interventions could, for instance, attempt to work with the difficulties of translating questions of quality into quantifiable, portable numbers in health care (see Zuiderent-Jerak 2015, in press).

[18] The notion of shaping agency alludes to the concept of *agencement* (Callon 2007), which captures the understanding of agency as an assemblage. This also permits the understanding that equipping agents gives them new capacities to be agents in the world.

CARE ABOUT VALUES: VISIBILITIES AND OTHERINGS

Care. Our second mode of intervention moves with the affective register. We can examine, and intervene in, the drawing of boundaries between cognitive and affective values. Following Bellacasa (de la Bellacasa 2011), such a mode of intervention would involve a commitment to values which are neglected. This is a mode of intervention that intertwines notions of nurturing care to give visibility to values that are oppressed, othered, made invisible, and made pathological. The intertwining of agency, affects, and values becomes highlighted.

In this mode of intervening, the social scientist might ask questions attuned to feminist discussions of standpoint and situated knowledges (Haraway 1988; Harding 1991): What values are we as social science researchers to nurture? What values need to be supported and cultivated? How is the enactment of values intertwined with the making of affect and cognition? How are values assigned to a cognitive or affective register of value?

INTERFERENCE IN STAKEMAKING: THE ACTIVIST MODE OF INTERVENING

Interference. Our third mode of intervention moves in the register of the activist. An intervention based on this mode would be an explicit call for action, a call to consciously and with commitment join the fray of enacting and ordering values in society. Intervening becomes a matter of joining in and participating in the enacting, ordering, and displacing of values in certain directions.

In this mode of intervening, along with Vasilis Galis and Anders Hansson (Galis and Hansson 2012), we might ask questions about different modes of activist intervention in value making. Questions of intervening in value making become matters of conscious interference—of action research. In the activist mode of intervening, consciousness of and commitment to value making become the crucial matters. What side should we take in controversies? How can we as social scientists and activists take sides in the committed creation of specific values?

INSPIRATION IN STAKEMAKING: THE PROMISSORY MODE OF INTERVENING

Inspiration. Our fourth mode of intervention involves galvanizing, energizing, stirring—so that things can indeed become otherwise. An intervention in this mode attempts to create value momentum, to participate in making the promissory moves of shaping the trajectories of assemblages of values, actors, and devices by devising provisional imaginaries of values. The sociology of expectations might be to inform a conscious effort to make certain futures possible.

In this mode of intervening we might ask questions about how to inspire to action, to imagine a different future. Rather than just analysing the promissory moves in life science, this mode of intervening could entail developing new promissory futures. The intervention might move ever closer to politics. What values do we want enacted in the making of future developments? What values can we imagine for a promissory and imagined future?

Coda

This book is based on the premise that debates and practices in the life sciences and medicine are saturated with values and that there is a pressing need to consider values as things to be explained and explored rather than accepted as given entities with explanatory power. In this final chapter we have provided some suggestions for making one's own valuography: we have outlined a number of approaches that can be used to examine the making of values, to examine the contingencies in their enactment, ordering, and displacement; we have proposed a few key concerns that can provide a direction for making an important valuographic inquiry; and we have, finally, briefly put forward a few ways in which a valuographic research programme can be made to matter in the world.

All of the above has been guided by our broad proposition that a developed pragmatic understanding of values can better address a set of crucial concerns in the dynamics of the life sciences and their place in society.

Our first such concern is related to the composition of values and the making of boundaries between them—be they economic, cultural, or epistemic. For the life sciences, one theme might be how something comes to be considered as both an economic and a non-economic value. Our argument is that understanding the processes by which these compositions and boundaries are made is crucial for understanding the dynamics of and challenges related to the life sciences.

Our second concern is related to the enactment and stabilization of values in relation to the epistemic, which, for instance, can be related to valuations performing what comes to be considered worth knowing and, equally crucially, according to what specific metrics. Again, we argue that understanding the processes by which this is done is crucial for understanding central dynamics and challenges in the life sciences and medicine.

Our third concern relates to how valuation practices are intertwined with the making of stakes in the life sciences and medicine. We claim that understanding the processes by which this is achieved is fundamental for understanding the dynamics of the life sciences, their practices of power, social order, subjectivities, and affects.

In this chapter, we have explored these themes, by way of identifying the contours of a valuographic research programme as well as the ways in which we, as scholars, on the basis of such a research programme, can take up its central challenges. Not only can we take an interest in how scientists, regulators, analysts, and publics regularly strive to define what is considered proper conduct in science and health care, economically and socially valuable, and worth knowing. Starting to outline a valuographic research programme also translates into exploring various ways we can aspire to make valuographies that matter.

We have, finally, argued that the life sciences and medicine are not the only arena that we can expect to find saturated with values in play in debates and practices. We hope this volume contains enough inspiration, currency, and leverage for both the study and making of values. There are enough challenges and concerns for us all—scholars, scientists, regulators, analysts, and publics—to address.

BIBLIOGRAPHY

Archive material for Chapter 3

Deutsch, Adam and Kallós, Paul (undated). 'Samtal med professor Axel Westman'. 2224: Leo Historical Archives, Helsingborg.

Deutsch, Adam and Kallós, Paul (1945). 'Rapport över besök hos Överläkare S. Genell, 4 April 1945, Malmö'. 2224: Leo Historical Archives, Helsingborg.

Leo Historical Archives (undated). 'Barnlös plötsligt gravid med nytt hormonpreparat'. 1368.

Leo Historical Archives (undated). 'Konto nr 304: Råvarukostnader Gonadex'. 2456.

Leo Historical Archives (1933). 'Till ledning för styrelsesammanträde i tredje kvartalet 1933, Bilaga 3'. 5097.

Leo Historical Archives (1943). 'Avtal mellan direktionen för Karolinska sjukhuset och aktiebolaget Leo angående bolagets hormonlaboratorium vid Karolinska sjukhuset'. 1169; signed in Hälsingborg 4 December 1943 and in Stockholm 21 December 1943.

Leo Historical Archives (1945). 'Protokoll vid sammanträdet den 6 april 1945'. 2224.

Leo Historical Archives (1948a). 'Rydin, Håkan to AB Leo, 3 December 1948'. 6029.

Leo Historical Archives (1948b). 'Rydin, Håkan to AB Leo, 8 October 1948'. 1368.

Leo Historical Archives (1948c). 'Professor Westmans anförande vid presskonferensen den 10 June 1948'. 1368.

Leo Historical Archives (1948d). 'Westman, Axel to AB Leo, 31 May 1948'. 1368.

Leo Historical Archives (1948e). 'Leo, AB to Axel Westman'. 1368.

Leo Historical Archives (1948f). 'Kalkyl över framställning av chorion-gonadotropin'. 6029.

Leo Historical Archives (1948g). 'P. M. för sammanträde i Hälsingborg lördagen den 17 januari 1948'. 1368.

Leo Historical Archives (1950a). 'Rapport från sammanträde den 14 juli 1950 beträffande urininsamling'. 2456.

Leo Historical Archives, Helsingborg (1950b). 'Redovisning för tiden den 11–16 december 1950'. 2456.

Stockholm County Council Archives (1944). 'Westman, Axel to Carl Holtman, 3 November 1944'. E1, 6.

Stockholm County Council Archives (1946). 'Westman, Axel to the Directorate of the Karolinska Hospital, 23 January 1946'. E1, 9.

References

Abi-Rached, Joelle M. (2012). 'From Brain to Neuro: The Brain Research Association and the Making of British Neuroscience, 1965–1996', *Journal of the History of the Neurosciences*, 21 (2): 189–213.

Acker, Caroline Jean (2010). 'How Crack Found a Niche in the American Ghetto: The Historical Epidemiology of Drug-related Harm', *BioSocieties*, 5: 70–88.

Adams, Vincanne, Murphy, Michelle, and Clarke, Adele E. (2009). 'Anticipation: Technoscience, Life, Affect, Temporality', *Subjectivity*, 28 (1): 246–65.

Affibody (2001). *Affibody Annual Report 2001*. Stockholm: Affibody.

Aftenposten (2012). 'En feit og fin og norsk en' [A nice, fat, Norwegian one] 12 May 2012, pp. 29–32.

Agaton, Charlotta et al. (2003). 'Affinity Proteomics for Systematic Protein Profiling of Chromosome 21 Gene Products in Human Tissues', *Molecular & Cellular Proteomics*, 2 (6): 405–14.

Ahlin, Jerker and Lundgren, Rolf (2002). *Från Leopiller till nicorette: En industri i Helsingborg under det tjugonde seklet*. Helsingborg: Pharmacia.

Ahmed, Sara (2010). *The Promise of Happiness*. Durham, NC & London: Duke University Press.

Ahrens, Thomas and Chapman, Christopher S. (2006). 'Doing Qualitative Field Research in Management Accounting: Positioning Data to Contribute to Theory', *Accounting, Organizations and Society*, 31 (8): 819–41.

Ainslie, George and Monterosso, John (2004). 'Behavior: A Marketplace in the Brain?', *Science*, 306 (5695): 421–3.

Aitken, David A. et al. (1996). 'Dimeric inhibin A as a Marker for Down's Syndrome in Early Pregnancy', *New England Journal of Medicine*, 334 (19): 1231–6.

Anderson, Stuart (ed.) (2005). *Making Medicines: A Brief History of Pharmacy and Pharmaceuticals*. London: Pharmaceutical Press.

Anonymous (1948). 'Den bakteriefria fabriken', *Teknik för alla*, 14: 3–5.

Anonymous (2010). 'Fundamental Pleasure Questions', in Morten L. Kringelbach and Kent C. Berridge (eds), *Pleasures of the Brain*. New York: Oxford University Press, 7–23.

Anonymous (2012). 'Post-Chernobyl disaster sheep controls lifted on last UK farms', (updated 1 June 2012) <http://www.bbc.co.uk/news/uk-england-cumbria-18299228>, accessed 4 June 2013.

Anonymous (2013a). 'Bases biologiques du plaisir: le système de récompense', *SVT féminin masculin: Un site pour les enseignants et les élèves* <http://svtfemininmasculin.com>, accessed February 11, 2013.

Anonymous (2013b). 'SVT féminin masculin: Un site pour les enseignants et les élèves', <http://svtfemininmasculin.com>, accessed February 11, 2013.

Anteby, Michel (2010). 'Markets, Morals, and Practices of Trade: Jurisdictional Disputes in the U.S. Commerce in Cadavers', *Administrative Science Quarterly*, 55 (4): 606–38.

Appadurai, Arjun (1986). 'Introduction: Commodities and the Politics of Value', in Arjun Appadurai (ed.), *The Social Life of Things: Commodities in Cultural Perspective*. Cambridge: Cambridge University Press, 3–63.

Araujo, Luis, Finch, John and Kjellberg, Hans (2010). *Reconnecting Marketing to Markets*. Oxford: Oxford University Press.

Asdal, Kristin (2004). *Politikkens teknologier*, Oslo: Det hist.-filos. fakultet, Univ. i Oslo.

Asdal, Kristin (2005). 'Returning the Kingdom to the King A Post-constructivist Response to the Critique of Positivism', *Acta Sociologica*, 48 (3): 253–61.

Asdal, Kristin (2008). 'On Politics and the Little Tools of Democracy: A Down-to-Earth Approach', *Distinktion: Scandinavian Journal of Social Theory*, 9 (1): 11–26.

Asdal, Kristin (2011). 'The Office: The Weakness of Numbers and the Production of Non-authority', *Accounting, Organizations and Society*, 36 (1): 1–9.

Asdal, Kristin (2014). 'From Climate Issue to Oil Issue: Offices of Public Administration, Versions of Economics and the Ordinary Technologies of Politics', *Environment and Planning A*, 46 (9): 2110–24.

Ashmore, Malcolm, Mulkay, Michael, and Pinch, Trevor J. (1989). *Health and Efficiency: A Sociology of Health Economics.* Milton Keynes: Open University Press.

Aspers, Patrik and Beckert, Jens (2011). 'Value in Markets', in Jens Beckert and Patrik Aspers (eds), *The Worth of Goods: Valuation and Pricing in the Economy.* Oxford: Oxford University Press, 3–38.

Atkinson-Grosjean, Janet and Fairley, Cory (2009). 'Moral Economies in Science: From Ideal to Pragmatic', *Minerva*, 47: 147–70.

Auden, W. H. (1970). 'Freedom and Necessity in Poetry', in Arne Tiselius and Sam Nilsson (eds), *The Place of Value in a World of Facts: Proceedings of the Fourteenth Nobel Symposium, Stockholm, September 15–20, 1969.* New York: John Wiley & Sons, 135–42.

Baccarani, Michele et al. (2013). 'European LeukemiaNet Recommendations for the Management of Chronic Myeloid Leukemia: 2013', *Blood*, 122 (6): 872–84.

Baily, Mary Ann and Murray, Thomas H. (2008). 'Ethics, Evidence, and Cost in Newborn Screening', *Hastings Center Report*, 38 (3): 23–31.

Bajos, Nathalie and Ferrand, Michèle (2011). 'De l'interdiction au contrôle: Les enjeux contemporains de la légalisation de l'avortement', *Revue française des affaires sociales*, (1): 42–60.

Baker, Wayne E. (1984). 'The Social Structure of a National Securities Market', *American Journal of Sociology*, 89: 775–81.

Bandelj, Nina (2012). 'Relational Work and Economic Sociology', *Politics & Society*, 40 (2): 175–201.

Bang, Jens and Northeved, Allan (1972). 'A New Ultrasonic Method for Transabdominal Amniocentesis', *American Journal of Obstetrics and Gynecology*, 114 (5): 599.

Barry, Andrew and Slater, Don (2002). 'Technology, Politics and the Market; An Interview with Michel Callon', *Economy and Society*, 31 (2): 285–306.

Bayer (2006). 'Sustainability Development Report', <http://www.sustainability2006.bayer.com/en/good-corporate-citizen.aspx>, accessed 10 October 2010.

Beck, Stephen and Niewöhner, Jörg (2009). 'Localising Genetic Testing and Screening in Cyprus and Germany: Contingencies, Continuities, Ordering Effects and Bio-cultural Intimacy', in Paul Atkinson, Peter Glaser and Margaret Lock (eds), *Handbook of Genetics & Society.* London: Routledge, 76–93.

Beckert, Jens (2011). 'The Transcending Power of Goods: Imaginative Value in the Economy', in Jens Beckert and Patrik Aspers (eds), *The Worth of Goods: Valuation and Pricing in the Economy.* Oxford: Oxford University Press.

Beckert, Jens and Aspers, Patrik (eds) (2011). *The Worth of Goods: Valuation and Pricing in the Economy.* Oxford: Oxford University Press.

Beebee, Trevor J. C. and Griffiths, Richard A. (2005). 'The Amphibian Decline Crisis: A Watershed for Conservation Biology?', *Biological Conservation*, 125: 271–85.

Ben-David, Joseph (1971). *The Scientist's Role in Society: A Comparative Study.* Englewood Cliffs, NJ: Prentice-Hall.

Berenson, Alex (2006). 'End of Drug Trial is a Big Loss for Pfizer', *New York Times*, <http://www.nytimes.com/2006/12/04/health/04pfizer.html?pagewanted=all&_r=0>, accessed 23 October 2014.

Berg, Marc, de Brantes, Francois, and Schellekens, Wim (2006). 'The Right Incentives for High-quality, Affordable Care; A New Form of Regulated Competition', *International Journal for Quality in Health Care,* 18 (4): 261–3.

Bermel, Joyce (1983). 'Update on Genetic Screening: An FDA Approved Kit, Views on Early Diagnosis', *The Hastings Center Report,* 13 (5): 3–4.

Berridge, Kent C. (2003). 'Pleasures of the Brain', *Brain and Cognition,* 52 (1): 106–28.

Berridge, Kent C. and Kringelbach, Morten L. (2008). 'Affective Neuroscience of Pleasure: Reward in Humans and Animals', *Psychopharmacology,* 199: 457–80.

Beunza, Daniel, Hardie, Iain and MacKenzie, Donald (2006). 'A Price Is a Social Thing: Towards a Material Sociology of Arbitrage', *Organization Studies,* 27 (5): 721–45.

Bevis, D. C. A. (1952). 'The Antenatal Prediction of Haemolytic Disease of the Newborn', *Lancet,* 1 (6704): 395–8.

Bhopal, Raj et al. (1997). 'The Vexed Question of Authorship: Views of Researchers in a British Medical Faculty', *BMJ: British Medical Journal,* 314 (7086): 1009.

Biagioli, Mario and Galison, Peter (2003). 'Introduction', in Mario Biagioli and Peter Galison (eds), *Scientific Authorship: Credit and Intellectual Property in Science.* New York: Routledge, 1–9.

Birch, Kean and Tyfield, David (2012). 'Theorizing the Bioeconomy: Biovalue, Biocapital, Bioeconomics or... What?', *Science, Technology & Human Values,* 38 (3): 299–332.

Black, John, Hashimzade, Nigar, and Myles, Gareth (2012). *A Dictionary of Economics.* 4th edn. Oxford: Oxford University Press.

Blackburn, Simon (2008). *The Oxford Dictionary of Philosophy.* 2nd rev. edn. Oxford: Oxford University Press.

Bleier, Ruth (1984). *Science and Gender: A Critique of Biology and Its Theories on Women.* New York: Pergamon Press.

Bloor, David (1976). *Knowledge and Social Imagery.* London: Routledge.

Blume, Stuart S. (1992). *Insight and Industry: On the Dynamics of Technological Change in Medicine.* Cambridge, MA: MIT Press.

Boeije, Hennie (2002). 'A Purposeful Approach to the Constant Comparative Method in the Analysis of Qualitative Interviews', *Quality & Quantity,* 36 (4): 391–409.

Boltanski, Luc (2004). *La condition fœtale: Une sociologie de l'engendrement et de l'avortement.* Paris: Gallimard.

Boltanski, Luc (2013). 'A Journey through French-style Critique', in Paul du Gay and Glenn Morgan (eds), *New Spirits of Capitalism? Crises, Justifications, and Dynamics.* Oxford: Oxford University Press, 43–59.

Boltanski, Luc and Thévenot, Laurent (1991). *De la justification: Les économies de la grandeur.* Paris: Gallimard.

Boltanski, Luc and Thévenot, Laurent (1999). 'The Sociology of Critical Capacity', *European Journal of Social Theory,* 2 (3): 359–77.

Boltanski, Luc and Thévenot, Laurent (2006). *On Justification: Economies of Worth.* Princeton, NJ: Princeton University Press.

Borell, Ulf (1994). 'Pionjärer inom svensk obstetrik och gynekologi: Axel Westman', *SFOGSs medlemsblad,* 1 (Jan): 25–7.

Borrell, Antoni et al. (2004). 'First-trimester Screening for Trisomy 21 Combining Biochemistry and Ultrasound at Individually Optimal Gestational Ages. An Interventional Study', *Prenatal Diagnosis,* 24 (7): 541–5.

Borup, Mads, Brown, Nik, Konrad, Kornelia, and Van Lente, Harro (2006). 'The Sociology of Expectations in Science and Technology', *Technology Analysis & Strategic Management*, 18 (3–4): 285–98.

Bouissac, Paul (1998). *Encyclopedia of Semiotics.* Oxford: Oxford University Press.

Bowker, Geoffrey C. and Star, Susan Leigh (1999). *Sorting Things Out: Classification and its Consequences.* Cambridge, MA: MIT Press.

Boyd, Patricia, DeVigan, Catherine, and Garne, Ester (2005). *Prenatal Screening Policies in Europe.* EUROCAT.

Boyd, Patricia A. et al. (2008). 'Survey of Prenatal Screening Policies in Europe for Structural Malformations and Chromosome Anomalies, and Their Impact on Detection and Termination Rates for Neural Tube Defects and Down's Syndrome', *BJOG: An International Journal of Obstetrics & Gynaecology,* 115 (6): 689–96.

Brahme, Leonard (1951). 'Framsteg och publicitet', *Svenska Läkartidningen,* 48: 415.

Braverman, Irus (2011). 'Looking at Zoos', *Cultural Studies,* 25 (6): 809–42.

Braverman, Irus (2012). *Zooland: The Institution of Captivity.* Palo Alto, CA: Stanford University Press.

Breiter, Hans C. et al. (2001). 'Functional Imaging of Neural Responses to Expectancy and Experience of Monetary Gains and Losses', *Neuron,* 30 (2): 619–39.

Brock, David J. H. (1982). *Early Diagnosis of Foetal Defects.* Edinburgh & London: Churchill Livingstone.

Brock, David J. H. and Sutcliffe, Roger G. (1972). 'Alpha-fetoprotein in the Antenatal Diagnosis of Anencephaly and Spina Bifida', *The Lancet,* 300 (7770): 197–9.

Brown, Nik (2003). 'Hope against Hype: Accountability in Biopasts, Presents and Futures', *Science Studies,* 16 (2): 3–21.

Brown, Nik (2009). 'Shifting Tenses: From Regimes of truth to Regimes of Hope', *Configurations,* 13 (3): 331–55.

Brown, Nik (2013). 'Vital Value: Dialectics and Contradictions', paper given at Dimensions of Value and Values in Science Technology and Innovation Studies, Edinburgh, 17–18 April 2013.

Brown, Nik and Michael, Mike (2003). 'A Sociology of Expectations: Retrospecting Prospects and Prospecting Retrospects', *Technology Analysis & Strategic Management,* 15 (1): 3–18.

Brown, Nik, Rappert, Brian, and Webster, Andrew (2000). *Contested Futures: A Sociology of Prospective Techno-science.* Aldershot: Ashgate.

Brown, Nik, Machin, Laura and McLeod, Danae (2011). 'Immunitary Bioeconomy: The Economisation of Life in the International Cord Blood Market', *Social Science & Medicine,* 72 (7): 1115–22.

Brown, Richard E. and Milner, Peter M. (2003). 'The Legacy of Donald O. Hebb: More Than the Hebb Synapse', *Nature Reviews Neuroscience,* 4 (12): 1013–19.

Buchanan, Ian (2010). *A Dictionary of Critical Theory.* Oxford: Oxford University Press.

Bud, Robert (2008). 'Upheaval in the Moral Economy of Science? Patenting, Teamwork and the World War II Experience of Penicillin', *History and Technology,* 24 (2): 173–90.

Bull, Malcolm (2012). 'What is the Rational Response? A Review of A Perfect Moral Storm: The Ethical Tragedy of Climate Change. Oxford, July 2011', *London Review of Books 34,* 10 (3–6).

Burke, Peter (1990). *The French Historical Revolution: The Annales School, 1929–89.* Cambridge: Polity Press.

Cabanac, Michel (2010). 'The Dialectics of Pleasure', in Morten L. Kringelbach and Kent C. Berridge (eds), *Pleasures of the Brain*. New York: Oxford University Press, 113–24.

Calhoun, Craig (2002). *Dictionary of the Social Sciences*. Oxford: Oxford University Press.

Callon, Michel (1986). 'Some Elements of a Sociology of Translation: Domestication of the Scallops and the Fishermen of St Brieuc Bay', in John Law (ed), *Power, Action and Belief: A New Sociology of Knowledge*. London: Routledge, 196–233.

Callon, Michel (1987). 'Society in the Making; The Study of Technology as a Tool for Sociological Analysis', in Wiebe Bijker, Thomas Hughes and Trevor Pinch (eds), *The Social Construction of Technological Systems: New Directions in the Sociology and History of Technology*. Cambridge, MA: MIT Press, 83–103.

Callon, Michel (ed.) (1998). *The Laws of the Markets*. Oxford: Blackwell.

Callon, Michel (2005). 'Why Virtualism Paves the Way to Political Impotence. Callon Replies to Miller', *Economic Sociology: European Electronic Newsletter*, 6 (2): 3–20.

Callon, Michel (2007). 'What Does It Mean to Say That Economics is Performative?', in Donald MacKenzie, Fabian Muniesa and Lucia Siu (eds), *Do Economists Make Markets? On the Performativity of Economics*. Princeton, NJ: Princeton University Press, 311–57.

Callon, Michel (2009). 'Civilizing Markets: Carbon Trading between in vitro and in vivo Experiments', *Accounting, Organizations and Society*, 34:535–48.

Callon, Michel and Latour, Bruno (1981). 'Unscrewing the Big Leviathan: How Actors Macro-structure Reality and How Sociologists Help Them to Do So', in Karin Knorr-Cetina and Aron V. Cicourel (eds), *Advances in Social Theory and Methodology: Toward an Integration of Micro- and Macro-Sociologies*. London: Routledge & Kegan Paul, 277–303.

Callon, Michel and Rabeharisoa, Vololona (2003). 'Research "in the Wild" and the Shaping of New Social Identities', *Technology in Society*, 25 (2): 193–204.

Callon, Michel and Rabeharisoa, Vololona (2008). 'The Growing Engagement of Emergent Concerned Groups in Political and Economic Life', *Science, Technology, & Human Values*, 33 (2): 230–61.

Callon, Michel, Méadel, Cécile, and Rabeharisoa, Vololona (2002). 'The Economy of Qualities', *Economy and Society*, 31 (2): 194–217.

Callon, Michel, Millo, Yuval and Muniesa, Fabian (eds) (2007). *Market Devices*. Oxford: Blackwell.

Cambrosio, Alberto and Keating, Peter (1992). 'A Matter of FACS: Constituting Novel Entities in Immunology', *Medical Anthropology Quarterly*, 6 (4): 362–84.

Cambrosio, Alberto and Keating, Peter (1995). *Exquisite Specficity: The Monoclonal Antibody Revolution*. Oxford: Oxford University Press.

Camerer, Colin, Loewenstein, George and Prelec, Drazen (2005). 'Neuroeconomics: How Neuroscience Can Inform Economics', *Journal of Economic Literature*, 43 (1): 9–64.

Carlat, Daniel (2007). 'Dr. drug rep', *New York Times*, 25 November 2007.

Castilla, Eduardo E. and Orioli, Iêda M. (2004). 'ECLAMC: The Latin-American Collaborative Study of Congenital Malformations', *Public Health Genomics*, 7 (2–3): 76–94.

Ceci, Stephen J. (1988). 'Scientists' Attitudes toward Data Sharing', *Science, Technology, & Human Values*, 13 (1/2): 45–52.

Centre national de la recherche scientifique (CNRS) (2010). 'Sex, money... Specific brain areas for each pleasure', <http://www2.cnrs.fr/en/1787.htm>, accessed 28 March 2013.

Chazan, Lilian Krakowski (2008). '"É...tá grávida mesmo! E ele é lindo!": A construção de 'verdades' na ultra-sonografia obstétrica', *História, Ciências, Saúde-Manguinhos*, 15 (1): 99–116.

Cherry, Mark J. (2005). *Kidney for Sale by Owner.* Washington DC: Georgetown University Press.

Childress, James F. (1989). 'Ethical Criteria for Procuring and Distributing Organs for Transplantation', *Journal of Health Politics Policy and Law*, 14 (1): 87–113.

Chitty, Lyn S. et al. (2012). 'Non-invasive Prenatal Testing for Aneuploidy–Ready for Prime Time?', *American Journal of Obstetrics and Gynecology*, 206 (4): 269–75.

Christie, Daphne and Tansey, Elizabeth (2003). *Genetic Testing.* 17: London: Wellcome Trust Centre for the History of Medicine at UCL.

Christie, Daphne and Zallen, Doris (2003). 'Genetic Testing', paper given at Witness Seminar in 20th Century Medicine, London, 13 July 2001.

Chrulew, Matthew (2011). 'Managing Love and Death at the Zoo: The Biopolitics of Endangered Species Preservation', *Australian Humanities Review*, 50:137–57.

Claesson, L., Högberg, B., Rosenberg, Th., and Westman, A. (1948). 'Crystalline Human Chorionic Gonadotrophin and Its Biological Action', *Acta Endocrinologica*, 1(1): 1–18.

Clarke, Adele E. and Montini, Theresa (1993). 'The Many Faces of RU486: Tales of Situated Knowledges and Technological Contestations', *Science, Technology and Human Values*, 18 (1): 42–78.

Clarke, Adele E. et al. (2003). 'Biomedicalization: Technoscientific Transformations of Health, Illness, and US Biomedicine', *American Sociological Review*, 68 (April): 161–94.

Claxton, Larry D. (2005a). 'Scientific Authorship. Part 1. A Window into Scientific Fraud', *Mutat Research*, 589 (1): 17–30.

Claxton, Larry D. (2005b). 'Scientific Authorship: Part 2. History, Recurring Issues, Practices, and Guidelines', *Mutation Research. Reviews in Mutation Research*, 589 (1): 31–45.

Cochoy, Franck (2007). 'A Sociology of Market-Things: On Tending the Garden of Choices in Mass Retailing', in Michel Callon, Yuval Millo and Fabian Muniesa (eds), *Market Devices.* Malden, MA: Blackwell, 109–29.

Cochoy, Franck (2009). 'Driving a Shopping Cart from STS to Business, and the Other Way Round: On the Introduction of Shopping Carts in American Grocery Stores (1936–1959)', *Organization*, 16 (1): 31–55.

Cockburn, W. Chas (1991). 'The International Contribution to the Standardization of Biological Substances I. Biological Standards and the League of Nations 1921–1946', *Biologicals*, 19 (3): 161–9.

Cockburn, W. Chas et al. (1991). 'The International Contribution to the Standardization of Biological Substances II. Biological Standards and the World Health Organization 1947–1990: General Considerations', *Biologicals*, 19 (4): 257–64.

Cohen, Lawrence (2005). 'Operability, Bioavailability, and Exception', in Aihwa Ong and Steven Collier (eds), *Global Assemblages: Technology, Politics, and Ethics as Anthropological Problems.* Malden, MA: Blackwell, 124–43.

Cole, R. David (1996). 'CHOH HAO LI', *Biographical Memoirs*, 70: 221–39.

Coleman, James Samuel, Katz, Elihu, and Menzel, Herbert (1966). *Medical Innovation: A Diffusion Study.* New York: Bobbs-Merrill.

Collins, Harry M. (1981). 'Son of Seven Sexes: The Social Destruction of a Physical Phenomenon', *Social Studies of Science,* 11 (1): 33–62.

Collins, Harry M. (1985). *Changing Order: Replication and Induction in Scientific Practice.* Beverly Hills, CA: Sage.

Collins, Harry M. and Pinch, Trevor (1998). *The Golem: What You Should Know about Science.* Cambridge: Cambridge University Press.

Cooper, Melinda (2007). 'Life, Autopoiesis, Debt: Inventing the Bioeconomy', *Distinktion,* 14:25–43.

Cooper, Melinda (2008). *Life as Surplus: Biotechnology and Capitalism in the Neoliberal Era.* Seattle, WA: University of Washington.

Cowan, Ruth Schwartz (2008). *Heredity and Hope: The Case for Genetic Screening.* Cambridge, MA: Harvard University Press.

Cuckle, Howard S., Wald, Nicholas J., and Lindenbaum, Richard H. (1984). 'Maternal Serum Alpha-Fetoprotein Measurement: A Screening Test for Down Syndrome', *The Lancet,* 323 (8383): 926–9.

Cuckle, Howard S., Wald, Nicholas J. and Thompson, S. G. (1987). 'Estimating a Woman's Risk of Having a Pregnancy Associated with Down's Syndrome Using Her Age and Serum Alpha-Fetoprotein Level', *BJOG: An International Journal of Obstetrics & Gynaecology,* 94 (5): 387–402.

Cunningham, George C. and Tompkinson, D. Gwynne (1999). 'Cost and Effectiveness of the California Triple Marker Prenatal Screening Program', *Genetics in Medicine,* 1:199–206.

Cutting Edge Information (2009) 'Thought leader fair-market value: Compensation benchmarks and procedures', <http://www.cuttingedgeinfo.com/thought-leader-fmv/>, accessed 17 August 2014.

Dally, Ann (1998). 'Thalidomide: Was the Tragedy Preventable?', *The Lancet,* 351 (9110): 1197–9.

Daston, Lorraine (1995). 'The Moral Economy of Science', *Osiris,* 10: 2–24.

Daston, Lorraine and Galison, Peter (2007). *Objectivity.* Brooklyn, NY: Zone Books.

Davidson, Ronald G. and Rattazzi, Mario C. (1972). 'Prenatal Diagnosis of Genetic Disorders: Trials and Tribulations', *Clinical Chemistry,* 18 (3): 179–87.

de Chadarevian, Soraya (2002). *Designs for Life: Molecular Biology after World War II.* Cambridge: Cambridge University Press.

de la Bellacasa, Maria (2011). 'Matters of Care in Technoscience: Assembling Neglected Things', *Social Studies of Science,* 41 (1): 85–106.

de Quervain, Dominique J.-F. et al. (2004). 'The Neural Basis of Altruistic Punishment', *Science,* 305 (5688): 1254–8.

Dear, Peter (2001). 'Science Studies as Epistemography', in Jay A. Labinger and Harry M. Collins (eds), *The One Culture? A Conversation about Science.* Chicago, IL: University of Chicago Press, 128–41.

Delgado, Richard (1991). 'Norms and Normal Science: Toward a Critique of Normativity in Legal Thought', *University of Pennsylvania Law Review,* 139 (4): 933–62.

Derry, Margaret E. (2003). *Bred for Perfection: Shorthorn Cattle, Collies, and Arabian Horses since 1800.* Baltimore, MD & London: Johns Hopkins University Press.

Devaney, Stephanie A., Palomaki, Glenn E., Scott, Joan A. and Bianchi, Diana W. (2011). 'Noninvasive Fetal Sex Determination Using Cell-free Fetal DNA', *JAMA (Journal of the American Medical Association),* 306 (6): 627.

Dewey, John (1913). 'The Problem of Values', *Journal of Philosophy, Psychology and Scientific Methods,* 10 (10): 268–9.

Dewey, John (1939). *Theory of Valuation. International Encyclopedia of Unified Science.* Chicago, IL: University of Chicago Press.

Dickinson, Anthony and Balleine, Bernard (2010). 'Hedonics: The Cognitive–Motivational Interface', in Morten L. Kringelbach and Kent C. Berridge (eds), *Pleasures of the Brain.* New York: Oxford University Press, 74–84.

Ditlefsen, Anne (2007). 'Tilfeldig oppdagelse med kjempegevinst' [Enormous gain from an accidental finding] <http://forskning.no/fisk-fiskehelse-oppdrett-landbruk-stub/2008/02/tilfeldig-oppdagelse-med-kjempegevinst>, accessed 22 August 2014.

DNDi (2006). <http://www.dndi.org/>, accessed 1 April 2011.

Donald, Ian (1969). 'Ultrasonics in Diagnosis (Sonar)', *Proceedings of the Royal Society of Medicine,* 62 (5): 442.

Doyle, Charles (2011). *A Dictionary of Marketing.* 3rd edn. Oxford: Oxford University Press.

Dr F 'Comments on Dollars for Docs', <http://www.propublica.org/article/profiles-of-the-top-earners-in-dollar-for-docs>, accessed 31 March 2011.

Dreger, Alice, Feder, Ellen K., and Tamar-Mattis, Anne (2012). 'Prenatal Dexamethasone for Congenital Adrenal Hyperplasia', *Journal of Bioethical Inquiry,* 9 (3): 277–94.

Duden, Barbara (1993). *Disembodying Women: Perspectives on Pregnancy and the Unborn.* Cambridge, MA: Harvard University Press.

Dupré, John (2007). 'Fact and Value', in Harold Kincaid, John Dupré and Alison Wylie (eds), *Value-Free Science? Ideals and Illusions.* Oxford: Oxford University Press, 27–41.

Dussauge, Isabelle (2010). 'Sex, lögner och evolutionens förvrängda löften [Sex, lies and evolution's distorted promises]', *Tidsskrift for Kjønnsforskning,* (4): 433–45.

Dussauge, Isabelle (2013). 'The Experimental Neuro-Framing of Sexuality', *Graduate Journal of Social Science,* 10 (1): 124–51.

Dussauge, Isabelle (2014). 'Sex, Cash, and Neuromodels of Desire', *BioSocieties,* advance online publication 11 August, doi: 10.1057/biosoc.2014.23.

The Economist (2006). 'When the drugs don't work', *The Economist* <http://www.economist.com/node/8375604>, accessed 23 October 2014.

Egmond, Stans van and Teun Zuiderent-Jerak (2010). Analysing Pocily Change: the performative role of economics in the constitution of a new policy programme in Dutch health care. In: Stans van Egmond (ed). *Science and Policy in Interaction: On practices of science policy interactions for policy-making in health care.* Rotterdam, Erasmus University Rotterdam: 133–54.

Ehrenreich, Barbara (2009). *Bright-sided: How the Relentless Promotion of Positive Thinking Has Undermined America.* New York: Metropolitan Books.

Elam, Mark (2010). 'Health as Life Bordering on Brain Disease: The Expanding Orbit of the Neurobiology of Addiction', paper given at Health—A New Religious Awakening in Western Societies? Copenhagen, 13–15 September.

Elliott, Carl (2010). *White Coat Black Hat: Adventures on the Dark Side of Medicine.* Boston, MA: Beacon Press.

Enthoven, Alain C. (1988). *Theory and Practice of Managed Competition in Healthcare Finance.* Amsterdam: Elsevier.

Enthoven, Alain C. and van de Ven, Wynand P. M. M. (2007). 'Going Dutch – Managed-Competition Health Insurance in the Netherlands', *New England Journal of Medicine,* 357 (24): 2421–3.

Epstein, Steven (1996). *Impure Science: AIDS, Activism, and the Politics of Knowledge,* 7: Berkeley, CA: University of California Press.

Eriksson, Gunnar (1987). 'Vetenskapens roller i läkemedelsframställningen: Från den galeniska eran till den svenska läkemedelsindustrins framväxt', in Tore Frängsmyr (ed.), *Vetenskap och läkemedel: Ett historiskt perspektiv*. Stockholm: Almqvist & Wiksell, 39–75.

Espeland, Wendy Nelson and Stevens, Mitchell L. (1998). 'Commensuration as a Social Process', *Annual Review of Sociology*, 24: 313–43.

Etzkowitz, Henry and Leydesdorff, Loet (2000). 'The Dynamics of Innovation: From National Systems and "Mode 2" to a Triple Helix of University–Industry–Government Relations', *Research Policy*, 29 (2): 109–23.

Ezzell, Carol (2002). 'Proteins Rule', *Scientific American*, 286 (4): 40.

Fan, H. Christina et al. (2012). 'Non-invasive Prenatal Measurement of the Fetal Genome', *Nature*, 487 (7407): 320–4.

Farlow, Andrew (2004). 'Over the Rainbow: The Pot of Gold for Neglected Diseases', *The Lancet*, 364 (9450): 2011–12.

Farlow, Andrew (2005). *The Global HIV Vaccine Enterprise, Malaria Vaccines, and Purchase Commitments: What is the Fit?* Geneva: WHO.

Farlow, Andrew (2006). *The Science, Economics, and Politics of Malaria Vaccine Policy*, paper given at a submission to UK Department for International Development and the Malaria Vaccine Technology Roadmap, London, 5 March.

Farlow, Andrew, Light, Donald, Mahoney, Richard, and Widdus, Roy (2005). *Concerns Regarding the Centre for Global Development Report 'Making Markets for Vaccines'*, paper given at submission to Commission on Intellectual Property Rights, Innovation and Public Health, WHO, Geneva, 29 April.

Fassin, Didier (2004). 'Le corps exposé: Essai d'économie morale de l'illégitimité', in Didier Fassin and Dominique Memmi (eds), *Le gouvernement des corps*. Paris: Editions de l'Ecole des Hautes Etudes en Sciences Sociales, 235–66.

Fassin, Didier (2005). 'Compassion and Repression: The Moral Economy of Immigration Policies in France', *Cultural Anthropology*, 20 (3): 362–87.

Fassin, Didier (2007). 'Economie morale du traumatisme', in D. Fassin and R. Rechtman (eds), *L'empire du traumatisme. Enquête sur la condition de victime*. Paris: Flammarion, 403–17.

Fassin, Didier (2009a). 'Les économies morales revisitées', in *Annales. Histoire, sciences sociales*, 64. Paris: Editions de l'EHESS, 1237–66.

Fassin, Didier (2009b). 'In the Heart of Humaneness: The Moral Economy of Humanitarianism', in Didier Fassin and Mariella Pandolfi (eds), *Contemporary States of Emergency: The Politics of Military and Humanitarian Intervention*. New York: Zone Books.

Feenberg, Andrew (1999). *Questioning Technology*. New York: Routledge.

Feenberg, Andrew (2002). *Transforming Technology: A Critical Theory Revisited*, 8. Oxford: Oxford University Press.

Feero, W. Gregory, Guttmacher, Alan E., Bodurtha, Joann and Strauss III, Jerome F. (2012). 'Genomics and Perinatal Care', *New England Journal of Medicine*, 366 (1): 64–73.

Finholt, Thomas A. (2002). 'Collaboratories', *Annual Review of Information Science and Technology*, 36 (1): 73–107.

Fishman, Jennifer R. (2004). 'Manufacturing Desire: The Commodification of Female Sexual Dysfunction', *Social Studies of Science*, 34 (2): 187–218.

Fleck, Ludwik (1979 [1935]). *Genesis and Development of a Scientific Fact*, ed. Thaddeus J. Trenn and Robert K. Merton. Chicago, IL: University of Chicago Press.

Forsythe, John (2009). *Transplantation: A Companion to Specialist Surgical Practice*. London: Saunders.

Fortun, Mike (1999). 'Projecting Speed Genomics', in Mike Fortun and Everret Mendelsohn (eds), *The Practices of Human Genetics. Sociology of the Sciences Yearbook 21* Dordrecht: Kluwer, 25–48.

Foucault, Michel (1970). *The Order of Things: An Archaeology of the Human Sciences.* London: Routledge.

Foucault, Michel (2003). *Society Must Be Defended. Lectures at the College de France 1975–1976*, eds Mauro Bertani et al., trans. David Macey. New York: Picador.

Foucault, Michel (2004). *Sécurité, territoire, population: cours au Collège de France (1977–1978).* Paris: Gallimard.

Fourcade, Marion (2011a). 'Price and Prejudice: On Economics and the Enchantment (and Disenchantment) of Nature', in J. Beckert and P. Aspers (eds), *The Worth of Goods: Valuation and Pricing in the Economy.* Oxford: Oxford University Press.

Fourcade, Marion (2011b). 'Cents and Sensibility: Economic Valuation and the Nature of "Nature"', *American Journal of Sociology,* 116 (6): 1721–77.

Fox News/Opinion Dynamics Poll (2002). 'Do you think it is acceptable to use cloning?' <http://www.pollingreport.com/science.htm>, accessed 4 June 2007.

Fradelos, Christina Kathryn (2008). *The Last Desperate Cure: Electrical Brain Stimulation and its Controversial Beginnings.* Chicago, IL: University of Chicago.

Franklin, Sarah (2007). *Dolly Mixtures: The Remaking of Genealogy.* Durham, NC: Duke University Press.

Franklin, Sarah and Lock, Margaret (eds) (2003). *Remaking Life & Death: Toward an Anthropology of the Biosciences.* Santa Fe, NM: School of American Research Press.

Franklin, Sarah and Roberts, Celia (2006). *Born and Made: An Ethnography of Preimplantation Genetic Diagnosis.* Princeton, NJ: Princeton University Press.

Freeman, Richard B. (2007). 'Survival Benefit: Quality Versus Quantity and Trade-offs in Developing New Renal Allocation Systems', *American Journal of Transplantation,* 7 (5): 1043–6.

Friese, Carrie (2009). 'Models of Cloning, Models for the Zoo: Rethinking the Sociological Significance of Cloned Animals', *BioSocieties,* 4 (4): 367–90.

Friese, Carrie (2010). 'Classification Conundrums: Categorizing Chimeras and Enacting Species Preservation', *Theory and Society,* 39 (2): 145–72.

Friese, Carrie (2013a). *Cloning Wild Life: Zoos, Captivity and the Future of Endangered Animals.* New York: New York University Press.

Friese, Carrie (2013b). 'Realizing Potential in Translational Medicine: The Uncanny Emergence of Care as Science', *Current Anthropology,* 54 (S7): S129–S138.

Frijda, Nico H. (2010). 'On the Nature and Function of Pleasure', in Morten L. Kringelbach and Kent C. Berridge (eds), *Pleasures of the Brain.* New York: Oxford University Press, 99–112.

Fujimura, Joan H. (2013). 'Confounded Categories: Is the HapMap a Race-based Project?', paper given at 4S/EASST Design and Displacement—Social Studies of Science and Technology, Copenhagen, 17–20 October.

Fukushima, Masato (2012). 'The Politics of Natural Products', in *Asian Biopoleis II: Crossing Boundaries*, 1–27.

Galis, Vasilis (2006). *From Shrieks to Technical Reports: Technology, Disability and Political Processes in Building Athens Metro.* Linköping: Linköping University.

Galis, Vasilis and Hansson, Anders (2012). 'Partisan Scholarship in Technoscientific Controversies: Reflections on Research Experience', *Science as Culture,* 21 (3): 335–64.

Galis, Vasilis and Lee, Francis (2014). 'A Sociology of Exclusion', *Science, Technology, & Human Values,* 39 (1): 154–79.

Gårdlund, Waldemar (1948). 'Vetenskaplig reklam och praktisk erfarenhet: Ett bidrag till problemet om kvinnans sterilitet', *Svenska Läkartidningen,* 45:1733–46.

Gascon, Claude et al. (2007). *Amphibian Conservation Action Plan.* Gland, Switzerland and Cambridge: IUCN/SSC Amphibian Conservation Summit.

Gaskell, George (2000). 'Agricultural Biotechnology and Public Attitudes in the European Union', *AgBioForum,* 3 (2 & 3): 87–96.

Gates Malaria Partnership (2006). *Gates Malaria Partnership Report,* <http://www.gatesmalariapartnership.org/sites/www.gatesmalariapartnership.org/files/reports/attachments/Annual%20Report%202001-2006.pdf>, accessed 4 September 2014.

Gaudillière, Jean-Paul (2005). 'Better Prepared Than Synthesized: Adolf Butenandt, Schering Ag and the Transformation of Sex Steroids into Drugs (1930–1946)', *Studies in History and Philosophy of Biological and Biomedical Sciences,* 36 (4): 612–44.

Gaudillière, Jean-Paul and Löwy, Ilana (1998). 'General Introduction', in Jean-Paul Gaudillière and Ilana Löwy (eds) *The Invisible Industrialist: Manufactures and the Production of Scientific Knowledge.* London and New York: St Martin's Press, 3–15.

Gaynor, Martin and Vogt, William B. (1999). *Antitrust and Competition in Health Care Markets.* Cambridge, MA: National Bureau of Economic Research.

Gaynor, Martin, Moreno-Serra, Rodrigo, and Propper, Carol (2010). 'Death by Market Power; Reform, Competition and Patient Outcomes in the National Health Service', *NBER working paper series,* Working Paper 16164. Cambridge, MA: National Bureau of Economic Research.

Georgiadis, Janniko R. and Kortekaas, Rudie (2010). 'The Sweetest Taboo: Functional Neurobiology of Human Sexuality in Relation to Pleasure', in Morten L. Kringelbach and Kent C. Berridge (eds), *Pleasures of the Brain.* New York: Oxford University Press, 178–201.

Geschwind, Irving I, Li, Choh Hao, and Barnafi, Livio (1956). 'Isolation and Structure of Melanocyte-Stimulating Hormone from Porcine Pituitary Glands 1', *Journal of the American Chemical Society,* 78 (17): 4494–5.

Gieryn, Thomas F. (1983). 'Boundary-work and the Demarcation of Science from Non-science: Strains and Interests in Professional Ideologies of Scientists', *American Sociological Review,* 48 (6): 781–95.

Gieryn, Thomas F. (1995). 'Boundaries of Science', in Sheila Jasanoff et al. (eds), *Handbook of Science and Technology Studies.* London: Sage, 393–443.

Gieryn, Thomas F. (1999). *Cultural Boundaries of Science: Credibility on the Line.* Chicago, IL: University of Chicago Press.

GlaxoSmithKline (GSK) (2005). *Corporate Responsibility Report,* <http://www.gsk.com/responsibility/cr_report_2005/index.htm>, accessed 10 October 2010.

Glennerster, Rachel, Kremer, Michael, and Williams, Heidi (2006). 'Creating Markets for Vaccines', *Innovations: Technology, Governance, Globalization,* 1 (1): 67–79.

Glimcher, Paul W. (2008). 'Neuroeconomics', *Scholarpedia,* 10.

Global Health Program (GHP) (2011). *Gates Foundation Malaria Strategy,* <http://www.gatesfoundation.org/malaria/Documents/malaria-strategy.pdf>, accessed 25 April 2012.

Go, Attie T. J. I., van Vugt, John M. G. and Oudejans, Cees B. M. (2011). 'Non-invasive Aneuploidy Detection Using Free Fetal DNA and RNA in Maternal Plasma: Recent Progress and Future Possibilities', *Human Reproduction Update*, 17 (3): 372–82.

Godin, Benoît (2006). 'The Linear Model of Innovation: The Historical Construction of an Analytical Framework', *Science, Technology, & Human Values*, 31 (6): 639–67.

Godlee, Fiona (2012). 'Why Markets Don't Work in Healthcare', *British Medical Journal*, 344: e3300.

Gold, Rachel Benson (2003). 'Lessons from before Roe: Will Past be Prologue?', *Guttmacher Report on Public Policy*, 6 (1).

Golinski, Jan (1998). *Making Natural Knowledge: Constructivism and the History of Science*. Cambridge: Cambridge University Press.

Gomez, Martha C. et al. (2003). 'Nuclear Transfer of Synchronized African Wild Cat Somatic Cells into Enucleated Domestic Cat Oocytes', *Biology and Reproduction*, 69 (3): 1032–41.

Gomez, Martha C. et al. (2004). 'Birth of African Wildcat Cloned Kittens Born from Domestic Cats', *Cloning and Stem Cells*, 6:247–58.

Gomez, Martha C. et al. (2008). 'Nuclear Transfer of Sand Cat Cells into Enucleated Domestic Cat Oocytes is Affected by Cryopreservation of Donor Cells', *Cloning and Stem Cells*, 10 (4): 469–83.

Goodman, Jordan (1998). 'Can it Ever Be Pure Science? Pharmaceuticals, the Pharmaceutical Industry and Biomedical Research in the Twentieth Century', in John Paul Gaudillière and Ilana Löwy (eds), *The Invisible Industrialist. Manufacturers and the Production of Scientific Knowledge*. London and New York: St Martin's Press, 143–66.

Goodwin, Charles (1994). 'Professional Vision', *American Anthropologist*, 96 (3): 606–33.

Gøtzsche, Peter C. et al. (2007). 'Ghost Authorship in Industry-initiated Randomised Trials', *PLoS medicine*, 4 (1): e19.

Gould, Stephen Jay (2000). 'More Things in Heaven and Earth', in Hilary Rose and Steven Rose (eds), *Alas, Poor Darwin: Arguments against Evolutionary Psychology*. London: Jonathan Cape.

Graeber, David (2001). *Toward an Anthropological Theory of Value: The False Coin of Our Own Dreams*. New York: Palgrave.

Grattan, Sean (2011). 'Ahmed, Sara. The Promise of Happiness', *Social Text*, <http://socialtextjournal.org/the-promise-of-happiness/>, accessed August 2014.

Greely, Henry T. (2011). 'Get Ready for the Flood of Fetal Gene Screening', *Nature*, 469 (7330): 289–91.

Greely, Henry T. and King, Jaime S. (2010). 'The Coming Revolution in Prenatal Genetic Testing', *AAAS Professional Ethics Report*, 23 (2): 1–4.

Greener, Ian (2003). 'Patient Choice in the NHS: The View from Economic Sociology', *Social Theory & Health*, 1 (1): 72–89.

Greenhoouse, Linda and Siegel, Reva (2010). *Before Roe v. Wade: Voices that Shaped the Abortion Debate before the Supreme Court Ruling*. New York: Kaplan.

Grinten, Tom van der (2006). *Zorgen om beleid; Over blijvende afhankelijkheden en veranderende bestuurlijke verhoudingen in de gezondheidszorg*. Rotterdam: Department of Health Policy and Management.

Grob, Rachel (2011). *Testing Baby: The Transformation of Newborn Screening, Parenting, and Policymaking*. Piscataway, NJ: Rutgers University Press.

Gross, Alan G., Harmon, Joseph E., and Reidy, Michael (2007). *Communicating Science: The Scientific Article from the 17th Century to the Present*. Chicago, IL: University of Chicago Press.

Gross, Sky E. and Shuval, Judith T. (2008). 'On Knowing and Believing: Prenatal Genetic Screening and Resistance to 'Risk-Medicine'', *Health, Risk & Society*, 10 (6): 549–64.

Guggenheim, Michael and Potthast, Joerg (2011). 'Symmetrical Twins. On the Relationship between ANT and Critical Capacities', <http://tuberlin.academia.edu/JoergPotthast/Papers/697242/Symmetrical_twins._On_the_relationship_between_ANT_and_the_sociology_of_critical_capacities_2012_>, accessed 25 April 2012.

Guston, David H. and Keniston, Kenneth (1994). 'Introduction: The Social Contract for Science', in *The Fragile Contract. University Science and the Federal Government.* Cambridge MA and London: MIT Press, 1–41.

Hacking, Ian (1999). *The Social Construction of What?* Cambridge, MA: Harvard University Press.

Hagstrom, Warren O. (1975). *The Scientific Community.* Carbondale, IL: Southern Illinois University Press

Hahn, Sinuhe et al. (2011). 'Determination of Fetal Chromosome Aberrations from Fetal DNA in Maternal Blood: Has the Challenge Finally Been Met?', *Expert Reviews in Molecular Medicine,* 13 (1).

Hamburger, Christian, Pedersen-Bjergaard, Kaj, and Wijnbladh, Hjalmar (1960). 'Axel Westman 29.12.1894–29.5.1960', *Acta Endocrinologica,* 34:XIX–XXXIII.

Hammer, Peter J. (1999). 'Questioning Traditional Antitrust Presumptions: Price and Non-price Competition in Hospital Markets', *U. Mich. JL Reform,* 32: 727–1119.

Hanson, Elizabeth (2002). *Animal Attractions: Nature on Display in American Zoos.* Princeton, NJ: Princeton University Press.

Haraway, Donna J. (1988). 'Situated Knowledges: The Science Question in Feminism and the Privilege of Partial Perspective', *Feminist Studies,* 14 (3): 575–99.

Haraway, Donna J. (1992). 'The Promises of Monsters: A Regenerative Politics for Inappropriate/d Others', in Lawrence Grossberg, Cary Nelson and Paula A. Treichler (eds), *Cultural Studies.* New York: Routledge.

Haraway, Donna J. (1997). *Modest_Witness@Second_Millennium. FemaleMan©_Meets_OncoMouse: Feminism and Technoscience.* New York: Routledge.

Haraway, Donna J. (2008). *When Species Meet.* Minneapolis, MN: Minnesota University Press.

Harding, Sandra (1991). 'Feminist Standpoint Epistemology', in Ingrid Bartsch and Muriel Lederman (eds), *The Gender and Science Reader.* London: Routledge, 145–68.

Harper, Peter S. (2006). *First Years of Human Chromosomes: The Beginnings of Human Cytogenetics.* Banbury: Scion Publishing.

Harris, Ieuan J. (1960). 'The Chemistry of Melanocyte-stimulating Hormones', in Arthur James Rook (ed.), *Progress in the Biological Sciences in Relation to Dermatology.* Cambridge: Cambridge University Press, 28–37.

Harris, Marvin (1976). 'History and Significance of the Emic/Etic Distinction', *Annual Review of Anthropology,* 5: 329–50.

Harris, R. and Andrews, T. (1988). 'Prenatal Screening for Down's Syndrome', *Archives of Disease in Childhood,* 63: 705–6.

Harrison, Glenn W. (2008). 'Neuroeconomics: A Critical Reconsideration', *Economics and Philosophy,* 24 (Special Issue 03): 303–44.

Hayden, Cori (2003). *When Nature Goes Public: The Making and Unmaking of Bioprospecting in Mexico.* Princeton, NJ: Princeton University Press.

Healy, David and Cattell, Dinah (2003). 'Interface between Authorship, Industry and Science in the Domain tf Therapeutics', *British Journal of Psychiatry,* 183 (1): 22–7.

Helderman, Jan-Kees, Schut, Erik, van der Grinten, Tom, and van de Ven, Wynand (2005). 'Market-Oriented Health Care Reforms and Policy Learning in the Netherlands', *Journal of Health Politics, Policy and Law,* 30 (1–2): 189–209.

Helgesson, Claes-Fredrik (2010). 'From Dirty Data to Credible Scientific Evidence: Some Practices Used to Clean Data in Large Randomised Clinical Trials', in Catherine Will and Tiago Moreira (eds), *Medical Proofs, Social experiments: Clinical Trials in Context.* Aldershot: Ashgate, 49–66.

Helgesson, Claes-Fredrik and Kjellberg, Hans (2005). 'Macro-actors and the Sounds of the Silenced', in Barbara Czarniawska and Tor Hernes (eds), *Actor-Network Theory and Organizing.* Malmö & Copenhagen: Liber & Copenhagen Business School Press, 145–64.

Helgesson, Claes-Fredrik and Lee, Francis (2012). 'Valuations of Experimental Designs in Proteomic Biomarker Experiments and Traditional RCTs', paper given at What Price Creativity? A workshop on the valuing of social/public goods, University of St Andrews, 12–14 December.

Helmreich, Stefan (2008). 'Species of Biocapital', *Science as Culture,* 17 (4): 463–78.

Hensley, Scott and Martinez, Barbara (2005). 'To sell their drugs, companies increasingly rely on doctors', *Wall Street Journal (Eastern ed.),* 15 July: A1, A2.

Herzenberg, Leonard A. et al. (1979). 'Fetal Cells in the Blood of Pregnant Women: Detection and Enrichment by Fluorescence-Activated Cell Sorting', *Proceedings of the National Academy of Sciences,* 76 (3): 1453–5.

Heuts, Frank and Mol, Annemarie (2013). 'What Is a Good Tomato? A Case of Valuing in Practice', *Valuation Studies,* 1 (2): 125–46.

Holt, William V., Pickard, Amanda R., and Prather, Randall S. (2004). 'Wildlife Conservation and Reproductive Cloning', *Society for Reproduction and Fertility,* 127: 317–24.

Hook, Ernest B. (1981). 'Rates of Chromosome Abnormalities at Different Maternal Ages', *Obstetrics & Gynecology,* 58 (3): 282–5.

Hopwood, Anthony G. (2009). 'Accounting and the Environment', *Accounting, Organizations and Society,* 34 (3): 433–9.

Ideland, Malin (2002). *Dagens gennyheter: Hur massmedier berättar om genetik och genteknik.* Lund: Nordic Academic Press.

InsiteResearch (2011). 'Can KOL Management Generate a Return on Investment?', *Next Generation Pharmaceutical 14* <http://www.ngpharma.com/>, accessed 28 March 2011.

Ivry, Tsipy (2009a). *Embodying Culture: Pregnancy in Japan and Israel.* New Brunswick, NJ & London: Rutgers University Press.

Ivry, Tsipy (2009b). 'The Ultrasonic Picture Show and the Politics of Threatened Life', *Medical Anthropology Quarterly,* 23 (3): 189–211.

Ivry, Tsipy, Teman, Elly, and Frumkin, Ayala (2011). 'God-sent Ordeals and Their Discontents: Ultra-Orthodox Jewish Women Negotiate Prenatal Testing', *Social Science & Medicine,* 72 (9): 1527–33.

Jain, S. Lochlann (2010). 'The Mortality Effect: Counting the Dead in the Cancer Trial', *Public Culture,* 22 (1): 89–117.

Jefferson, T. et al. (2009). 'Relation of Study Quality, Concordance, Take Home Message, Funding, and Impact in Studies of Influenza Vaccines: Systematic Review', *BMJ: British Medical Journal,* 338.

Jensen, Arne et al. (1985). *Å dyrke havet. Perspektivanalyse på norsk havbruk.* Trondheim: Tapir.

Joas, Hans (2000). *The Genesis of Values*, trans. Gregory Moore. Manchester: Polity.

Johansson Krafve, Linus (2012). 'To Design Free Choice and Competitive Neutrality: The Construction of a Market in Primary Care', *Scandinavian Journal of Public Administration*, 15 (4): 45–66.

Johri, Mira and Ubel, Peter A. (2003). 'Setting Organ Allocation Priorities: Should We Care What the Public Cares about?', *Liver Transplantation*, 9 (8): 878–80.

Jost, Timothy S. (2007). *Health Care at Risk: A Critique of the Consumer-Driven Movement*. Durham, NC: Duke University Press.

Journal Officiel (France) (2009). 'Arrête du 23 Juin 2009: Les règles de bonnes pratiqes en matière de dépistage et de diagnostique prénatals avec utilisation des marqueurs sériques maternels de la trisomie 21'.

Kahneman, Daniel, Diener, Ed, and Schwarz, Norbert (1999). *Well-Being: The Foundations of Hedonic Psychology*. New York: Russell Sage Foundation.

Kajii, T., Kida, M., and Takahashi, K. (1973). 'The Effect of Thalidomide Intake during 113 Human Pregnancies', *Teratology*, 8 (2): 163–6.

Karpik, Lucien (2010). *Valuing the Unique: The Economics of Singularities*. Princeton, NJ: Princeton University Press.

Katz, Elihu and Lazarsfeld, Paul Felix (2006). *Personal Influence: The Part Played by People in the Flow of Mass Communications*. New Brunswick, NJ: Transaction.

Kaufman, Leslie (2012). 'Zoos' bitter choice: To save some species, letting others die', *The New York Times*, sec. News, p. A1.

Keating, Peter and Cambrosio, Alberto (2012). *Cancer on Trial: Oncology as a New Style of Practice*. Chicago, IL: University of Chicago Press.

Kekwick, R. A. and Pedersen, Kai O. (1974). 'Arne Tiselius. 1902–1971', *Biographical Memoirs of Fellows of the Royal Society*, 20: 401–28.

Kelly, Michael (1998). *Encyclopedia of Aesthetics*. Oxford: Oxford University Press.

Kerr, Anne (2009). 'Reproductive Genetics. From Choice to Ambivalence and Back Again', in Paul Atkinson, Peter Glaser and Margaret Lock (eds), *The Handbook of Genetics & Society: Mapping the New Genomic Era*. London: Routledge, 59–75.

Kerr, Anne and Cunningham-Burley, Sarah (2000). 'On Ambivalence and Risk: Reflexive Modernity and the New Human Genetics', *Sociology*, 34 (2): 283–304.

Kevles, Daniel J. (1998). *The Baltimore Case: A Trial of Politics, Science, and Character*. New York: Norton.

King, Jaime S. (2011). 'And Genetic Testing For All . . . The Coming Revolution in Non-Invasive Prenatal Genetic Testing', *Rutgers LJ*, 42: 599–819.

Kjellberg, Hans and Helgesson, Claes-Fredrik (2007). 'The Mode of Exchange and Shaping of Markets: Distributor Influence in the Swedish Post-war Food Industry', *Industrial Marketing Management*, 36 (7): 861–78.

Kjellberg, Hans and Mallard, Alexandre et al. (2013). 'Valuation Studies? Our Collective Two Cents', *Valuation Studies*, 1 (1): 11–30.

Klingberg, Marcus (2010). 'An Epidemiologist's Journey from Typhus to Thalidomide, and from the Soviet Union to Seveso', *JRSM*, 103 (10): 418–23.

Knorr Cetina, Karin (1995). 'Laboratory Studies: The Cultural Approach to the Study of Science', in S. Jasanoff et al. (eds), *Handbook of Science and Technology Studies*. London: Sage.

Knorr Cetina, Karin and Bruegger, Urs (2002). 'Global Microstructures: The Virtual Societies of Financial Markets', *American Journal of Sociology*, 107 (4): 905–50.

KnowledgePoint360 (2010). 'Promotional brochure'.

Knut och Alice Wallenbergs Stiftelse (2007). *Verksamheten 2006*. Stockholm: Knut & Alice Wallenbergs Stiftelse.

Knutson, Brian, Wimmer, G. Elliott, Kuhnen, Camelia M., and Winkielman, Piotr (2008). 'Nucleus Accumbens Activation Mediates the Influence of Reward Cues on Financial Risk Taking', *Neuroreport*, 19 (5): 509–13.

Kohler, Robert E. (1994). *Lords of the Fly: Drosophila Genetics and the Experimental Life*. Chicago, IL: University of Chicago Press.

Kolata, Gina Bari (1980). 'Mass Screening for Neural Tube Defects', *Hastings Center Report*: 8–10.

Kole, Adriaan et al. (2003). 'The Effects of Different Types of Product Information on the Consumer Product Evaluation for Fresh Cod in Real Life Settings', in *RIVO Report*. Yerseke, Netherlands: Netherlands Institute for Fisheries Research (RIVO).

Komisaruk, Barry R., Whipple, Beverly and Beyer, Carlos (2010). 'Sexual Pleasure', in Morten L. Kringelbach and Kent C. Berridge (eds), *Pleasures of the Brain*. New York: Oxford University Press, 169–77.

Kottmeier, Hans-Ludvig (1961). 'Axel Westman 29.XII.1894–29.V.1960', *Nordisk Medicin*, 65: 549–51.

Kringelbach, Morten L. (2010). 'The Hedonic Brain: A Functional Neuroanatomy of Human Pleasure', in Morten L. Kringelbach and Kent C. Berridge (eds), *Pleasures of the Brain*. New York: Oxford University Press, 202–21.

Kringelbach, Morten L. and Berridge, Kent C. (2010). 'Introduction: The Many Faces of Pleasure', in Morten L. Kringelbach and Kent C. Berridge (eds), *Pleasures of the Brain*. New York: Oxford University Press, 3–6.

Kurlansky, Mark. (1997). *Cod: A Biography of the Fish that Changed the World*. New York: Walker.

Kurunmäki, Liisa (1999). 'Professional vs Financial Capital in the Field of Health Care – Struggles for the Redistribution of Power and Control', *Accounting, Organizations and Society*, 24: 95–124.

Kurunmäki, Liisa and Miller, Peter (2008). 'Counting the Costs: The Risks Of Regulating and Accounting for Health Care Provision', *Health, Risk & Society*, 10 (1): 9–21.

Lachmann, H. J. et al. (2013). 'OR10-005 – Treatment Responses in TRAPS: Eurofever/Eurotraps', *Pediatric Rheumatology*, 11 (Suppl 1): A188.

Lamont, Michèle (2009). *How Professors Think: Inside the Curious World of Academic Judgment*. Cambridge, MA: Harvard University Press.

Lamont, Michèle (2012). 'Toward a Comparative Sociology of Valuation and Evaluation', *Annual Review of Sociology*, 38 (1): 201–21.

Landecker, Hannah (1999). 'Between Beneficence and Chattel: The Human Biological in Law and Science', *Science in Context*, 12 (1): 203–26.

Lanza, Robert P., Dresser, Betsy L., and Damiani, Philip (2000a). 'Cloning Noah's Ark: Biotechnology Might Offer the Best Way to Keep Some Endangered Species from Disappearing from the Planet', *Scientific American*, November:85–9.

Lanza, Robert P. et al. (2000b). 'Cloning of an Endangered Species Using Interspecies Nuclear Transfer', *Cloning*, 2 (2): 79–90.

Larsen, Lars Thorup (2007). 'Speaking Truth to Biopower: On the Genealogy of Bioeconomy', *Distinktion,* 14:9–24.

Larsson, Larsåke (2005). *Upplysning och propaganda: Utveckling av svensk PR och information.* Lund: Studentlitteratur.

Latour, Bruno (1987). *Science in Action: How to Follow Scientists and Engineers through Society.* Cambridge, MA: Harvard University Press.

Latour, Bruno (1999a). *Pandora's Hope.* Cambridge, MA: Harvard University Press.

Latour, Bruno (1999b). 'Morale et technique; La fin des moyens', *Réseaux,* 100: 39–58.

Latour, Bruno (2002). 'Morality and Technology: The End of the Means', *Theory, Culture and Society,* 19 (5/6): 247–60.

Latour, Bruno (2004). 'Why Has Critique Run Out of Steam? From Matters of Fact to Matters of Concern', *Critical Inquiry,* 30 (2): 225–48.

Latour, Bruno (2005). *Reassembling the Social: An Introduction to Actor-Network-Theory,* Clarendon Lectures in Management Studies. Oxford: Oxford University Press.

Latour, Bruno (2007). 'How to Think Like a State', in *The Thinking State?* The Hague: WRR/ Scientific Council for Government Policy, 19–32.

Latour, Bruno (2010). 'An Attempt at a "Compositionist Manifesto"', *New Literary History,* 41 (3): 471–90.

Latour, Bruno and Lépinay, Vincent Antonin (2009). *The Science of Passionate Interests: An Introduction to Gabriel Tarde's Economic Anthropology.* Chicago, IL: Prickly Paradigm Press.

Latour, Bruno and Woolgar, Steve (1979). *Laboratory Life: The Social Construction of Scientific Facts.* Princeton, NJ: Princeton University Press.

Law, John (1996). 'Organising Account Ethics: Ontology and the Mode of Accounting', in R Munro and J Mouritsen (eds), *Accountability: Power, Ethos and the Technologies of Managing.* London: International Thomson Business Press.

Law, John (2004). 'Matter-ing, Or How Might STS Contribute?', <http://www.lancs.ac.uk/fss/sociology/papers/law-matter-ing.pdf>, accessed 3 March 2004.

Lazarsfeld, Paul F (1944). 'The election is over', *Public Opinion Quarterly,* 8 (3): 317–30.

Le Grand, Julian (2007). *The other invisible hand: Delivering public services through choice and competition.* Princeton: Princeton University Press.

Lee, Nick (1999). 'The Challenge of Childhood Distributions of Childhood's Ambiguity in Adult Institutions', *Childhood,* 6 (4): 455–74.

Lerner, Aaron B. (1993). 'The discovery of the Melanotropins: A History of Pituitary Endocrinology', *Annals of the New York Academy of Sciences,* 680: 1–12.

Levidow, Les and Young, Robert Maxwell (1981). *Science, technology, and the labour process: Marxist studies. Volume 1.* London: CSE Books.

Levidow, Les and Young, Robert Maxwell (1985). *Science, technology, and the labour process: Marxist studies. Volume 2.* London: CSE Books.

Lewis, Celine, Hill, Melissa, Skirton, Heather, and Chitty, Lyn S (2012). 'Non-invasive prenatal diagnosis for fetal sex determination: benefits and disadvantages from the service users' perspective', *European Journal of Human Genetics,* 20 (11): 1127–33.

Liebenau, Jonathan (1987). *Medical science and medical industry: The formation of the American pharmaceutical industry.* Baltimore: Johns Hopkins University Press.

Light, Donald W. (2000). 'The Medical Profession and Organizational Change: From Professional Dominance to Countervailing Power', in Chloe E. Bird, Peter Conrad and Allen M. Fremont (eds), *Handbook of Medical Sociology.* Upper Saddle River, NJ: Prentice Hall, 201–16.

Light, Donald W. (2010). 'GAVI's Advance Market Commitment', *The Lancet*, 375: 638.

Lippman, Abby (1991). 'Prenatal Genetic Testing and Screening: Constructing Needs and Reinforcing Inequities', *American Journal of Law and Medicine*, 17: 15–50.

Lippman, Abby (1999). 'Choice as a Risk to Women's Health', *Health, Risk & Society*, 1 (3): 281–91.

Lloyd, Elisabeth A. (2005). *The Case of the Female Orgasm: Bias in the Science of Evolution.* Cambridge, MA: Harvard University Press.

Lock, Margaret (2001). 'The Tempering of Medical Anthropology: Troubling Natural Categories', *Medical Anthropology Quarterly*, 15 (4): 478–92.

Lock, Margaret (2002). *Twice Dead: Organ Transplants and the Reinvention of Death.* Berkeley, CA: University of California Press.

Lönngren, Rune et al. (1999). *Svensk farmaci under 1900-talet: Utvecklingen över tiden.* Stockholm: Apotekarsocieteten.

López-Beltrán, Carlos (2007). '5 The Medical Origins of Heredity', in *Heredity Produced: At the Crossroads of Biology, Politics, and Culture, 1500–1870*, 105.

Lorence, Daniel P. and Richards, Michael (2002). 'Variation in Coding Influence across the USA: Risk and Reward in Reimbursement Optimization', *Journal of Management in Medicine*, 16 (6): 422–35.

Löwy, Ilana (2000). 'Trustworthy Knowledge and Desperate Patients: Clinical Test from New Drugs from Cancer to AIDS', in Margaret Lock, Allan Young and Alberto Cambrosio (eds), *Living and Working with New Medical Technologies: Intersections of Inquiry.* Cambridge: Cambridge University Press, 49–81.

Löwy, Ilana (2010a). 'Über die gesellschiftiche Function des Wissenschaftlers heute', in D. Gugerli et al. (eds), *Nach Feierabend: special issue, Universität.* Zurich: Diaphanes, 35–54.

Löwy, Ilana (2010b). 'Carlo Ginzburg: Le genre caché de la micro-histoire', in Danielle Chabaud-Rychter et al. (eds), *Sous les sciences sociales, le genre: Relectures critiques, de Max Weber à Bruno Latour.* Paris: Éditions La Découverte, 177–89.

Lundgren, Anders (1998). 'The Development of Chemical Industry in Sweden and the Contribution of Academic Chemistry after 1900', in Anthony S. Travis, Harm G. Schröter, Ernst Homburg, and Peter J. T. Morris (eds), *Determinants in the Evolution of the European Chemical Industry, 1900–1939: New Technologies, Political Frameworks, Markets and Companies.* Dordrecht: Kluwer, 123–42.

Lundh, Andreas et al. (2012). 'Industry Sponsorship and Research Outcome', *Cochrane Database Syst Rev*, 12.

Lundvall, Bengt-Åke (1992). *National Systems of Innovation: Towards a Theory of Innovation and Interactive Learning.* London: Pinter.

Luten, Joop et al. (2002). 'Evaluation of Wild Cod versus Wild Caught, Farmed Raised Cod from Norway by Dutch Consumers', *Økonomisk Fiskeriforskning*, 12.

Lynch, Michael (1993). *Scientific Practice and Ordinary Action: Ethnomethodology and Social Studies Of Science.* Cambridge: Cambridge University Press.

Lynch, Michael (1998). 'The Discursive Production of Uncertainty: The OJ Simpson 'Dream Team' and the Sociology of Knowledge Machine', *Social Studies of Science* 28 (5/6): 829–68.

Lynch, Michael (2013). 'Postscript Ontography: Investigating the Production of Things, Deflating Ontology', *Social Studies of Science.*

Maasen, Sabine (2010). 'Neuroeconomics – A Marriage of Giants Thanks to Neoliberal Knowledge Society?!', paper given at the Neurosociety: What Is it with the Brain These Days?, Oxford, 7–8 December.

MacKenzie, Donald (1990). *Inventing Accuracy: A Historical Sociology of Nuclear Missile Guidance.* Cambridge, MA: MIT Press.

MacKenzie, Donald (2004). 'The Big, Bad Wolf and the Rational Market; Portfolio Insurance, the 1987 Crash and the Performativity of Economics', *Economy and Society,* 33 (3): 303–34.

MacKenzie, Donald (2006). *An Engine, Not a Camera: How Financial Models Shape Markets.* Cambridge, MA: MIT Press.

MacKenzie, Donald (2007). 'Is Economics Performative? Option Theory and the Construction of Derivatives Markets', in Donald MacKenzie, Fabian Muniesa, and Lucia Siu (eds), *Do Economists Make Markets? On the Performativity of Economics.* Princeton, NJ: Princeton University Press.

MacKenzie, Donald (2009a). 'Making Things the Same: Gases, Emission Rights and the Politics of Carbon Markets', *Accounting, Organizations and Society,* 34 (3-4): 440–55.

MacKenzie, Donald (2009b). *Material Markets: How Economic Agents Are Constructed.* Oxford: Oxford University Press.

MacKenzie, Donald and Millo, Yuval (2003). 'Constructing a Market, Performing Theory: The Historical Sociology of a Financial Derivatives Exchange', *American Journal of Sociology,* 109 (1): 107–45.

MacKenzie, Donald and Wajcman, Judy (1999). *The Social Shaping of Technology.* Buckingham & Philadelphia, PA: Open University Press.

MacKenzie, Donald, Muniesa, Fabian, and Siu, Lucia (2007). *Do Economists Make Markets? On the Performativity of Economics.* Princeton, NJ: Princeton University Press.

McCaughey, Martha (2008). *The Caveman Mystique: Pop-Darwinism and the Debates over Sex, Violence, and Science.* London & New York: Routledge.

McClure, Samuel M., Laibson, David I., Loewenstein, George, and Cohen, Jonathan D. (2004). 'Separate Neural Systems Value Immediate and Delayed Monetary Rewards', *Science,* 306 (5695): 503–7.

McCray, W. Patrick (2000). 'Large Telescopes and the Moral Economy of Recent Astronomy', *Social Studies of Science,* 30 (5): 685.

McNeil, David (2008). 'We Can Work with Pharma', *Clinical Psychiatry News, February 2008* <http://www.clinicalpsychiatrynews.com/>, accessed 31 March 2011.

Martin, Emily (2000). 'Mind-Body Problems', *American Ethnologist,* 27 (3): 569–90.

Martin, Emily (2006). 'Pharmaceutical Virtue', *Culture, Medicine and Psychiatry,* 30 (2): 157–74.

Martin, Emily (2010). 'Self-making and the Brain', *Subjectivities,* 3 (4): 366–81.

Marx, Karl ([1867] 1990). *Capital Volume I.* London: Penguin.

Melander, Hans et al. (2003). 'Evidence B(i)ased Medicine—Selective Reporting from Studies Sponsored by Pharmaceutical Industry: Review of Studies in New Drug Applications', *BMJ: British Medical Journal,* 326 (7400): 1171.

Merton, Robert K. (1957). 'Priorities in Scientific Discovery: A Chapter in the Sociology of Science', *American Sociological Review*, 22 (6): 635–59.

Merton, Robert K. (1973a). 'The Normative Structure of Science', in *The Sociology of Science: Theoretical and Empirical Investigations*. Chicago, IL: Chicago University Press, 267, 275–7.

Merton, Robert K. (1973b). *The Sociology of Science: Theoretical and Empirical Investigations*. Chicago, IL: University of Chicago Press.

Meskus, Mianna (2012). 'Personalized Ethics: The Emergence and the Effects in Prenatal Testing', *BioSocieties*, 7 (4): 373–92.

Miller, Carolyn R. (1984). 'Genre as Social Action', *Quarterly Journal of Speech*, 70 (2): 151–67.

Miller, Daniel (2002). 'Turning Callon the Right Way up', *Economy and Society*, 31 (2): 218–33.

Ministry of Health Welfare and Sport (2005). 'Better Faster', <www.snellerbeter.nl/english>, accessed 11 October 2005.

Mirowski, Philip (2004). *The Effortless Economy of Science?*, Science and Cultural Theory. Durham, NC: Duke University Press.

Mirowski, Philip (2011). *Science-mart: Privatizing American Science*. Cambridge, MA: Harvard University Press.

Mitchell, Robert and Waldby, Catherine (2010). 'National Biobanks: Clinical Labor, Risk Production, and the Creation of Biovalue', *Science, Technology and Human Values*, 35 (3): 330–55.

Mol, Annemarie (1999). 'Ontological Politics. A Word and Some Questions', in John Law and John Hassard (eds), *Actor Network Theory and After*. Oxford: Blackwell, 74–89.

Mol, Annemarie (2002). *The Body Multiple: Ontology in Medical Practice*. Durham, NC & London: Duke University Press.

Mol, Annemarie (2008). *The Logic of Care: Health and the Problem of Patient Choice*. Oxon Routledge.

Mol, Annemarie (2013). 'Mind Your Plate! The Ontonorms of Dutch Dieting', *Social Studies of Science*, 43 (3): 379–96.

Moreira, Tiago (2012). *The Transformation of Contemporary Health Care: The Market, the Laboratory, and the Forum*. New York: Routledge.

Morgan, Lynn (2009). *Icons of Life: A Cultural History of Human Embryos*. Berkeley, CA: University of California Press.

Moynihan, Ray (2008). 'Drug Marketing: Key Opinion Leaders: Independent Experts or Drug Representatives in Disguise?', *BMJ: British Medical Journal*, 336 (7658): 1402.

Muniesa, Fabian (2012). 'A Flank Movement in the Understanding of Valuation', in Lisa Adkins and Celia Lury (eds), *Measure and Value*, Sociological Review. Oxford: Wiley-Blackwell, 24–38.

Muniesa, Fabian and Helgesson, Claes-Fredrik (2013). 'Valuation Studies and the Spectacle of Valuation', *Valuation Studies*, 1 (2): 119–23.

Muniesa, Fabian and Trébuchet-Breitwiller, Anne-Sophie (2010). 'Becoming a Measuring Instrument: An Ethnography of Perfume Consumer Testing', *Journal of Cultural Economy*, 3 (3): 321–37.

Muniesa, Fabian, Millo, Yuval, and Callon, Michel (2007). 'An Introduction to Market Devices', in Michel Callon, Yuval Millo and Fabian Muniesa (eds), *Market Devices*. Oxford: Blackwell, 1–12.

Nadler, Henry L. (1968). 'Antenatal Detection of Hereditary Disorders', *Pediatrics,* 42 (6): 912–18.

Nakata, Naomi, Wang, Yuemei, and Bhatt, Sucheta (2010). 'Trends in Prenatal Screening and Diagnostic Testing among Women Referred for Advanced Maternal Age', *Prenatal Diagnosis,* 30 (3): 198–206.

National Geographic (2007). 'Stopping a Global Killer', <http://ngm.nationalgeographic.com/2007/07/malaria/finkel-text>, accessed 10 October 2010.

Nelkin, Dorothy (1995). *Selling Science: How the Press Covers Science and Technology (1987).* rev. edn. New York: Freeman.

Nelkin, Dorothy and Lindee, M. Susan (1995). *The DNA Mystique: The Gene as a Cultural Icon.* New York: Freeman.

Neuberger, James and Thorburn, Douglas (2005). 'MELD – Moving Steadily towards Equality, Equity, and Fairness', *Liver Transplantation,* 11 (5): 585–7.

Neuberger, James et al. (1998). Assessing Priorities for Allocation of Donor Liver Grafts: Survey of Public and Clinicians', *BMJ: British Medical Journal,* 317(18 July).

Nevéus, Torgny (2000). *Honoris causa: Promotioner, hedersdoktorer och hedersmedlemmar vid Uppsala universitet 1800–2000: förteckningar*. Stockholm: Uppsala University.

Neyland, Daniel (2012). 'Parasitic Accountability', *Organization,* 19 (6): 845–63.

Neyland, Daniel and Simakova, Elena (2012). 'Managing Electronic Waste: A Study of Market Failure', *New Technology, Work and Employment,* 27 (1): 36–51.

Nicholson, Malcolm and Fleming, John (2000). 'Looking at the Unborn: Historical Aspects of Obstetrics Ultrasound', paper given at Wellcome to Witness Seminars to Twentieth Century Medicine, London, 10 March 1998.

Nicolaides, Kypros H. et al. (1992). 'Fetal Nuchal Translucency: Ultrasound Screening for Chromosomal Defects in First Trimester of Pregnancy', *BMJ: British Medical Journal,* 304 (6831): 867.

Nilsson, Mikael (2012). 'Science as Propaganda: Swedish Scientists and the Co-production of American Hegemony in Sweden during the Cold War, 1953–68', *European Review of History,* 19:275–302.

Norberg, Birgitta (2002). 'Project Description: Cod Puberty – Age at First Maturation in Relation to Season, Growth and Energy Acquisition during the First Year of Life', The Research Council of Norway <https://www.forskningsradet.no/prosjektbanken_beta/#/project/146677>, accessed 22 August 2014.

Nordlund, Christer (2007). 'Hormones for Life? Behind the rise and fall of a hormone remedy (Gonadex) against sterility in the Swedish welfare state', *Studies in History and Philosophy of Biological and Biomedical Science* 38: 191–216.

Nordlund, Christer (2011). *Hormones of Life: Endocrinology, the Pharmaceutical Industry and the Dream of a Remedy for Sterility, 1930–1970.* Sagamore Beach, MA: Science History Publications.

Norgren, Lennart (1989). *Kunskapsöverföring från universitet till företag: en studie av universitetsforskningens betydelse för de svenska läkemedelsföretagens produktlanseringar 1945–1984.* PhD thesis, Uppsala University: Allmänna förlaget.

O'Connell, Joseph (1993). 'Metrology: The Creation of Universality by the Circulation of Particulars', *Social Studies of Science,* 23(1): 129–73.

Olds, James and Milner, Peter M. (1954). 'Positive Reinforcement Produced by Electrical Stimulation of Septal Area and Other Regions of Rat Brain', *Journal of Comparative and Physiological Psychology,* 47: 419–27.

Oniscu, Gabriel C. et al. (2003). 'Equity of Access to Renal Transplant Waiting List and Renal Transplantation in Scotland: Cohort Study', *BMJ: British Medical Journal*, 327 (7426): 1261.

Orenstein, Peggy (2008). *Waiting for Daisy: A Tale of Two Continents, Three Religions, Five Infertility Doctors, an Oscar, an Atomic Bomb, a Rom.* London & New York: Bloomsbury.

Ostrom, Elinor (2000). 'Collective Action and the Evolution of Social Norms', *Journal of Economic Perspectives*, 14 (3): 137–58.

Oudshoorn, Nelly (1993). 'United We Stand: The Pharmaceutical Industry, Laboratory, and Clinic in the Development of Sex Hormones into Scientific Drugs, 1920–1940', *Science, Technology & Human Values*, 18 (1): 5–24.

Oxenford, Kerry, Silcock, Caroline, Hill, Melissa, and Chitty, Lyn (2013). 'Routine Testing of Fetal Rhesus *D* Status in Rhesus D Negative Women Using Cell Free Fetal DNA: An Investigation into the Preferences and Information Needs of Women', *Prenatal Diagnosis*, 33 (7): 688–94.

Palm, Ineke (2005). *De zorg is geen markt; een kritische analyse van de marktwerking in de zorg vanuit verschillende perspectieven.* Rotterdam: Wetenschappelijk Bureau SP.

Palmblad, Eva (1997). *Sanningens gränser: Kvacksalveriet, läkarna och samhället, Sverige 1880–1990.* Stockholm: Carlsson.

Palmer, Julie (2009). 'The Placental Body in 4D: Everyday Practices of Non-diagnostic Sonography', *Feminist Review*, 93 (1): 64–80.

Palomaki, Glenn E. et al. (2010). 'Four Years' Experience with an Interlaboratory Comparison Program Involving First-trimester Markers of Down Syndrome', *Archives of Pathology and Laboratory Medicine*, 134 (11): 1685–91.

Parascandola, John (1990). '"The Preposterous Provision": The American Society for Pharmacology and Experimental Therapeutics' Ban on Industrial Pharmacologists 1908–1941', in Jonathan Liebenau, Gregory J. Higby and Elaine C. Stroud (eds), *Pill Peddlers: Essays on the History of the Pharmaceutical Industry.* Madison, WI: American Institute of the History of Pharmacy, 29–48.

Paul, Diane B. (1995). *Controlling Human Heredity: 1865 to the Present.* Atlantic Highlands, NJ: Humanities Press.

Paul, Diane B. and Spencer, Hamish G. (1995). 'The Hidden Science of Eugenics', *Nature*, 374 (6520): 302.

Persad, Govind, Wertheimer, Alan, and Emanuel, Ezekiel J. (2009). 'Principles for Allocation of Scarce Medical Interventions', *The Lancet*, 373 (9661): 423–31.

Pestre, Dominique (2013). 'Les études sur les sciences: Épistémologies, ontologies et attitudes politiques', in *À Contre-science: Politiques et savoirs des sociétés contemporaines.* Paris: Éditions du Seuil, 193–219.

Petchesky, Rosalind Pollack (1987). 'Fetal Images: The Power of Visual Culture in the Politics of Reproduction', *Feminist Studies*, 13 (2): 263–92.

Phillips, David P. et al. (1991). 'Importance of the Lay Press in the Transmission of Medical Knowledge to the Scientific Community', *New England Journal of Medicine*, 325:1180–83.

Pickering, Andrew (1995). *The Mangle of Practice: Time, Agency, and Science.* Chicago, IL: University of Chicago Press.

Pieters, Toine (1998). 'Managing Differences in Biomedical Research: The Case of Standardizing Interferons', *Studies in History and Philosophy of Science Part C: Studies in History and Philosophy of Biological and Biomedical Sciences*, 29 (1): 31–79.

Pieters, Toine (2002). 'About Media, Audiences and Marketing Medicines: The Interferons', in Godelieve van Heteren, Marijke Gijswijt-Hofstra and E.M. Tansey (eds), *Biographies of*

Remedies: Drugs, Medicines and Contraceptives in Dutch and Anglo–American Healing Cultures. Clio Medica 66. Amsterdam: Rodopi.

Pieters, Toine (2005). *Interferon: The Science and Selling of a Miracle Drug.* London: Routledge.

Pinch, Trevor and Swedberg, Richard (2008). *Living in a Material World: Economic Sociology Meets Science and Technology Studies.* Cambridge, MA: MIT Press.

Podolsky, Scott H. and Greene, Jeremy A. (2008). 'A Historical Perspective of Pharmaceutical Promotion and Physician Education', *JAMA: Journal of the American Medical Association,* 300 (7): 831–3.

Pollit, Christopher and Bouckaert, Geert (2000). *Public Management Reform.* Oxford: Oxford University Press.

Pope, Gregory C. (1989). 'Hospital Nonprice Competition and Medicare Reimbursement Policy* 1', *Journal of Health Economics,* 8 (2): 147–72.

Porter, Michael E. and Teisberg, Elizabeth Olmsted (2006). *Redefining Health Care: Creating Value-based Competition on Results.* Boston, MA: Harvard Business School Press.

Porter, Theodore M. (2000). 'Life Insurance, Medical Testing, and the Management of Mortality', in Lorraine Daston (ed.), *Biographies of Scientific Objects.* Chicago, IL: University of Chicago Press, 226–46.

POST (2005). Fighting Diseases of Developing Countries. <http://www.parliament.uk/business/publications/research/briefing-papers/POST-PN-241/fighting-diseases-of-developing-countries-june-2005>, accessed 4 September 2014.

Powledge, Tabitha M. (1979). 'Prenatal Diagnosis: New Techniques, New Questions', *Hastings Center Report,* 9 (3): 16–17.

Powledge, Tabitha M. and Sollitto, Sharmon (1974). 'Prenatal Diagnosis: The Past and the Future', *Hastings Center Report,* 4 (5): 11–13.

Priest, Susanna Hornig (2000). 'US Public Opinion Divided over Biotechnology?', *Nature Biotechnology,* 18: 989–42.

Pronk, Evert (2006). 'Darmkanker Gestroomlijnd', *Medisch Contact,* 61 (4): 150–2.

ProPublica (2011). 'Dollars for Docs', <http://projects.propublica.org/docdollars/>, accessed 15 August 2011.

Puri, Sunita, Adams, Vincanne, Ivey, Susan, and Nachtigall, Robert D (2011). '"There is Such a Thing as Too Many Daughters, But Not Too Many Sons": A Qualitative Study of Son Preference and Fetal Sex Selection among Indian Immigrants in the United States', *Social Science and Medicine (1982),* 72 (7): 1169–76.

Putnam, Hilary (1982). 'Beyond the Fact-value Dichotomy', *Crítica: Revista Hispanoamericana de Filosofía,* 14 (41): 3–12.

Randle, Philip (1990). 'Frank George Young. 25 March 1908–20 September 1988', *Biographical Memoirs of Fellows of the Royal Society,* 36: 583–99.

Rapp, Rayna and Ginsbourg, Faye (1999). 'Fetal Reflections: Confessions of Two Feminist Anthropologists as Mutual Informants', in Lynne Marie Morgan and Meredith V. Michaels (eds), *Fetal Positions, Feminist Practices.* Philadelphia, PA: University of Pennsylvania Press, 279–95.

Rasmussen, Nicolas (2004). 'The Moral Economy of the Drug Company-Medical Scientist Collaboration in Interwar America', *Social Studies of Science,* 34: 161–85.

Reagan, Leslie J. (2010). *Dangerous Pregnancies: Mothers, Disabilities, and Abortion in Modern America.* Berkeley, CA: University of California Press.

Redouté, Jérôme et al. (2000). 'Brain Processing of Visual Sexual Stimuli in Human Males', *Human Brain Mapping,* 11 (3): 162–77.

Reinecke, Juliane (2010). 'Beyond a Subjective Theory of Value and towards a "Fair Price": An Organizational Perspective on Fairtrade Minimum Price Setting', *Organization*, 17 (5): 563–81.

Remennick, Larissa (2006). 'The Quest for the Perfect Baby: Why Do Israeli Women Seek Prenatal Genetic Testing?', *Sociology of Health & Illness*, 28 (1): 21–53.

Research Council of Norway and Innovation Norway (RCN/IN) (2006). 'Plan for a Coordinated Investment in Cod. Farming and Capture-based Aquaculture 2001–2010'. Updated April 2006 [Plan for koordinert satsing på torsk. Oppdrett og fangsbasert akvakultur. 2001–2010. Oppdatert april 2006.]

Research Council of Norway and Norwegian Industrial and Regional Development Fund (RCN/NIRDF) (2001). 'Farming of Cod. Strategy for a Coordinated Investment by SND and the Norwegian Research Council 2001–2010' [Oppdrett av torsk. Strategi for koordinert satsing fra SND og Norges forskningsråd 2001–2010.]

Research Council of Norway and Norwegian Industrial and Regional Development Fund (RCN/NIRDF) (2003). 'Farming of Cod. Strategy for a Coordinated Investment by SND and The Norwegian Research Council 2001–2010'. Updated July 2003. [Oppdrett av torsk. Strategi for koordinert satsing fra SND og Norges forskningsråd 2001–2010. Oppdatert juli 2003.]

Research Council of Norway, Innovation Norway and Norwegian Seafood Research Fund (RCN/IN/NSRF) (2009). 'Plan for a Coordinated Investment in Cod: Farming and Capture-based Aqaculture 2010–2020'. [Plan for koordinert satsing på torsk. Oppdrett og fangsbasert akvakultur 2010-2020.]

Resta, Robert G. (2002). 'Historical Aspects of Genetic Counseling: Why Was Maternal Age 35 Chosen as The Cut-off for Offering Amniocentesis?', *Medicina Nei Secoli*, 14 (3): 793–811.

Riis, Povl and Fuchs, Fritz (1960). 'Antenatal Determination of Fœtal Sex: In Prevention of Hereditary Diseases', *The Lancet*, 276 (7143): 180–2.

Ritvo, Harriet (1995). 'Possessing Mother Nature: Genetic Capital in Eighteenth-century Britain', in John Brewer and Susan Staves (eds), *Early Modern Conceptions of Property*. London: Routledge, 413–26.

Rolls, Edmund T. (1999). *The Brain and Emotion*. Oxford & New York: Oxford University Press.

Rose, Hilary (2000a). 'Colonising the Social Sciences?', in Hilary Rose and Steven Rose (eds), *Alas, Poor Darwin: Arguments against Evolutionary Psychology*. London: Jonathan Cape.

Rose, Steven (2000b). 'Escaping Evolutionary Psychology', in Hilary Rose and Steven Rose (eds), *Alas, Poor Darwin: Arguments against Evolutionary Psychology*. London: Jonathan Cape.

Rose, Hilary and Rose, Steven (2000). *Alas, Poor Darwin: Arguments against Evolutionary Psychology*. London: Jonathan Cape.

Rose, Nikolas (2007). *The Politics of Life Itself: Biomedicine, Power, and Subjectivity in the Twenty-first Century*. Princeton, NJ: Princeton University Press.

Rosemberg, Eugenia (ed.) (1968). *Gonadotropins: Proceedings of the Workshop Conference held at Vista Hermosa, Mor., Mexico, June 24–26, 1968*. Los Altos, CA: Geron-X.

Roth, Alvin E. (2007). 'Repugnance as a Constraint on Markets', *Journal of Economic Perspectives*, 21 (3): 37–58.

Roth, Alvin E., Sönmez, Tayfun, and Ünver, M. Utku (2005). 'A Kidney Exchange Clearing House in New England', *American Economic Review*, 95 (2): 376–80.

Rothfels, Nigel (2002). *Savages and Beasts: The Birth of the Modern Zoo*. Baltimore, MD: Johns Hopkins University Press.

Rothman, Barbara Katz (1986). *The Tentative Pregnancy: Prenatal Diagnosis and the Future of Motherhood.* New York: Viking Books.

Rothschild, Joan (2005). *The Dream of the Perfect Child.* Bloomington, IN: Indiana University Press.

Royal College of Obstetricians and Gynaecologists (RCOG) (1993). *Report of the RCOG Working Party on Biochemical Markers and the Detection of Down's Syndrome.* London: ROCS Press.

Ryan, Alan (1987). 'Introduction', in Alan Ryan (ed.), *John Stuart Mill and Jeremy Bentham: Utilitarianism and Other Essays.* London: Penguin.

Ryder, Oliver A. and Benirschke, Kurt (1997). 'The Potential Use of "Cloning" in the Conservation Effort', *Zoo Biology,* 16:295–300.

Saldaña, Johnny (2009). *The Coding Manual for Qualitative Researchers.* Los Angeles, CA: Sage.

Samerski, Silja (2009). 'Genetic Counseling and the Fiction of Choice: Taught Self-determination as a New Technique of Social Engineering', *Signs,* 34 (4): 735–61.

Sandelowski, Margarete and Barroso, Julie (2005). 'The Travesty of Choosing after Positive Prenatal Diagnosis', *Journal of Obstetric, Gynecologic, & Neonatal Nursing,* 34 (3): 307–18.

Schafer, Arthur (2004). 'Biomedical Conflicts of Interest: A Defence of The Sequestration Thesis—Learning from the Cases of Nancy Olivieri and David Healy', *Journal of Medical Ethics,* 30 (1): 8–24.

Schaubel, Douglas et al. (2009). 'Survival Benefit-based Deceased-donor Liver Allocation', *American Journal of Transplantation,* 9 (4.2): 970–81.

Schüll, Natascha Dow and Zaloom, Caitlin (2011). 'The Shortsighted Brain: Neuroeconomics and the Governance of Choice in Time', *Social Studies of Science,* 41 (4): 515–38.

Schultz, Wolfram (2002). 'Getting Formal with Dopamine and Reward', *Neuron,* 36 (2): 241–63.

Schut, Erik (2009). 'Is de marktwerking in de zorg doorgeschoten?', *Socialisme & Democratie,* 7/8:68–80.

Schwennesen, Nete and Koch, Lene (2012). 'Representing and Intervening: "Doing" Good Care in First Trimester Prenatal Knowledge Production and Decision-making', *Sociology of Health & Illness,* 34 (2): 283–98.

Schwennesen, Nete, Svendsen, Mette N., and Koch, Lene (2010). 'Beyond Informed Choice: Prenatal Risk Assessment, Decision-Making and Trust', *Clinical Ethics,* 5 (4): 207–16.

Scott, James C. (1977). *The Moral Economy of the Peasant: Rebellion and Subsistence in Southeast Asia.* New Haven, CT: Yale University Press.

Scott, John and Marshall, Gordon (2009). *A Dictionary of Sociology,* 3rd edn. Oxford: Oxford University Press.

Scott-Lichter D. et al. (2012). *CSE's White Paper on Promoting Integrity in Scientific Journal Publications, 2012,* 3rd rev. edn. Wheat Ridge, CO: Council of Science Editors.

Service, Robert F. (2001a). 'Proteomics: Proteomics 2.0: The View Ahead', *Science,* 294 (5549): 2076.

Service, Robert F. (2001b). 'High-speed Biologists Search for Gold in Proteins', *Science,* 294 (7 December): 2074–7.

Service, Robert F. (2001c). 'Gene and Protein Patents Get Ready to Go Head to Head', *Science,* 294(5549), 2082–3 doi: 10.1126/science.294.5549.2082.

Sescousse, Guillaume, Redouté, Jérôme, and Dreher, Jean-Claude (2010). 'The Architecture of Reward Value Coding in the Human Orbitofrontal Cortex', *J. Neurosci.,* 30 (39): 13095–104.

Shapin, Steven (1984). 'Pump and Circumstance: Robert Boyle's Literary Technology', *Social Studies of Science,* 14 (4): 481–520.

Shapin, Steven (1994). *A Social History of Truth: Civility and Science in Seventeenth-century England.* Chicago, IL: University of Chicago Press.

Shapin, Steven (2008). *The Scientific Life: A Moral History of a Late Modern Vocation.* Chicago, IL: University of Chicago Press.

Shapiro, Ron (2007). 'Kidney Allocation and the Perception of Fairness', *American Journal of Transplantation,* 7 (5): 1041–2.

Sharp, Lesley (2006). *Strange Harvest: Organ Transplants, Denatured Bodies, and the Transformed Self.* Berkeley, CA: University of California Press.

Sheldon, Trevor A and Simpson, John (1991). 'Appraisal of a New Scheme for Prenatal Screening for Down's Syndrome', *BMJ: British Medical Journal,* 302 (6785): 1133.

Shiva, Vandana (1997). *Biopiracy: The Plunder of Nature and Knowledge.* Boston, MA: South End Press.

Shrum, Wesley, Genuth, Joel and Chompalov, Ivan (2007). *Structures of Scientific Collaboration.* Cambridge, MA: MIT Press.

Shukin, Nicole (2009). *Animal Capital: Rendering Life in Biopolitical Times.* Minneapolis, MN: University of Minnesota Press.

Silvester, Kate et al. (2004). 'Reducing Waiting Times in the NHS; Is Lack of Capacity the Problem?', *Clinician in Management,* 12:105–11.

Simakova, Elena (2012). *Marketing Technologies: Corporate Cultures and Technological Change.* London: Routledge.

Simakova, Elena and Coenen, Christopher (2013). 'Visions, Hype, and Expectations: A Place for Responsibility', in Richard Owen, Maggy Heintz, and John Bessant (eds), *Responsible Innovation,* 241–66.

Simakova, Elena and Neyland, Daniel (2008). 'Marketing Mobile Futures: Assembling Constituencies and Creating Compelling Stories for an Emerging Technology', *Marketing Theory,* 8 (1): 91–116.

Simborg, Donald W. (1981). 'DRG Creep: A New Hospital-acquired Disease', *New England Journal of Medicine,* 304 (26): 1602–4.

Sinding, Christiane (2002). 'Making the Unit of Insulin: Standards, Clinical Work, and Industry, 1920–1925', *Bulletin of the History of Medicine,* 76 (2): 231–70.

Sismondo, Sergio (2007). 'Ghost Management: How Much of the Medical Literature is Shaped behind the Scenes by the Pharmaceutical Industry?', *PLoS Medicine,* 4 (9): e286.

Sismondo, Sergio (2009). 'Ghosts in the Machine Publication Planning in the Medical Sciences', *Social Studies of Science,* 39 (2): 171–98.

Sjögren, Ebba and Helgesson, Claes-Fredrik (2007). 'The Q(u)ALYfying hand: Health Economics and Medicine in the Shaping of Swedish Markets for Subsidised Pharmaceuticals', in Michel Callon, Yuval Millo and Fabian Muniesa (eds), *Market Devices.* Oxford: Blackwell, 215–40.

Skaerbaek, Peter (2009). 'Public Sector Auditor Identities in Making Efficiency Auditable: The National Audit Office of Denmark as Independent Auditor and Modernizer', *Accounting, Organizations and Society,* 34: 971–87.

Skeggs, Beverly and Loveday, Vik (2012). 'Struggles for Value: Value Practices, Injustice, Judgment, Affect and the Idea of Class', *British Journal of Sociology,* 63 (3): 472–90.

Smith, Charles W. (1990). *Auctions: The Social Construction of Value.* Berkeley and Los Angeles, CA: University of California Press.

Smith, Kyle S., Mahler, Stephen V., Peciña, Susana, and Berridge, Kent C. (2010). 'Hedonic Hotspots: Generating Sensory Pleasure in the Brain', in Morten L. Kringelbach and Kent C. Berridge (eds), *Pleasures of the Brain.* New York: Oxford University Press, 27–49.

Sonderholm, Jorn (2010). 'A Theoretical Flaw in the Advance Market Commitment Idea', *Journal of Medical Ethics,* 36 (6): 339–43.

Star, Susan Leigh (1991). 'Power, Technologies and the Phenomenology of Conventions: On Being Allergic to Onions', in John Law (ed.), *A Sociology of Monsters: Essays on Power, Technology and Domination.* London: Routledge, 26–56.

Star, Susan Leigh and Griesemer, James R. (1989). 'Institutional Ecology, "Translations" and Boundary Objects: Amateurs and Professionals in Berkeley's Museum of Vertebrate Zoology, 1907–39', *Social Studies of Science,* 19: 387–420.

Star, Susan Leigh and Strauss, Anselm (1999). 'Layers of Silence, Arenas of Voice: The Ecology of Visible and Invisible Work', *Computer Supported Cooperative Work,* 8(1–2): 9–30.

Stark, David (2000). 'For a Sociology of Worth'. *Working Paper Series, Center on Organizational Innovation.* New York: Columbia University.

Stark, David (2011). 'What's Valuable?', in Jens Beckert and Patrik Aspers (eds), *The Worth of Goods: Valuation and Pricing in the Economy.* Oxford: Oxford University Press, 319–38.

Statham, Helen (2002). 'Prenatal Diagnosis of Fetal Abnormality: The Decision to Terminate the Pregnancy and the Psychological Consequences', *Fetal and Maternal Medicine Review,* 13 (4): 213–47.

Stein, Zena and Susser, Mervyn (1971). 'The Preventability of Down's Syndrome', *HSMHA Health Reports,* 86 (7): 650.

Stein, Zena, Susser, Mervyn, and Guterman, Andrea V. (1973). 'Screening Programme for Prevention of Down's Syndrome', *The Lancet,* 301 (7798): 305–10.

Steinbusch, Paul, Oostenbrink, Jan, Zuurbier, Joost and Schaepkens, J. (2007). 'The Risk of Upcoding in Casemix Systems: A Comparative Study', *Health Policy,* 81(2–3): 289–99.

Strasser, Bruno J. (2011). 'The Experimenter's Museum: GenBank, Natural History, and the Moral Economies of Biomedicine', *ISIS,* 102 (1): 60–96.

Strathern, Marilyn (1992). *After Nature: English Kinship in the Late Twentieth Century,* 1989. Cambridge: Cambridge University Press.

Stuart, Simon N. et al. (2004). 'Status and Trends of Amphibian Declines and Extinctions Worldwide', *Science,* 306: 1783–6.

Sunder Rajan, Kaushik (2003). 'Genomic Capital: Public Cultures and Market Logics of Corporate Biotechnology', *Science as Culture,* 12 (1): 87–121.

Sunder Rajan, Kaushik (2006). *Biocapital: The Constitution of Postgenomic Life.* Durham, NC: Duke University Press.

Sunder Rajan, Kaushik (ed.) (2012). *Lively Capital: Biotechnologies, Ethics, and Governance in Global Markets. Experimental Futures.* Durham, NC: Duke University Press.

Svensk Farmaceutisk Tidskrift (1901). 'En ny svensk industri', 5: 293–5.

Svensk Författningssamling (1934). 'Kungl. Maj:ts kungörelse angående handel med farmaceutiska specialiteter'. [SFS; Swedish Statute Book.]

Swann, John Patrick (1988). *Academic Scientists and the Pharmaceutical Industry: Cooperative Research in Twentieth-century America.* Baltimore, MD: Johns Hopkins University Press.

Talkowski, Michael E. et al. (2012). 'Clinical Diagnosis by Whole-genome Sequencing of a Prenatal Sample', *New England Journal of Medicine,* 367 (23): 2226–32.

Taranger, Geir Lasse (2001). 'Project Description: Use of continuous light to delay puberty in farmed Atlantic cod', http://www.regionaleforskningsfond.no/servlet/Satellite?c=Prosjekt&cid=1193731566079&pagename=agder/Hovedsidemal&p=1226994216935, accessed 22 August 2014.

Taranger, Geir Lasse et al. (2005). 'Project Summary. Photoperiod Control of Puberty in Farmed Fish: Development of New Techniques and Research into Underlying Physiological Mechanisms (PUBERTIMING)', http://www.imr.no/pubertiming/__data/page/7985/Summary_Pubertiming_final.pdf, accessed 28 May 2012.

Taranger, Geir Lasse et al. (2010). 'Control of Puberty in Farmed Fish', *General and Comparative Endocrinology,* 165 (3): 483–515.

Tarde, Gabriel (1902). *Psychologie économique.* Paris: Félix Alcan.

Taylor, James (2005). *Stakes and Kidneys.* Aldershot: Ashgate.

Thacker, Eugene (2005). *The Global Genome: Biotechnology, Politics, and Culture.* Cambridge, MA: MIT Press.

Thévenot, Laurent (2007). 'Which Road to Follow? The Moral Complexity of an "Equipped" Humanity', in John Law and Annemarie Mol (eds), *Complexities: Social Studies of Knowledge Practices.* Durham, NC and London: Duke University Press, 53–87.

Thompson, Charis (2005). *Making Parents: The Ontological Choreography of Reproductive Technologies.* Cambridge, MA: MIT Press.

Thompson, Edward P. (1971). 'The Moral Economy of the English Crowd in the Eighteenth Century', *Past & Present,* 50: 76–136.

Tiefer, Leonore (2002). 'Arriving at a "New View" of Women's Sexual Problems', *Women & Therapy,* 24 (1–2): 63–98.

Tiselius, Arne (1970). 'Opening Address', in Arne Tiselius and Sam Nilsson (eds), *The Place of Value in a World of Facts: Proceedings of the Fourteenth Nobel Symposium.* New York: John Wiley and Sons, 11–15.

Tiselius, Arne, Long, C. N. H., Thomson, David L., and Young, Frank G. (1957). 'MSH Structure and Isolation', *Chemical and Engineering News,* 10–12.

Tong, Alison et al. (2010). 'Community Preferences for the Allocation of Solid Organs for Transplantation: A Systematic Review', *Transplantation,* 89 (7): 796–805.

Toplak, Natasa et al. (2012). 'An International Registry on Autoinflammatory Diseases: The Eurofever Experience', *Annals of the Rheumatic Diseases,* 71 (7): 1177–82.

Turnbull, David (1989). 'The Push for a Malaria Vaccine', *Social Studies of Science,* 19(2): 283–300.

Tutton, Richard (2011). 'Promising Pessimism: Reading the Futures to be Avoided in Biotech', *Social Studies of Science,* 41 (3): 411–29.

Ubel, Peter and Loewenstein, George (1996). 'Distributing Scarce Livers: The Moral Reasoning of the General Public', *Social Science and Medicine,* 42: 1049–55.

Vassy, Carine (2005). 'How Prenatal Diagnosis became Acceptable in France', *Trends in Biotechnology*, 23 (5): 246–9.

Vassy, Carine (2006). 'From a Genetic Innovation to Mass Health Programmes: The Diffusion of Down's Syndrome Prenatal Screening and Diagnostic Techniques in France', *Social Science & Medicine*, 63 (8): 2041–51.

Vertesi, Janet and Dourish, Paul (2011). 'The Value of Data: Considering the Context of Production in Data Economies', in *Proceedings of the ACM 2011 Conference on Computer Supported Cooperative Work*, 533–42.

Vidal, Fernando and Ortega, Francisco (2011). 'Approaching the Neurocultural Spectrum: An Introduction', in Francisco Ortega and Fernando Vidal (eds), *Neurocultures: Glimpses into an Expanding Universe.* Frankfurt am Main: Peter Lang.

Ville, Isabelle (2011). 'Politiques du handicap et périnatalité: La difficile conciliation de deux champs d'intervention sur le handicap', *ALTER—European Journal of Disability Research/ Revue Européenne de Recherche sur le Handicap*, 5 (1): 16–25.

VOA (2005). 'Malaria Vaccine May be on the Way', <http://www.voanews.com/english/ news/a-13-2005-04-23-voa2.html>, accessed 10 October 2010.

Wadmann, Sarah (2014) *Governing Drug Testing and Prescription in Preventive Cardiovascular Medicine in Denmark.* PhD. thesis, University of Copenhagen.

Wald, Nicholas J. and Hackshaw, Allan K. (1997). 'Combining Ultrasound and Biochemistry in First-trimester Screening for Down's Syndrome', *Prenatal Diagnosis*, 17 (9): 821–9.

Wald, Nicholas J. et al. (1988). 'Maternal Serum Screening for Down's Syndrome in Early Pregnancy', *BMJ: British Medical Journal*, 297 (6653): 883.

Wald, Nicholas J. et al. (1996). 'Serum Screening for Down's Syndrome between 8 and 14 Weeks of Pregnancy', *BJOG: An International Journal of Obstetrics & Gynaecology*, 103 (5): 407–12.

Waldby, Catherine (2000). *The Visible Human Project: Informatic Bodies and Posthuman Medicine.* London and New York: Routledge.

Waldby, Catherine (2002). 'Stem Cells, Tissue Cultures and the Production of Biovalue', *Health: An Interdisciplinary Journal for the Social Study of Health, Illness and Medicine*, 6 (3): 305–23.

Waldby, Catherine and Mitchell, Robert (2006). *Tissue Economies: Blood, Organs, and Cell Lines in Late Capitalism.* Durham, NC: Duke University Press.

Walker, W., Fairweather, D. V. I., and Jones, P. (1964). 'Examination of Liquor Amnii as a Method of Predicting Severity of Haemolytic Disease of Newborn', *BMJ: British Medical Journal*, 2 (5402): 141.

Walknowska, Janina, Conte, Felix A. and Grumbach, Melvin M. (1969). 'Practical and Theoretical Implications of Fetal/Maternal Lymphocyte Transfer', *The Lancet*, 293 (7606): 1119–22.

Watermeadow (2007). 'Rethinking the 'KOL Culture', *Next Generation Pharmaceutical 4* <http:// www.ngpharma.com>, accessed 29 March 2011.

Watson, Karli K., Shepherd, Stephen V., and Platt, Michael L. (2010). 'Neuroethology of Pleasure', in Morten L. Kringelbach and Kent C. Berridge (eds), *Pleasures of the Brain.* New York: Oxford University Press, 85–95.

Watson, William J., Chescheir, Nancy C., Katz, Vern L., and Seeds, John W. (1991). 'The Role of Ultrasound in Evaluation of Patients with Elevated Maternal Serum Alpha-fetoprotein: A Review', *Obstetrics & Gynecology*, 78 (1): 123–8.

Watts, Jonathan (2011). 'Fukushima disaster: it's not over yet', (updated Friday 9 September 2011 23.01 BST) <http://www.guardian.co.uk/world/2011/sep/09/fukushima-japan-nuclear-disaster-aftermath>, accessed 4 June 2013.

Wave Healthcare (2011) 'KOL Training', <http://www.wavehealthcare.co.uk/en/1/koltraining.html>, accessed 25 March 2011.

Weber, Tracy and Ornstein, Charles (2010). 'Dollars for Docs: Who's on Pharma's Top-paid List?', *ProPublica* <http://www.propublica.org/article/profiles-of-the-top-earners-in-dollar-for-docs>, accessed 29 March 2011.

Weir, Robert (1995). 'The Issue of Fairness in the Allocationo of Organs', *Journal of Corporate Law,* 20:91–108.

Wennerholm, Staffan (2004). '"Unga forskare sökes": Att tilltala vetenskapens påläggskalvar i teknikerbristens tidevarv', in Anders Ekström (ed.), *Den mediala vetenskapen.* Nora: Nya Doxa, 141–62.

Wertz, Dorothy C. and Fletcher, John C. (1993). 'A Critique of Some Feminist Challenges to Prenatal Diagnosis', *Journal of Women's Health,* 2 (2): 173–88.

Westman, Axel (1933). 'Den moderna hormonforskningens betydelse för gynekologien', *Uppsala Läkareförenings Förhandlingar,* 38: 183–91.

Westman, Axel (1936). 'Den ofrivilliga steriliteten', in Swedish Government Official Reports (ed.), *Betänkande i sexualfrågan,* Swedish Government Official Reports 1936:59. Stockholm: Befolkningskommissionen, 159–66.

Westman, Axel (1937a). 'Experimentella undersökningar över hypofysens produktion av gonadotropt hormon', *Nordisk Medicin,* 14: 1209–14.

Westman, Axel (1937b). 'Untersuchungen über die Wirkung des gonadotropen Hypophysenvorderlappenhormones, Antex (Leo), auf die Ovarien der Frau', *Acta Obstetricia et Gynecologica Scandinavica,* 17 (4): 492–515.

Westman, Axel (1938). 'Ueber den Luteinisierungseffekt des gonadotropen Chorionhormons auf die Ovarien der Frau', *Acta Pathologica et Microbiologica Scandinavica,* 15 (suppl. 37): 560–77.

Westman, Axel (1940a). *Barnlös: en bok om steriliteten och sterilitetsbekämpandet.* Stockholm: Wahlström & Widstrand.

Westman, Axel (1940b). *Hormoner.* Stockholm: Wahlström & Widstrand.

Westman, Axel (1948). 'Klinisk prövning av choriongonadotropinet "Gonadex Leo"', *Svenska Läkartidningen,* 45: 1156–60.

Widmalm, Sven (2007). 'Introduction: Science and the Creation of Value', *Minerva,* 45:115–20.

Widmalm, Sven (2014). 'United in Separation: The Inventions of Gel Filtration and the Moral Economy of Research in Swedish Biochemistry *ca*1950–1970', *Science in Context,* 27 (2): 249–74.

Wiesner, Russell et al. (2003). 'Model for End-stage Liver Disease (MELD) and Allocation of Donor Livers', *Gastroenterology,* 124 (1): 91–6.

Wigzell, Hans (2012). 'The Nobel Prize in Physiology or Medicine 1984', <http://www.nobelprize.org/nobel_prizes/medicine/laureates/1984/presentation-speech.html>, accessed 14 Dec 2013.

Wilkinson, Steven (2003). *Bodies for Sale: Ethics and Exploitation in the Human Body Trade.* New York: Routledge.

Williams, Claire, Alderson, Priscilla, and Farsides, Bobbie (2002). 'Too Many Choices? Hospital and Community Staff Reflect on the Future of Prenatal Screening', *Social Science & Medicine,* 55 (5): 743–53.

Wilmot, Stephen and Ratcliffe, Julie (2002). 'Principles of Distributive Justice Used by Members of the General Public in the Allocation of Donor Liver Grafts for Transplantation: A Qualitative Study', *Health Expectations,* 5 (3): 199–209.

Winner, Langdon (1977). *Autonomous Technology: Techniques Out of Control as a Theme in Political Thought.* Cambridge, MA: MIT Press.

Woolgar, Steve (1981). 'Interests and Explanation in the Social Study of Science', *Social Studies of Science,* 11 (3): 365–94.

Woolgar, Steve (1988). *Science: The Very Idea.* Ellis Horwood Series in Key Ideas. Chichester: Ellis Horwood.

Woolgar, Steve (1998). 'A New Theory of Innovation?', *Prometheus,* 16 (4): 441–53.

Woolgar, Steve (2012). 'Ontological Child Consumption', in Bengt Sandin, Johanna Sjöberg and Anna Sparrman (eds), *Situating Child Consumption: Rethinking Values and Notions of Children, Childhood and Consumption.* Lund: Nordic Academic Press, 33–51.

Woolgar, Steve and Lezaun, Javier (2013). 'The Wrong Bin Bag: A Turn to Ontology in Science and Technology Studies', *Social Studies of Science,* 43 (3): 321–40.

Woolgar, Steve and Pawluch, Dorothy (1985). 'Ontological Gerrymandering: The Anatomy of Social Problems Explanations', *Social Problems*: 214–27.

Woolgar, Steve, Coopmans, Catelijne, and Neyland, Daniel (2009). 'Does STS Mean Business?', *Organization,* 16 (1): 5–30.

World Health Organization (WHO) (2008). *Malaria Report.* <http://www.who.int/malaria/publications/atoz/9789241563697/en/>, accessed 4 September 2014.

World Medical Association International Code of Medical Ethics <http://www.wma.net/en/30publications/10policies/c8/>, accessed 28 January 2013.

Wray, K. Brad (2007). 'Evaluating Scientists: Examining the Effects of Sexism and Nepotism', in Harold Kincaid, John Dupré, and Alison Wylie (eds), *Value-Free Science? Ideals and Illusions.* Oxford: Oxford University Press, 87–106.

WRR (2000). *Het borgen van publiek belang.* Den Haag: SDU.

Wynne, Brian (1989). 'Sheepfarming after Chernobyl: A Case Study in Communicating Scientific Information', *Environment: Science and Policy for Sustainable Development,* 31 (2): 10–39.

Young, Bob (1985). 'Is Nature a Labour Process?', in L. Levidow and B. Young (eds), *Science, Technology and the Labour Process. Marxist Studies. Volume 1.* London: CSE Books.

Yoxen, Edward (1981). 'Life as a Productive Force: Capitalising the Science and Technology of Molecular Biology', in L. Levidow and R. Young (eds), *Science, Technology and the Labour Process.* London: Blackrose Press, 66–122.

Zelizer, Viviana A. (1978). 'Human Values and the Market: The Case of Life Insurance and Death in 19th-Century America', *American Journal of Sociology,* 84 (3): 591–610.

Zelizer, Viviana A. (1981). 'The Price and Value of Children: The Case of Children's Insurance', *American Journal of Sociology,* 86 (5): 1036–56.

Zelizer, Viviana A. (1989). 'The Social Meaning of Money: "Special Monies"', *American Journal of Sociology,* 95 (2): 342–77.

Zelizer, Viviana A. (2005). *The Purchase of Intimacy.* Princeton, NJ: Princeton University Press.

Zelizer, Viviana A. (2012). 'How I Became a Relational Economic Sociologist and What Does That Mean?', *Politics & Society,* 40 (2): 145–74.

Zuckerman, Harriet (1996). *Scientific Elite: Nobel Laureates in the United States,* 1st edn. published in 1977. New Brunswick, NJ & London: Transaction.

Zuffoletti, Jim and Freire, Octavio (2006). 'Marketing to Professionals: Key Opinion Control', *Pharmaceutical Executive* (updated 1 October 2006) <http://www.pharmexec.com>, accessed 28 March 2011.

Zuiderent-Jerak, Teun (2009). 'Competition in the Wild; Configuring Healthcare Markets', *Social Studies of Science,* 39 (5): 765–92.

Zuiderent-Jerak, Teun (2015, in press). *Situated Intervention; Sociological Experiments in Healthcare.* Cambridge, MA: MIT Press.

Zuiderent-Jerak, Teun and van der Grinten, Tom E. D. (2009). 'Zorg voor medische technologie; Over het ontwikkelen van zorgtechnologie, vormgeven van technologiebeleid en behartigen van publieke belangen', in Lotte Asveld and Michiel Besters (eds), *Medische technologie: ook geschikt voor thuisgebruik.* Den Haag: Rathenau Instituut, 61–108.

Zulueta, Benjamin C. (2009). 'Master of the Master Gland: Choh Hao Li, the University of California, and Science, Migration, and Race', *Historical Studies in the Natural Sciences,* 39 (2): 129–70.

INDEX